A Farewell to Entropy: Statistical Thermodynamics Based on Information

A Farewell to Entropy: Statistical Thermodynamics Based on Information

Arieh Ben-Naim
The Hebrew University, Israel

NEW JERSEY · LONDON · SINGAPORE · BEIJING · SHANGHAI · HONG KONG · TAIPEI · CHENNAI

Published by

World Scientific Publishing Co. Pte. Ltd.

5 Toh Tuck Link, Singapore 596224

USA office: 27 Warren Street, Suite 401-402, Hackensack, NJ 07601

UK office: 57 Shelton Street, Covent Garden, London WC2H 9HE

British Library Cataloguing-in-Publication Data
A catalogue record for this book is available from the British Library.

A FAREWELL TO ENTROPY
Statistical Thermodynamics Based on Information

ISBN-13 978-981-270-706-2
ISBN-10 981-270-706-9
ISBN-13 978-981-270-707-9 (pbk)
ISBN-10 981-270-707-7 (pbk)

Typeset by Stallion Press
Email: enquiries@stallionpress.com

Printed by FuIsland Offset Printing (S) Pte Ltd, Singapore

This book is dedicated to Ruby

Contents

List of Abbreviations

BE	Bose-Einstein
FD	Fermi-Dirac
GF	Generating function
GP	Gibbs paradox
ID	Indistinguishable
lhs	Left hand side
MB	Maxwell-Boltzmann
MI	Missing information
PF	Partition function
rhs	Right hand side
rv	Random variable

Preface

Nowadays, it is difficult to justify the writing of a new textbook on statistical thermodynamics. A quick glance at the bibliography at the end of this book shows an abundance of such textbooks. There are also a few books on statistical thermodynamics that use information theory such as those by Jaynes, Katz, and Tribus. These books use the principle of maximum entropy to "guess" the "best" or the least biased probability distributions of statistical mechanics.

The main purpose of this book is to go one step forward, not only to use the principle of maximum entropy in predicting probability distributions, but to replace altogether the concept of entropy with the more suitable concept of information, or better yet, the missing information (MI).

I believe that the time is ripe to acknowledge that the term *entropy*, as originally coined by Clausius, is an unfortunate choice. Moreover, it is also a misleading term both in its meaning in ancient and in contemporary Greek.[1] On this matter, I cannot do any better

[1] In the Merriam-Webster Collegiate Dictionary (2003), "Entropy is defined as: "change," literary "turn," a measure of the "unavailable energy" in a closed thermodynamic system... a measure of the system's degree of order..."
From Merriam-Webster online:

1. A measure of the unavailable energy in a closed thermodynamic system.
2. A measure of the system's disorder.
3. The degradation of matter and energy in the universe to an ultimate state of inert uniformity. $\varepsilon\nu$ = *in*, $\tau\rho o\pi\eta$ = *trope* = *transformation* $\varepsilon\nu\tau\rho o\pi\iota\alpha$ = entropy = transformation inwards in modern usage, entropy: turn into, or turn to be or evolves into, the way something will turn out, will change; could be *evolves*.

than Leon Cooper (1968). Cooper cites the original passage from Clausius: in choosing the word "Entropy," Clausius wrote:

> *"I prefer going to the ancient languages for the names of important scientific quantities, so that they mean the same thing in all living tongues. I propose, accordingly, to call S the* ***entropy*** *of a body, after the Greek word "****transformation****." I have designedly coined the word entropy to be similar to* ***energy****, for these two quantities are so analogous in their physical significance, that an analogy of denominations seems to be helpful."*

Right after quoting Clausius' explanation on his reasons for the choice of the word "Entropy," Cooper commented:

> *"By doing this, rather than extracting a name from the body of the current language (say:* ***lost heat****), he succeeded in coining a word that meant the same thing to everybody:* ***nothing****."*

I fully agree with Cooper's comment; however, I have two additional comments, and contrary to Cooper, I venture into taking the inevitable conclusion:

First, I agree that "entropy means the same thing to everybody: *nothing*." But more than that, entropy is also a misleading term. The two quantities "energy" and "entropy" are not analogous in their physical significance; hence, there is no reason for using analogous denominations.

Second, I do not believe that Cooper's apparently casual suggestion that "lost heat" might be a better choice, is a good idea, as much as the more common "unavailable energy" interpretation attached to "entropy" in most dictionaries.

As we shall discuss in Chapter 1, both the "heat loss" and "unavailable energy" may be applied under certain conditions to $T\Delta S$ but not to entropy. The reason it is applied to S rather than to $T\Delta S$, is that S, as presently defined, contains the units of energy and temperature. This is unfortunate. If entropy had been recognized from the outset as a measure of information, or of uncertainty, then it would be dimensionless, and the burden of carrying the units of energy would be transferred to the temperature T.

Finally, I believe that the time has come to reach the inevitable conclusion that "entropy" is a misnomer and should be replaced by either *missing information* or *uncertainty*. These are more appropriate terms for what is now referred to as "entropy."

Unfortunately, there is a vigorous ongoing debate on the very *interpretation* of entropy as information, let alone the replacement of entropy by information. This aspect will be discussed at length in Chapter 1. In Chapter 1, I shall also discuss the more common interpretation of entropy as disorder, mixed-upness, disorganization, chaos and the like. In my opinion all these terms are also inappropriate interpretations of entropy.

Today, the concept of entropy is used in many fields far from, and unrelated to, thermodynamics. Even in a superficial survey of the applications of the term entropy in various fields, from communications to economics, sociology to psychology, linguistics, arts and many more, one immediately realizes that the *concept* that is used is *information*, and not entropy as defined by Clausius and as is used in thermodynamics.

I can understand the continual application of the term "entropy" by practitioners in thermodynamics and statistical mechanics. It is a tradition that is hard to change.

I fail to understand why so many authors use the term entropy where in fact what they really mean is *information or uncertainty* (that includes Shannon himself who, as the story goes, renamed his measure for information as entropy; see below). To me, the usage of the term entropy is a corruption of the meaningful concept of information. Entropy does not mean what it meant in ancient Greek, does not mean what it presently means in modern Greek, and does not mean what it was supposed to mean when Clausius made this unfortunate choice. To use a concept that means "nothing," to replace a simple, familiar and meaningful concept such as information, by entropy, is at best, a perversive practice. The origin of this practice is found in Tribus'[2] story:

> "*What's in a name? In the case of Shannon's measure the naming was not accidental. In 1961 one of us (Tribus) asked Shannon*

[2] Tribus (1971).

what he had thought about when he had finally confirmed his famous measure. Shannon replied: "My greatest concern was what to call it. I thought of calling it 'information,' but the word was overly used, so I decided to call it 'uncertainty.' When I discussed it with John von Neumann, he had a better idea. Von Neumann told me, 'You should call it entropy, for two reasons. In the first place your uncertainty function has been used in statistical mechanics under that name. In the second place, and more important, no one knows what entropy really is, so in a debate you will always have the advantage.'"

On von Neumann's suggestion, Denbigh (1981) comments:

"*In my view von Neumann did science a disservice!*" adding "*there are, of course, good mathematical reasons why information theory and statistical mechanics both require functions having the same formal structure. They have a common origin in probability theory, and they also need to satisfy certain common requirements such as additivity. Yet, this formal similarity does not imply that the functions necessarily signify or represent the same concepts. The term 'entropy' had already been given a well-established physical meaning in thermodynamics, and it remains to be seen under what conditions, if any, thermodynamic entropy and information are mutually inconvertible.*"

I agree with the first sentence by Denbigh and indeed: *In my view von Neumann did science a disservice.*

My reason for embracing Denbigh's statement is that *information* (or choice, or uncertainty) is a simple, familiar, meaningful and well-defined concept. Renaming it *entropy* merely corrupts the term information. I shall delve into this more in Chapter 1.

The term "information" is indeed a far more general, and "overly used," concept than the quantity defined by Shannon. This is probably the reason why Shannon sought the advice of von Neumann for an appropriate term for the quantity he defined to measure information (or choice, or uncertainty). It would have been helpful if an appropriate term could have been found that had retained the qualitative properties of information on the one hand, yet was restricted to usage for that specific quantity as defined by Shannon, on the other. Unfortunately, the choice of "entropy" certainly does not fulfill this requirement, and in my opinion is an inadequate term. The choice made in this book is "missing information," or MI, or better yet, the amount of MI. This term captures the meaning of the concept sought by Shannon as a measure of the missing information, and therefore it is an appropriate term to replace the traditional term entropy. Perhaps, in the future, a term that combines the concepts of "measure" and of "information" (or of "measure" and of "uncertainty") should be coined. After all, Shannon himself did not discuss the *information* itself, but some measure of the size of a message that may or may not carry information. Perhaps a term like "**enformetry**," which has a part from **en**tropy, a part the root word of in**form**ation, and **metry** that indicates a measure of the *size* of the message, would serve better than entropy. In Section 3.3, we shall see that the most appropriate interpretation of "entropy" is the number of binary questions that one needs to ask to acquire the missing information.

It seems to me that if the absolute temperature had been defined *after* the development of the kinetic theory of gases, and recognized as a measure of the average kinetic energy of the particles, it would have been bestowed with the units of energy.[3] These units are more "natural" and more appropriate units for the temperature. Having temperature in units of energy would automatically render the Boltzmann constant superfluous and the Boltzmann entropy a dimensionless quantity. Such a dimensionless quantity would still be a state function. In this case, it would be easier to accept the interpretation of entropy as information. Furthermore, it would ease the acceptance of information as a cornerstone of the fabric of reality among matter and energy. The replacement of "entropy" by "information," in itself, would not provide an explanation of the Second Law of Thermodynamics. However, being an intangible quantity, it would be easier to accept the fact that information, or rather the MI increases in one direction only; it would also remove much of the mystery associated with entropy.

The aim of this book is to develop statistical thermodynamics from the cornerstone concept of information, as defined by

[3]This is true for ideal gases. See also Section 1.1. Note that "*units* of *energy*", is not the same as the *energy* of the system.

Shannon, not only as a means of guessing the best (or the least biased) probability distributions, but to replace entropy altogether.

To achieve that, I have allotted more space to present the elements of probability and information theory than to the foundations of statistical mechanics.

In Chapter 1, I will present the pros and cons of the usage of information in statistical mechanics. I will also discuss briefly the usage of disorder and related concepts in connection with entropy. Chapter 2 presents the elements of probability theory, enough to convince the reader of its usefulness and its beauty, enough to understand what information, as defined by Shannon, is. Probability theory is a rich, fascinating and extremely useful branch of mathematics. I have chosen some examples as exercises (most worked out in detail). Some are simple and straightforward, while some are more difficult, or "brain-teasers." All are useful for understanding the rest of the book. In Chapter 3, I will present the concept of information, the definition, the meanings and the application of this concept in statistical mechanics. We shall not touch on the application of information in communication theory, but only on those aspects that are relevant and potentially useful to statistical thermodynamics.

Information as defined by Shannon is a real objective quantity as much as mass and energy in physics, a point or a circle in geometry. In fact, I believe that the measure of information as defined by Shannon is more "objective" than a "mass" in physics, or a "point" in geometry. Different persons would measure different masses of the same piece of matter; it is hard to claim an *exact* value of its mass.[4] Similarly, a point in geometry is conceived as an idealization, or as a mathematical limit, but not a real physical object. However, a coin hidden in one of eight identical boxes *defines* a Shannon measure of the MI, which is $\log_2 8 = 3$. It is a real, objective and exact quantity. It would be futile to argue that this measure is "subjective," or "all in the mind," as claimed by some authors (see also Chapter 1). Of course, information, as uncertainty, is always *about* something. That "something" could be chosen in different

[4]Denbigh and Denbigh (1985).

ways even for the same system. In information theory, that "something" is irrelevant to the theory. What matters is only the *size* of the message carrying the information. (See also Chapter 3.)

If one accepts the probabilistic interpretation of the entropy, and agrees on the meaning of Shannon's information, then the interpretation of the *thermodynamic entropy* as *thermodynamic information* becomes inevitable.

It is sometimes argued that information is not uncertainty. Indeed it is not. The information one has about an event is different from the uncertainty about that event. However, both of these concepts are valid and plausible interpretations of the quantity $-\sum p_i \log p_i$, which features in information theory. This, in itself, does not make the two concepts identical, but it allows us to use the two interchangeably whenever we refer to the quantity $-\sum p_i \log p_i$. Both of these concepts subscribe to the same requirements originally put forward by Shannon when constructing the measure of information, or uncertainty. It should be noted however that order and disorder do not subscribe to these requirements, and therefore should not be used to describe the quantity $-\sum p_i \log p_i$ (for more details, see Section 1.2).

Chapter 4 is a transition chapter. Instead of plunging directly into statistical mechanics, I decided to gradually transform the concept of the amount of missing information, denoted by H, as defined by Shannon, and as is used in many fields, into the more specific usage of information in thermodynamics. The transition from the general concept of MI to the specific thermodynamic MI is carried out along three (not necessarily independent) routes.

(i) From a small number of states (or events, or configuration) to a very large number of states.
(ii) From one type of information to two types.
(iii) From discrete to continuous information.

This transitional chapter will culminate in the re-derivation of the well known Sackur–Tetrode equation for the entropy. In contrast to the traditional derivation, the new one is based on information–theoretical arguments, and therefore deserves to be renamed the *equivalent* of the *Sackur–Tetrode equation for the thermodynamic amount of missing information of an ideal gas.*

In Chapter 5, we present the fundamental structure of statistical mechanics. This is a standard subject. Therefore, we shall refrain from any details that can be found in any textbook on statistical mechanics. We shall be brief and almost sketchy. The general framework of the structure of statistical thermodynamics, and some standard applications for ideal gases, will be mentioned briefly only if they shed new light on the entropy being replaced by MI.

Chapter 6 can be viewed either as a collection of simple applications of statistical thermodynamics to a few simple processes, or as exercises for practicing the calculation of informational changes in these elementary processes. Some of the processes are discussed in most textbooks, such as expansion, mixing or heat transfer. In this chapter, however, we shall emphasize the informational changes accompanying these processes.

By doing that, we shall stumble upon two important findings; the first is the illusion regarding the irreversibility of mixing ideal gases. This has to do with the common and almost universally accepted conclusion that mixing of ideal gases is an inherently irreversible process, and the interpretation of the quantity $-\sum x_i \ln x_i$ as "entropy of mixing." It will be shown that mixing, in an ideal gas system can be reversible or irreversible, as can be demixing. The association of the increase of entropy with mixing or with increase of disorder is only an illusion.

The second is the illusion associated with the sense of loss of information in the assimilation process. This is a deeper illusion which has its roots in Maxwell's and Gibbs' writings regarding the apparent loss of information due to the assimilation process. We shall discuss this in Chapter 6. We shall show that both Maxwell's statement as well as Gibbs' have originated from the intuitive feeling that in an assimilation process, there is a sense of loss of information due to the loss of *identity* of the particles. This illusion is a result of "thinking classically" about identical particles; it can be resolved by properly interpreting the role of indistinguishability of the particles. Both of these illusions have misled many authors, including Gibbs, to reach the wrong conclusions.

Chapter 6 ends with a formulation of the Second Law of Thermodynamics in terms of probabilities and missing information. In the over-a-hundred-years of the history of the Second Law, people were puzzled by the apparent conflict between the reversibility of the equations of motion, and the irreversibility associated with the Second Law. Boltzmann was the first to attempt to *derive* the Second Law from the dynamics of the particles. In my opinion, this, as well as other attempts, will inevitably fail in principle. First, because it is impractical to solve the equations of motion for some 10^{23} particles. Second, because one cannot get *probabilities* from the *deterministic* equations of motion. Third, and perhaps most important, because of the indistinguishability of the particles. It is well known that whenever we write the equation of motions of any number of particles, we must first *label* the particles. This is true for classical as well as for the quantum mechanical equations of motion. However, the very act of *labeling* the particles violates the principle of ID of the particles. Therefore, one cannot derive the second law of thermodynamics which applies to indistinguishable (or unlabel-able in principle; see also appendices J & M) particles from the equations of motion of labeled particles. To put it differently, if we are strictly forbidden from labeling the ID particles, not even temporarily, then we have no equations of motion, hence there exists no conflict.

I have pondered long and hard over the choice of notation. On the one hand, I wanted to use a new notation for the "entropy" and the "temperature," to stress the difference in the meaning and the units of these quantities. Yet, I did not want to change to a completely unfamiliar notation. During the time of writing the book, I was considering a sort of compromise to use $\bar{S}$ and $\bar{T}$ instead of S and T, but in different units. Finally, I came to the conclusion that the best way is to keep the same notation for S and T, but to emphasize, quite frequently that we mean S/k and kT for the new quantities S and T. There is one modified notation however. We use H, following Shannon, for the more general measure of information in the sense of $-\sum p_i \log p_i$, when applied to *any* distribution. Whenever we discuss a thermodynamic system at equilibrium, H becomes identical with S. This conserves the notations of S and T

on the one hand, and makes the distinction between the more general concept of information and the specific application of the same concept for thermodynamic systems, on the other. For instance, in Chapter 4, we shall discuss various problems of one or two particles (or coins) bound to M sites (or hidden in M boxes). These are purely informational problems, and we use H in this context. However, for $N \to \infty$ and $M \to \infty$, we get into the realm of a thermodynamic system of N ligands absorbed on M sites. In this case, we shall switch from the informational measure of information H to the informational measure that previously was referred to as entropy, and is denoted by S. Thus, the notation S is identical, both formally and conceptually, with Shannon's measure of information H, whenever it applies to a thermodynamic system at equilibrium.

As for the justification of this book: there are a few books which include in their titles both "statistical mechanics" and "information theory." In fact, all of these, including Jaynes' pioneering work use "information theory." They use the maximum-entropy principle to predict the most plausible distributions in statistical mechanics. None of these *base* statistical mechanics on the concept of *information*, which is the aim of this book. Moreover, all the examples given in Chapter 6 are presented, discussed and analyzed in terms of the changes in the amount of missing information in spontaneous processes. I found this point of view very fruitful, illuminating and worth publishing.

The book is written in textbook style but it is not a textbook of statistical thermodynamics.

If you are reading this preface and pondering whether or not to read the rest of the book, I suggest that you take a simple "test," the result of which should help you in making the decision.

Consider the following chain of reasoning:

(i) Mixing is conceived as a process that increases disorder in the system.
(ii) Increase in disorder is associated with increase in entropy. Therefore, from (i) and (ii), it follows that:
(iii) Mixing increases the entropy.

If you have no idea what I am talking about in this "test," you should first study one of the standard textbooks on statistical thermodynamics, then come back to read this book.

If you *do not agree* with conclusion (iii), i.e., if the result of the "test" is *negative*, then I am sure you can read, understand and hopefully enjoy the book.

If you *agree* with conclusion (iii), i.e., the result of the "test" is *positive*, then you have a problem. You should take medication, and read this book until you test negative! I hope that after reading the book, you will understand why I suggest this test.

ACKNOWLEDGMENT

I would like to thank my friends and colleagues Eshel Ben-Jacob, Max Berkowitz, Diego Casadei, João Paulo Ferreira, John Harland, Ken Harris, Art Henn, Harvey Leff, Azriel Levy, Robert Mazo, Mihaly Mezei and Andres Santos. They all did a wonderful job of reading the manuscript and detecting many errors and mistakes. I wish to thank once again the assistance and the inspiration I got from my wife Ruby. As always, without her help, writing this book would not have been possible.

This book was written while I was a visiting Scholar at the center of theoretical biological physics, the University of California, San Diego, CA.

I am very grateful to Peter Wolynes for his hospitality while visiting UCSD.

Arieh Ben-Naim
Department of Physical Chemistry
The Hebrew University of Jerusalem
Jerusalem, Israel

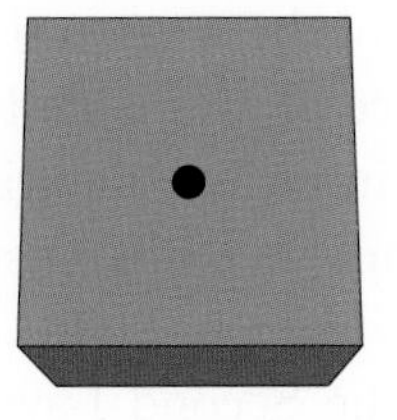

Chapter 1

Introduction

As stated in the preface, the aim of this book is to show that statistical thermodynamics will benefit from the replacement of the concept of entropy by "information" (or any of its equivalent terms; see also Section 1.3 and Chapter 3). This will not only simplify the interpretation of what entropy is, but will also bring an end to the mystery that has befogged the concept of entropy for over a hundred years. In the first section (1.1) of this chapter, we shall briefly survey the historical milestones in the understanding of the key concept of thermodynamics: temperature and entropy. I shall try to convince the reader that had the molecular theory of gases (the so-called kinetic theory of heat) preceded the definition of the absolute temperature, the entropy would have been defined as a dimensionless quantity which, in turn, would have eased the acceptance of the informational interpretation of the entropy. In the following two sections (1.2 and 1.3), we shall discuss the pros and cons of the two main groups of interpretations of the entropy — one based on the idea of order-disorder (or any related concept), and the second on the concept of information or missing information (or any related concept).

1.1 A Brief History of Temperature and Entropy

Temperature and entropy are the two fundamental quantities that make the entire field of thermodynamics deviate from other branches of physics. Both of these are statistical quantities, crucially dependent on the existence of atoms and their properties. Temperature, like pressure, is an intensive quantity; both can be felt by our senses. Entropy, like volume, is an extensive quantity depending on the size of the system. Unlike pressure and volume, temperature and entropy (as well as the Second Law), would not have existed had matter not been made up of an immense number of atoms and molecules. It is true that both temperature and entropy were *defined* and *measured* without any reference to the atomic constituency of matter. However, the *understanding* of these quantities and in fact, their very existence, *is* dependent on the atomic constituency of matter.

Perhaps, the first and the simplest quantity to be explained by the dynamics of the particles is the pressure of a gas. The pressure is defined as the force exerted on a unit area. This definition is valid not only without reference to the atomic constituency of matter, but also if matter were not atomistic at all, i.e., if matter were continuous. Not so for the temperature. Although temperature can be sensed and measured without reference to atoms, its very existence and certainly its explanation is intimately dependent on the atomistic constituency of matter. During the 19th century, it was widely believed that heat was a kind of substance that flows from a hot to a cold body. The association of the notion of temperature with motions of particles was one of the most important achievements of scientists of the late 19th century.

Robert Boyle (1660) found that at a given temperature of the gas, the product of the volume and pressure is constant. This is now known as the Boyle–Marriote law:

$$PV = constant. \tag{1.1.1}$$

Pressure could easily be explained by the particles incessantly bombarding the walls of the container. The pressure $\boldsymbol{P}$ is defined as force $\boldsymbol{F}$ per unit area, i.e.,

$$\boldsymbol{F} = \boldsymbol{P}A. \tag{1.1.2}$$

The force exerted by a moving particle of mass m and velocity v_x colliding with a wall perpendicular to the x-axis is equal to

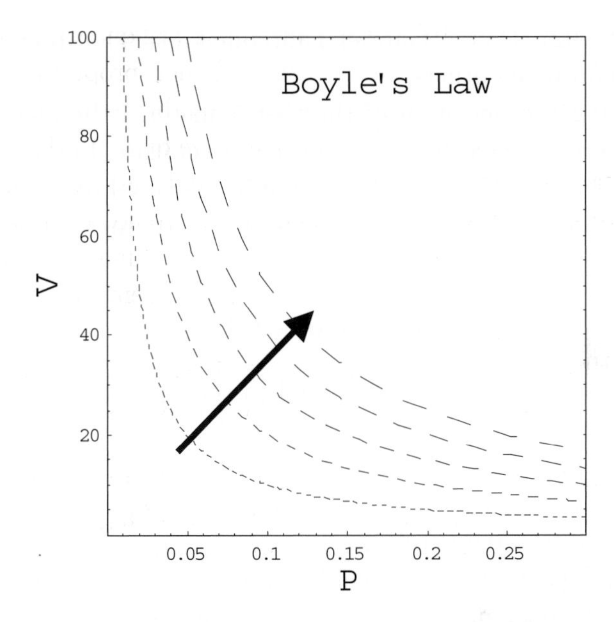

Figure 1.1. Volume as a function of pressure for a gas at different temperatures. Temperature increases in the direction of the arrow.

the change in the momentum per unit of time. An elementary argument[1] leads to the expression for the pressure in terms of the average square velocity of the particles[2]

$$P = \frac{\rho m \langle v^2 \rangle}{3}, \tag{1.1.3}$$

where ρ is the number density of the gas (i.e., the number of particles per unit volume), P is the (scalar) pressure, and

$$\langle v^2 \rangle = \langle v_x^2 \rangle + \langle v_y^2 \rangle + \langle v_z^2 \rangle = 3\langle v_x^2 \rangle. \tag{1.1.4}$$

This explanation was first given by Daniel Bernoulli (1732), long before the mechanical interpretation of the temperature.

Although the sense of hot and cold has been experienced since time immemorial, its measurement started only in the 17th and 18th centuries. Various thermometers were designed by Ole Christensen

[1] See, for example, Cooper (1968).

[2] We shall use both the notations $\langle x \rangle$ and $\bar{x}$ for the average of the quantity x.

Rømer (1702), Daniel Gabriel Fahrenheit (1714), Anders Celsius (1742), and many others.

Following the invention of the thermometer, it became possible to make precise measurements of the temperature. In the early 19th century, Jacques Charles and Joseph Louis Gay-Lussac discovered the experimental law that at constant pressure, the volume is linear in the temperature t (in °C), i.e.,

$$V = C(t + 273), \tag{1.1.5}$$

where C is a constant, proportional to the amount of gas.

From the graphs in Figure 1.2, it is clear that if we extrapolate to lower temperature all the curves converge to a single point. This led to the realization that there exists a minimal temperature, or an absolute zero temperature.

Combining the two empirical laws, one can write the equation of state of the ideal gas as:

$$PV = C \times T = nRT, \tag{1.1.6}$$

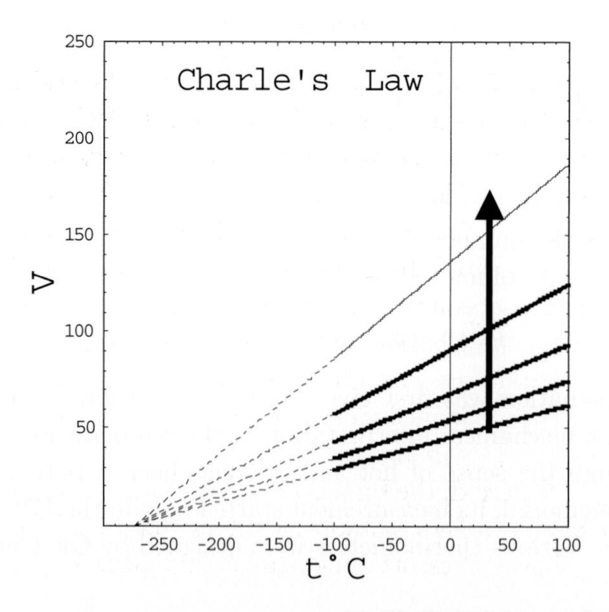

Figure 1.2. Volume as a function of temperature at different pressures. The pressure decreases in the direction of the arrow.

where T is the absolute temperature, n the number of moles, and the "constant," R, for one mole of gas is

$$R = 8.31 \times 10^7 \ \text{erg/mol K}. \tag{1.1.7}$$

Having a statistical mechanical theory of pressure, i.e., an interpretation of the pressure in terms of the average kinetic energy of the particles, on the one hand, and the equation of state of the gas on the other, one could derive the expression for the temperature in terms of the average kinetic energy of the particles. Thus, from (1.1.3) and (1.1.6), one can derive the equation

$$k_B T = \frac{2}{3} N \frac{m\langle v^2 \rangle}{2}, \tag{1.1.8}$$

where $k_B = 1.380 \times 10^{-23}$ J/K is now referred to as the Boltzmann constant, and N is the number of particles of the gas. In 1811, Amedeo Avogadro published his hypothesis that the average number of molecules in a given volume of *any* gas is the same (at a fixed temperature and pressure). With this finding, one can relate the Boltzmann constant k_B to the gas constant R by the relation

$$R = N_{AV} k_B \tag{1.1.9}$$

where $N_{AV} = 6.023 \times 10^{23}$ (particles/mol) is the Avogadro number.

Sadi Carnot published a seminal work on the ideal heat engine in 1824.[3] Carnot's prosaic style of writing was reformulated in mathematical terms by Clapeyron (1834). The latter laid the basis for the formulation of the Second Law by Clausius and by Kelvin.

William Thomson, later known as Lord Kelvin, established the existence of the absolute temperature (1854) and showed that by extrapolating from Charles' and Gay-Lussac's law (see Figure 1.2), there exists an absolute zero temperature at about −273°C (today the Kelvin scale is defined in such a way that its value at the triple point of water is, by definition, 273.16 K).

During that period, the First Law of Thermodynamics was established by many workers, notably by Julius Robert Mayer, William Thomsom, James Prescott Joule, and John James Waterston.

[3]Sadi Carnot (1824).

The Second Law was first formulated by Clausius in 1849. A second formulation by William Thomson (Lord Kelvin) followed in 1850. In 1854, Clausius studied the quantity dQ/T in connection with heat engines, where dQ is a small amount of heat transferred and T, the absolute temperature. However, the concept of entropy was introduced only in 1863.

James Clerk Maxwell published the dynamical theory of gases in 1860. This was extended by Ludwig Boltzmann in 1867. These works established both the existence and the form of the equilibrium velocity distribution of the molecules in the gas, known today as the Maxwell–Boltzmann distribution. This distribution established the connection between the average kinetic energy of the molecules with the absolute temperature.

During the late 19th and the early 20th centuries, the molecular interpretation of the entropy was developed by Ludwig Boltzmann and by Josiah Willard Gibbs. At that time, the molecular theory of matter was far from being universally accepted. A turning point in the acceptance of the atomistic theory of matter was the publication of the theory of Brownian motion by Albert Einstein (1905) followed by the experimental corroboration by Jean Perrin (1910).

As we have seen in this brief history of temperature and entropy, the concept of temperature, the various thermometers and the various scales came *before* the establishment of the Maxwell–Boltzmann (MB) distribution of velocities. The *acceptance* of the Maxwell–Boltzmann distribution as well as Boltzmann's expression for the entropy came later. It is clear from this sequence of events that the units of temperature were determined at least two decades before the recognition and acceptance of the molecular interpretation of the temperature.

When Clausius formulated the Second Law in terms of dQ/T, it was only natural to define the entropy in units of energy divided by temperature. From that, it followed that Boltzmann's entropy had to be defined with the constant k_B (later known as Boltzmann's constant) having the same units as Clausius' entropy, i.e., energy divided by temperature or Joule/Kelvin, or J/K.

Events could have unfolded differently had the identification of temperature with the average velocity of the atoms come earlier, in which case it would have been more natural to define temperature in units of energy,[4] i.e., one would have arrived at the identity

$$\frac{m\langle v^2\rangle}{2} = \frac{3}{2}T \tag{1.1.10}$$

(instead of $3k_B\, T/2$ on the right-hand side). Having this identification of the temperature would not have any effect on the formal expression of either the efficiency of Carnot's engine,

$$\eta = \frac{T_2 - T_1}{T_1}, \tag{1.1.11}$$

or on the definition of Clausius' entropy,

$$dS = \frac{dQ}{T}. \tag{1.1.12}$$

The only difference would have been that the entropy S would now be a dimensionless quantity. It would still be a state function, and it would still make no reference to the molecular constituency of matter. Clausius could have called this quantity entropy or whatever. It would not have changed anything in the formulation of the Second Law, in either the Clausius or in the Kelvin versions. The molecular interpretation of the entropy would have to wait however, until the close of the 19th century, when the foundations of statistical mechanics were laid down by Maxwell, Boltzmann and Gibbs.

What would have changed is the formal relationship between the entropy S and the number of states W. To relate the two, there will be no need to introduce a constant bearing the units of energy divided by the temperature. Boltzmann would have made the correspondence between the Clausius' thermodynamic entropy and the statistical entropy simply by choosing a dimensionless constant, or better, leaving the base of the logarithm in the relation $S = \log W$ unspecified (it will be shown in Chapter 3 that the choice of

---

[4]As actually is the practice in many branches of physics and in statistical mechanics. It should be noted that the relation between the temperature and the average kinetic energy of the particles is true for classical ideal gases. It is not true in general. For instance, ideal Fermi gas at T=0 K, has non-zero energy, see also Leff (1999) and Hill (1960).

base 2 has some advantages in connection with the interpretation of entropy in terms of missing information; see Section 3.5).

The removal of the units from Clausius' entropy would not have any effect on the interpretation of the entropy within the classical (or the non-atomistic) formulation of the Second Law. It would have a tremendous effect on the interpretation of Boltzmann's entropy, in terms of probability, and later in terms of the missing information.

The units of Clausius' entropy were, for a long time, the stumbling blocks that hindered the acceptance of Boltzmann's entropy. In one, we have a quantity that is *defined* by the ratio of energy and temperature; in the other, we have a quantity representing the number of states of the system. The two quantities seem to be totally unrelated. Had entropy been *defined* as a dimensionless quantity, the identification between the two entropies and eventually the interpretation of entropy as a measure of information would have become much easier.

The redefinition of the temperature in units of energy will also render the (erroneous) interpretation of the *entropy* as either "heat loss" or "unavailable energy" as superfluous. Consider for simplicity that we have an ideal mono-atomic gas so that all the internal energy consists of only the kinetic energy. We write the Helmholtz energy in the traditional forms as[5]

$$A = E - TS = \frac{3N}{2} k_B T - TS. \tag{1.1.13}$$

In this form, and under certain conditions, TS may be referred to as "heat loss" or "unavailable energy." In the conventional definition of the temperature, it is the *entropy* that bears the "energy" part of the units. Therefore, it is almost natural to ascribe the *energy*, associated with TS, to the entropy S. This leads to the common interpretation of the entropy as either "heat loss" or "unavailable energy" as it features in many dictionaries. If, on the other hand, one defines T in units of energy and S as a dimensionless quantity, this erroneous assignment to S can be avoided. In the example of an ideal gas, we would have written, instead of (1.1.13),

[5] See, for example, Denbigh (1966).

the equivalent relation

$$A = \frac{3}{2}NT - TS, \tag{1.1.14}$$

where T itself is related to the average kinetic energy of the particles by

$$\frac{3}{2}NT = N\frac{m\langle v^2\rangle}{2} \tag{1.1.15}$$

or equivalently

$$T = \frac{m\langle v^2\rangle}{3}. \tag{1.1.16}$$

Hence

$$\begin{aligned} A &= N\frac{m\langle v^2\rangle}{2} - N\frac{m\langle v^2\rangle}{3}S/N \\ &= N\frac{m\langle v^2\rangle}{2}\left[1 - \frac{2S}{3N}\right]. \end{aligned} \tag{1.1.17}$$

In this form, the squared brackets on the right-hand side of (1.1.17) includes the entropy, or rather the missing information (MI). Since S is an extensive quantity, S/N is the MI per particle in this system. The Helmholtz energy A is viewed as the *total* kinetic energy, which when multiplied by the factor in the squared brackets in Equation 1.1.17, is converted to the "free energy."

1.2 The Association of Entropy with Disorder

During over a hundred years of the history of entropy, there have been many attempts to interpret and understand entropy. We shall discuss the two main groups of such interpretations of entropy.

The earliest, and nowadays, the most common interpretation of the entropy is in terms of disorder, or any of the related concepts such as "disorganization," "mixed-upness," "spread of energy," "randomness," "chaos" and the like.

Perhaps, the earliest association of changes in entropy in spontaneous processes with increase of disorder is already implied in Boltzmann's writings. It is commonly believed that Bridgman (1953) was the first to spell out the association of entropy with disorder. Leff (1996, 2007) has recently advocated in favor of the idea of "spread of energy".

The association of entropy with disorder is at best a vague qualitative, and highly subjective one. It rests on observations that in some simple spontaneous processes, when viewed on a molecular level may be conceived as a disordering that takes place in the system. This is indeed true for many processes, but not for all.[6]

Consider the following processes.

Expansion of an ideal gas

Figure 1.3 shows schematically three stages in a process of expanding of an ideal gas.

On the left-hand side, we have N atoms in volume V. In the middle, some of the N atoms have moved to occupy a larger volume $2V$, and on the right-hand side, the atoms are spread evenly in the entire volume $2V$. Take a look. Can you tell which of the three systems is the more ordered? Well, one can argue that the system on the left, where the N atoms are gathered in one half of the volume, is more ordered than N atoms spread in the entire volume. That is plausible when we associate entropy with missing information (see below), but as for order, I personally do not see either of the systems in the figures to be more ordered, or disordered than the other.[7]

The mixing of two different gases

Consider the two systems (Figure 1.4). In the left system, we have N_A blue, and N_B red particles. In the right, we have all the particles

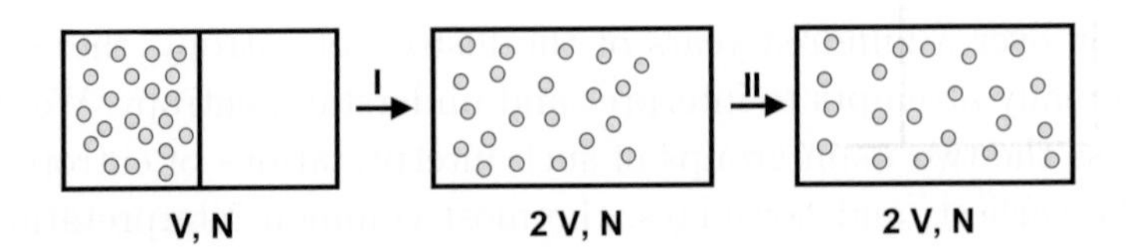

Figure 1.3. Three stages in the process of an expanding ideal gas.

[6]It is interesting to note that Landsberg (1978) not only contended that disorder is an ill-defined concept, but actually made the assertion that "it is reasonable to expect 'disorder' to be an intensive variable."

[7]This is certainly true if you ask people, who have never heard of the concept of entropy, to rate the degree of order in the three systems in Figure 1.3, as the author did.

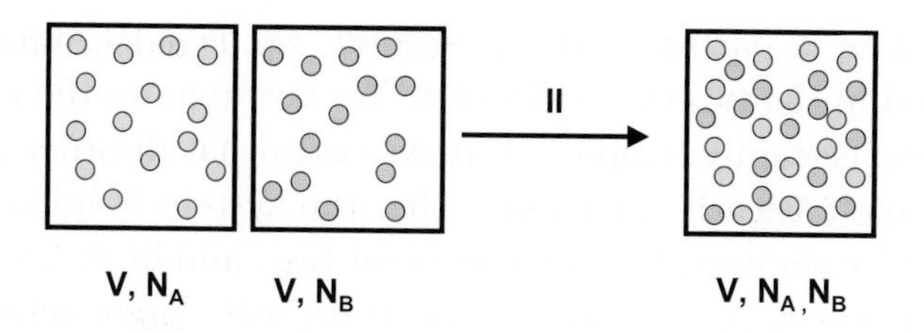

Figure 1.4. Mixing of two different gases.

mixed up in the *same* volume V. Now, which is more ordered? In my view, the left side is more ordered — all the blues and all the reds are separated in different boxes. On the right-hand side, they are mixed up. "Mixed-up" is certainly a disordered state, colloquially speaking. In fact, even Gibbs himself used the word "mix-upness" to describe entropy (see Sections 6.4–6.7). Yet, one can prove that the two systems have *equal* entropy. The association of mixing, with increase in disorder, and hence increase in entropy, is therefore only an illusion. The trouble with the concept of order and disorder is that they are not well-defined quantities — "order" as much as "structure" and "beauty" are in the eyes of the beholder!

"Mixing" of the same gas

Consider the following two processes. In the first (Figure 1.5a), we start with one atom of type A in each compartment of volume V,

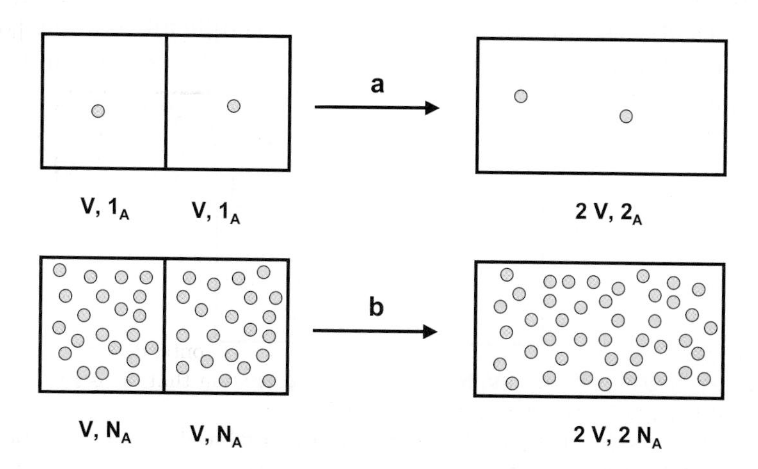

Figure 1.5. Assimilation, or "mixing" of two gases of the same kind.

separated by a partition. In the second, we have N atoms of the same kind in the two compartments. We now remove the partition. Something certainly happens. Can we claim that disorder increases or decreases? Clearly, in Figure 1.5a, the system is initially more "ordered," each particle is in a different box, and in the final stage, the two particles are mixed up. The same holds true for the case of N particles in each box, Figure 1.5b. Therefore, using the concept of "disorder," we should interpret the changes that take place in this process as *increase* in disorder. Yet in terms of thermodynamics, the entropy, in this process, should not be noticeably changed. As we shall see in Chapters 4 and 6, both processes involve an *increase* in MI; it is only in the macroscopic limit of very large N that the change in the MI in the process in Figure 1.5b becomes negligibly small and hence unmeasurable.

Extensivity of disorder?

Consider the two systems in Figure 1.6. We have two boxes of equal volume V and the same number of particles N in each box. Each of the boxes looks disordered. But can we claim that the "disorder" of the *combined* system (i.e., the two boxes in Figure 1.6) is twice the disorder of a single box? It is difficult to make a convincing argument for such an assertion since disorder is not well defined. But even as a qualitative concept, there is no reason to claim that the amount of disorder in the combined system is the sum of the disorder of each system. In other words, it is difficult to argue that disorder should have an additive property, as the entropy and the MI have.

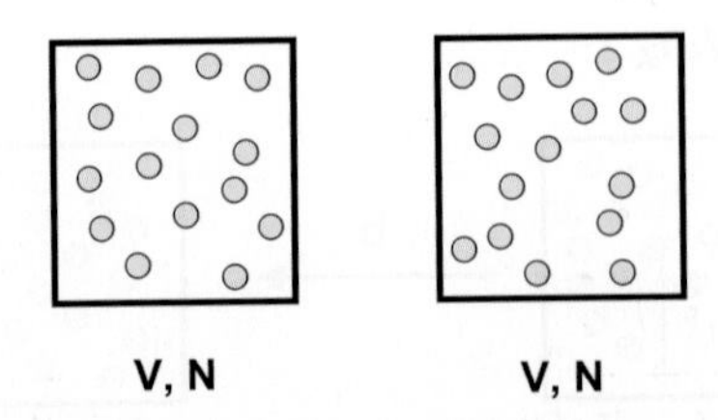

Figure 1.6. Is the extent of the disorder of the combined systems the sum of the disorder of each system?

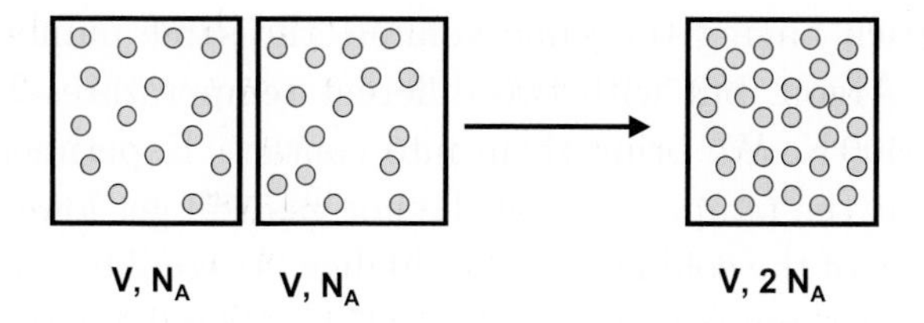

Figure 1.7. A pure assimilation process.

The assimilation process

Consider the process depicted in Figure 1.7. We start with two boxes of the same volume V and the same number of N particles in each. We bring all the particles into one box. In Section 6.3, we shall refer to this process as a pure assimilation process. Here, we are only interested in the question: Is the final state less or more disordered than the initial state? Again, one can argue either way. For instance, one can argue that having N particles in each different box is a more ordered state than having all the $2N$ particles mixed up in one box. However, it can be shown that, in fact, the entropy as well as the MI *decreases* in this process. Therefore, if we associate decrease in entropy with decrease in disorder, we should conclude that the final state is more ordered.

Heat transfer from a hot to a cold gas

As a final example, we discuss here an experiment involving change of temperatures. This experiment is important for several reasons. First, it is one of the classical processes in which entropy increases: in fact, this was the process for which the Second Law was first formulated. Second, it is a good example that demonstrates how difficult it is to argue that disorder is associated with entropy. The process is shown in Figure 1.8. Initially, we have two isolated

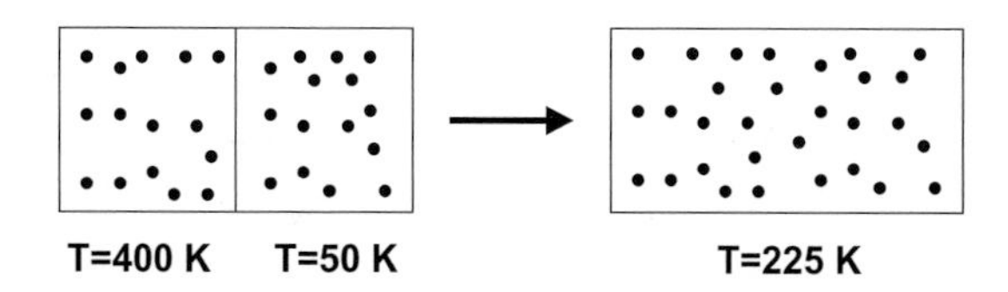

Figure 1.8. Heat transfer from a hot to a cold gas.

systems, each having the same volume, the same number of particles, say Argon, but with two different temperatures $T_1 = 50\,\mathrm{K}$ and $T_2 = 400\,\mathrm{K}$. We bring them into contact. Experimentally, we observe that the temperature of the hot gas will get lower, and the temperature of the cold gas will get higher. At equilibrium, we shall have a uniform temperature of $T = 225\,\mathrm{K}$ throughout the system.

Clearly, heat or thermal energy is transferred from the hot to the cold gas. But can we understand the changes that occur in the system in terms of disorder?

We know that temperature is associated with the distribution of molecular velocities. In Figure 1.9, we illustrate the distribution of velocities for the two gases in the initial state (the left-hand side of Figure 1.9). The distribution is sharper for the lower temperature gas, and is more dispersed for the higher temperature gas. At thermal equilibrium, the distribution is somewhat intermediate between the two extremes, and is shown as a dashed curve in Figure 1.9.

The heat transfer that we observed experimentally is interpreted on a molecular level as the change in the distribution of molecular velocities. Some of the kinetic energies of the hotter gas is transferred to the colder gas so that a new, intermediary distribution is attained at equilibrium.

Now look at the two curves on the right-hand side of Figure 1.9, where we plotted the velocity distribution of the entire system before and after the thermal contact. Can you tell which

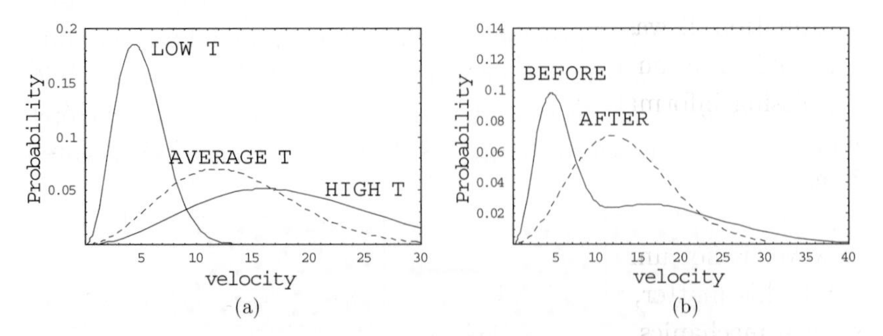

Figure 1.9. Velocity distributions for a gas at different temperatures. (a) The distributions corresponding to the left-hand side of Fig. 1.8. (b) The overall distribution of velocities, before and after the process of heat transfer.

distribution is more ordered or *disordered*? Can you tell in which distribution the *spread* of kinetic energy is more even, or over a larger range of velocities? To me, the final distribution (the dashed curve) looks more ordered, and the distribution looks less spread out over the range of velocities. Clearly, this is a highly subjective view. For this reason and some others discussed in Chapters 3 and 6, I believe that neither "disorder," nor "spread of energy" are adequate descriptions of entropy. On the other hand, information or MI is adequate. As we shall see in Chapter 6, the increase in the entropy in this process can be interpreted as an increase in the MI. It will be shown that the final distribution of velocities is that with the minimum Shannon's *information* or maximum MI. Although this result cannot be *seen* by looking directly at the system, nor by looking at the velocity distribution curves, it can be *proven* mathematically.

Order and disorder are vague and highly subjective concepts. It is true that in many cases, increase in entropy can be qualitatively correlated with increase in disorder, but that kind of correlation is not always valid. Furthermore, the commonly encountered statement that "nature's way is to go from order to disorder," is to say the same as "nature's way is to go from low to high entropy." It does not *explain* why disorder should increase in a spontaneous process. My objection to the association of entropy with disorder is mainly due to the fact that order and disorder are not well-defined and fuzzy concepts. They are very subjective, sometimes ambiguous, and at times totally misleading.

Ever since Shannon introduced the definition of the measure of information, it was found very useful in interpreting entropy (more on this in Section 1.3 and Chapter 3). In my opinion, the concept of missing information not only contributes to our understanding of what is the *thing* that changes and which is called entropy, but it also brings us closer to the last and final step in understanding entropy's behavior (more on this in Section 6.12). This view however is not universal.

On this matter, referring to the informational approach to statistical mechanics, Callen (1984), on page 384 writes:

> "*There is a school of thermodynamics who view thermodynamics as a subjective science of prediction.*"

In a paragraph preceding the discussion of entropy as disorder, Callen explains the origin of this subjectivity and writes:

> "*The concept of probability has two distinct interpretations in common usage. 'Objective probability' refers to a* ***frequency****, or a* ***fractional occurrence****; the assertion that 'the probability of newborn infants being male is slightly less than one half' is a statement about census data. 'Subjective probability' is a measure of* ***expectation based on less than optimum information****. The (subjective) probability of a* ***particular yet unborn*** *child being male,* ***as assessed by a physician****, depends upon that physician's knowledge of the parents' family histories, upon accumulating data on maternal hormone levels, upon the increasing clarity of ultrasound images, and finally upon an educated, but still subjective, guess.*"

I have quoted Callen's paragraph above to show that his argument in favoring "disorder" is essentially fallacious. I believe Callen has mis-applied a probabilistic argument to deem information as "subjective" and to advocate in favor of "disorder," which in his view is "objective" (more on this in Chapter 2).

I am not aware of any precise *definition* of order and disorder that can be used to validate the interpretation of the entropy in terms of the extent of disorder. There is one exception however. Callen (1985) writes on page 380:

> "*In fact, the conceptual framework of 'information theory' erected by Claude Shannon, in the late 1940, provides a basis for interpretation of the entropy in terms of Shannon's measure of* ***disorder****.*"

And further, on page 381, Callen concludes:

> "*For closed systems the entropy corresponds to Shannon's quantitative measure of the maximum possible disorder in the distribution of the system over its permissible microstates.*"

I have taught thermodynamics for many years and used Callen's book as a textbook. It is an excellent book. However, with all due respect to Callen and to his book, I must say that Callen misleads the reader with these statements. Shannon never defined nor referred to "disorder" in his writings. In my opinion, Callen is fudging with the definition of disorder in the quoted statement and in the rest of that chapter. What for? To "*legitimize*" the usage of *disorder* in interpreting entropy. That clearly does not jibe with Shannon's writings. What Callen refers to as Shannon's definition of *disorder* is, in fact, Shannon's definition of *information*. As we shall see in Chapter 3, the measure of "information" as defined by Shannon also retains some of the flavor of the meaning of information as we use in everyday life. This is not the case for disorder. Of course, one can *define* disorder as Callen does, precisely by using Shannon's definition of *information*. Unfortunately, this definition of "disorder" does not have, in general, the *meaning* of disorder as we use the word in our daily lives, as was demonstrated in the examples above.[8]

As we have seen above, even *mixing*, under certain conditions cannot be associated with an increase in disorder nor with the increase in entropy. In fact, we shall show in Chapter 6 that if mixing can be associated with increase in entropy, then also demixing should be associated with increase in entropy. However, mixing by itself has nothing to do with the increase in entropy.

Callen defined "*Shannon's disorder*" by (for more details, see Chapters 2 and 3)

$$\text{disorder} = -\sum p_i \log p_i. \tag{1.2.1}$$

This is clearly a distortion of Shannon's definition of information. Shannon never defined nor discussed disorder in connection with his work on information theory. I would make this statement even stronger by claiming that neither Shannon, nor anyone else *could have* arrived at such a definition of "disorder."

As we shall see in Chapter 3, Shannon did not start by defining information (or *choice* or *uncertainty*, certainly not *disorder*). What Shannon did was to start with the general concept of "information"

[8]McGlashan (1979) has also expressed reservations regarding the association of entropy with disorder. He excluded two cases however: "When, if ever, has the (number of states) anything to do with any of these words like mixed-upness." He then continues: "It does, but only for two very special cases. These are mixtures of perfect gases, and crystals at temperatures close to zero."

or rather the quantity of information transmitted along communication lines. Shannon then asked himself: if a quantitative measure of information exists, what properties must it have? After pronouncing several plausible properties of such a measure, he found that such a quantity must have the form $-\sum p_i \ln p_i$ (see Chapter 3). Had he set his aim to construct a measure of *disorder*, it is hard to believe that he, or anyone else, would have reached the same definition for *disorder*. In fact, it is not clear at all what requirement should a measure of disorder fulfill. It is not clear that the property of additivity required from information would also be required for disorder[9]; it might also be meaningless to require the "consistency property" (or the "independence on grouping" — see Chapter 3) from a quantity such as disorder.

Thus, even if Shannon would have pondered the plausible requirements for "disorder" to make it a quantitative measure, it is unlikely that he would have arrived at the definition (1.2.1), as claimed by Callen. Moreover, as we shall see in Chapter 3, many of the *derived* properties of information, such as conditional information, mutual information and the like would not and could not be assigned to disorder.

It is unfortunate and perhaps even ironic that Callen dismisses "information" as subjective, while at the same time embracing Shannon's definition of information, but renaming it as disorder. By doing that, he actually replaces a well-defined, quantitative and objective quantity by a more subjective concept of disorder. Had Callen not used *Shannon's definition of information*, the concept of disorder would have remained an undefined, qualitative and highly subjective quantity.

In my view, it does not make any difference if you refer to *information* or to *disorder*, as subjective or objective. What matters is that order and disorder are not well-defined scientific concepts. On the other hand, information is a well-defined scientific quantity, as

[9]See examples above. Note also that Callen in his 1987 edition, had made a list of requirements similar to those made by Shannon that "disorder" should subscribe to. In my opinion these requirements, although valid for information, are not even plausible for disorder. Similar reference to "Shannon's measure of disorder" may be found in Lee (2002).

much as a point or a line are *scientific* in geometry, or mass or charge of a particle are *scientific* in physics.

In concluding, we can say that increase in disorder (or any of the equivalent terms) can sometimes, but not always, be associated qualitatively with increase in entropy. On the other hand, "information" or rather MI can *always* be associated with entropy, and therefore it is superior to disorder.

1.3 The Association of Entropy with Missing Information

As in the case of order and disorder, the involvement of the notion of information is already implied in Boltzmann's expression for the entropy. It is also implicit in the writings of Maxwell. Maxwell was probably the first to use probabilistic ideas in the theory of heat. As we shall see in Chapter 3, probability is only a short step from information.

More explicitly, although not stated as such, the idea of Maxwell's demon, a being that can defy the Second Law, contains elements of the notion of information.

The demon was introduced by James Clerk Maxwell in the 1871 book, "Theory of Heat"[10]:

> "*Starting with a uniform temperature, let us suppose that such a vessel is divided into two portions or by a division in which there is a small hole, and that a being, who can see the individual molecules, opens and closes this hole so as to allow only the swifter molecules to pass from A to B, and only the slower ones pass from B to A. He will thus, without expenditure of work raise the temperature of B and lower that of A in contradiction to the second law of thermodynamics.*"[11]

[10]From Leff and Rex (1990), *Maxwell's Demon, Entropy, Information, Computing*, Adam-Hilger, Bristol, UK.

[11]It should be noted that in this statement Maxwell recognized the relation between the temperature and the average kinetic energies of the particles. Note also that intuitively, Maxwell felt that if such a Demon could have existed, he could use his *information* to achieve a process that violates the Second Law of Thermodynamics.

William Thomson (1874) referred to this imaginary being as "Maxwell's intelligent demon." Although Maxwell himself did not discuss the connection between the Second Law and information, this is implicitly implied by his description of what the demon is capable of doing, i.e., using information to lower the entropy.

Szilard (1929) in analyzing the implications of Maxwell's demon to the Second Law, referred to the "intervention of an intelligent being who uses information to defy the Second Law." Many articles have been written on Maxwell's demon and entropy.[12] However, the association of Maxwell's demon with information and with entropy was quite loose. The association of entropy with information was firmed up and quantified only after Shannon introduced the quantitative measure of information in 1948.

Previous to 1948, perhaps the most eloquent and explicit identification of entropy with information (albeit not the quantitative Shannon's measure) is the statement of G. N. Lewis in 1930:

> "*Gain in entropy always means loss of information, and nothing more. It is a subjective concept, but we can express it in a least subjective form, as follows. If, on a page, we read the description of a physico-chemical system, together with certain data which help to specify the system, the entropy of the system is determined by these specifications. If any of the essential data are erased, the entropy becomes greater; if any essential data are added, the entropy becomes less. Nothing further is needed to show that the irreversible process neither implies one-way time, nor has any other temporal implications. Time is not one of the variables of pure thermodynamics.*"

This is an almost prophetic statement made eighteen years before information theory was born. Lewis' statement left no doubt that he considered entropy as *conceptually* identical with information. Note however that Lewis claims that entropy is a subjective concept. This is probably a result of his usage of the *general* concept of information and not the specific measure of information as defined by Shannon.

[12] See Leff and Rex (1990).

The first to capitalize on Shannon's concept of information was Brillouin. In his book "Science and Information" (1956), Brillouin expressed the view that Clausius' entropy and Shannon's entropy are identical. Brillouin (1962) also suggested referring to information as neg-entropy. In my view, this amounts to replacing a simple, familiar and informative term with a vague and essentially misleading term. Instead, I would have suggested replacing entropy with either neg-information, missing information, or uncertainty.

Ilya Prigogine (1997) in his recent book "End of Certainty" quotes Murray Gell–Mann (1994) saying:

> "*Entropy and information are very closely related. In fact, entropy can be regarded as a measure of ignorance. When it is known only that a system is in a given macrostate, the entropy of the macrostate measures the degree of ignorance the microstate is in by counting the number of bits of additional information needed to specify it, with all the microstates treated as equally probable.*"[13]

I fully agree with the content of this quotation by Gell–Mann, yet Ilya Progogine, commenting on this very paragraph writes:

> "*We believe that these arguments are untenable. They imply that it is our own ignorance, our coarse graining, that leads to the second law.*"

Untenable? Why?

The reason for these two diametrically contradictory views by two great scientists has its sources in the confusion of the *general* concept of information, with the specific measure of information as defined by Shannon. I shall discuss this issue in greater detail below and in Chapter 3.

In my opinion, Gell–Mann is not only right in his statement, he is also careful to say "entropy *can* be regarded as a measure of ignorance ... Entropy ... measures the degree of ignorance." He does not say *"our own ignorance,"* as misinterpreted by Prigogine.

Indeed information, as we shall see in Chapter 3, is a measure that *is there* in the system. Within information theory, the term

[13] Gell–Mann (1994).

"information" is not a subjective quantity. Gell–Mann uses the term "ignorance" as a synonym to "lack of information." As such, ignorance is also an objective quantity that belongs to the system and is not the same as "*our own ignorance*," which might or might not be an objective quantity.

The misinterpretation of the information-theoretical entropy as a subjective information is quite common. Here is a paragraph from Atkins' preface from the book "The Second Law."[14]

> "*I have deliberately omitted reference to the relation between information theory and entropy. There is the danger, it seems to me, of giving the impression that entropy requires the existence of some cognizant entity capable of possessing 'information' or of being to some degree 'ignorant.' It is then only a small step to the presumption that entropy is all in the mind, and hence is an aspect of the observer. I have no time for this kind of muddleheadedness and intend to keep such metaphysical accretions at bay. For this reason I omit any discussion of the analogies between information theory and thermodynamics.*"

Atkins' comment and his rejection of the informational interpretation of entropy on the grounds that this "relation" might lead to the "presumption that entropy is all in the mind" is ironic. Instead, he uses the terms "disorder" and "disorganized," etc., which in my view are concepts that are far more "in the mind."

The fact is that there is not only an *analogy* between entropy and information, but an *identity* between the thermodynamic entropy and Shannon's measure of information.

The reason for the confusion is that the term information itself has numerous interpretations. We can identify at least three "levels" in which we can interpret the term "information." At the most general level, information is any knowledge that we can obtain by our senses. It is an abstract concept which may or may not be subjective. The information on "the weather conditions in New York state" might have different significance, meaning or value to different persons. This information is *not* the subject of interest of information theory. When Shannon sought a quantity to measure

[14] Atkins (1984).

information transmitted across communication lines, he was not interested in the *content* or the *meaning* of information, but in a quantity that measures the *amount* of information that is being transmitted.

Leaving aside the *content* or the meaning of information and focusing only on the *size* or the amount of information, we are still far from Shannon's measure of information. One can speak of different amounts of information, or information on different amounts of something. Consider the following two messages:

A: Each of the ten houses in the street costs one million dollars.
B: The first house in this street costs one million dollars, the second house costs one million dollars,... and the tenth house costs one million dollars.

Clearly, the two messages A and B carry the same information. The *size* of A is however much smaller than the size of B. Consider the next message:

C: Each of the houses in this town costs one million dollars.

Clearly, C contains more information than B. The message C tells us the price of more houses than the message B, yet it is much shorter than B (shorter in some sense that will be discussed in Chapter 3).
Consider the next message:

D: Each of the houses in this *country* costs one billion dollars.

This message conveys information on *more houses* and on *more money*, yet it is roughly of the same size as the messages C or A.

Information theory is neither concerned with the content of the message, nor with the amount of information that the message conveys. The only subject of interest is the *size* of the message itself. The message can carry small or large amounts of information, it can convey important or superfluous information; it may have different meanings or values to different persons or it can even be meaningless; it can be exact and reliable information or approximate and dubious information. All that matters is some measure of the size of the message. In Chapter 3, we shall make the last statement more quantitative. We shall also see that the size of the message can be expressed as the number of binary questions one needs to ask in order to retrieve the message. We shall be interested in the information *on* the message and not on the information carried by the message. We shall also see that entropy is nothing but the amount of missing information (MI).

Thus, neither the *entropy*, nor the Shannon measure of MI, are subjective quantities. In fact, no one has claimed that either of these is subjective. The subjectivity of information enters only when we apply the concept of information in its broader sense.

Jaynes pioneered the application of information theory to statistical mechanics. In this approach, the fundamental probabilities of statistical mechanics (see Chapter 5) are obtained by using the principle of maximum entropy. This principle states that the equilibrium distributions are obtained by maximizing the entropy (or the MI) with respect to all other distributions. The same principle can be applied to derive the most non-commital, or the least biased, distribution that is consistent with all the given information. This is a very general principle that has a far more general applicability. See also Chapters 4–6.

In his first paper on this subject, Jaynes wrote (1957)[15]:

> "*Information theory provides a constructive criterion for setting up probability distributions on the basis of partial knowledge and leads to a type of statistical inference which is called the maximum-entropy estimate.*"

And he added:

> "*Henceforth, we will consider the terms 'entropy' and 'uncertainty' as synonyms.*" "*The thermodynamic entropy is identical with information theory — entropy of the probability distribution except for the presence of Boltzmann's constant.*"
>
> "*...we accept the von-Neumann–Shannon expression for entropy, very literally as a measure of the amount of uncertainty represented by the probability distribution; thus entropy becomes the primitive concept... more fundamental than energy.*"

[15]It should be noted that Jaynes' approach was criticized by several authors, e.g., Friedman and Shimony (1971) and Diaz and Shimony (1981).

Regarding the question of the subjectivity of "information," Jaynes writes:

> "*In Euclidean geometry the coordinates of a point are 'subjective' in the sense that they depend on the orientation of the observer's coordinate system; while the distance between two points is 'objective' in the sense that it is independent of the observer's orientation.*"

In 1983 Jaynes writes[16]:

> "*The function H is called entropy, or better, information entropy of the distribution* $\{p_i\}$. *This is an unfortunate terminology which now seems impossible to correct. We must warn at the outset that the major occupational disease of this field is a persistent failure to distinguish between information entropy, which is a property of any probability distribution, and experimental entropy of thermodynamics, which is instead a property of a thermodynamics state as defined, for example, by such observed quantities as pressure volume, temperature, magnetization of some physical system. They should never have been called by the same name; the experimental entropy makes no reference to any probability distribution, and the information entropy makes no reference to thermodynamics. Many textbooks and research papers are fatally flawed by the author's failure to distinguish between these entirely different things, and in consequence proving nonsense theorems.*"
>
> "*The mere fact that the same mathematical expression* $-\sum p_i \log p_i$ *occurs both in statistical mechanics and information theory does not in itself establish any connection between these fields. This can be done only by finding new viewpoints from which thermodynamic entropy and information theory entropy appear as the same concept.*"

And later, Jaynes writes:

> "*It perhaps takes a moment of thought to see that the mere fact that a mathematical expression like*
>
> $$-\sum p_i \log p_i$$
>
> *shows up in two different fields, and that the same inequalities are used in two different fields does not in itself establish any*

[16] Jaynes (1983).

connection at all between the fields. Because, after all e^x, $\cos\theta$, $J_0(z)$ *are expressions that show up in every part of physics and engineering. Every place they show up... nobody interprets this as showing that there is some deep profound connection between, say, bridge building and meson theory.*"

I generally agree with the aforementioned statements by Jaynes. Indeed, the fact that $\cos\theta$ appears in two different fields, say in the propagation of an electromagnetic wave, and in the swinging of a pendulum, does not imply a deep and profound connection between the two fields. However, in both fields, the appearance of $\cos\theta$ indicates that the *phenomena* are *periodic*.

Likewise, the appearance of $-\sum p_i \log p_i$ in two different fields does not indicate that there exists a deep profound connection between the two fields. This is true not only between communication theory and thermodynamics. The measure of information $-\sum p_i \log p_i$ in linguistics makes no reference to the distribution of coins in boxes, or electrons in energy levels, and the measure of information $-\sum p_i \log p_i$ in thermodynamics makes no reference to the frequencies of the alphabet letters in a specific language; the two fields, or even the two subfields (say, in two different languages) are *indeed* different. The information is about different *things*, but all are measures of information nonetheless!

Thus, although it is clear that the types of information discussed in thermodynamics and in communication theory are different, they are all measures of information. Moreover, even in the same field, say in analyzing the information in two languages in linguistics, or even the locational and the momenta information of a thermodynamics system, the *types* of information are different. But in all cases that the expression $-\sum p_i \log p_i$ appears, it conveys the meaning of a measure of information. The information is *about* different things in each case, but conceptually they are measures of information in all cases.

To conclude, the concept of information is very general: it can be applied to many different fields, and in many subfields. The same is true of periodic phenomena which are very general and can occur in many fields. However, the significance of the concept of information and the significance of the periodic phenomena are the same in whatever fields they happen to appear. Specifically, the entropy, or better the thermodynamic entropy, or even better the thermodynamic missing information, is one particular example of the general notion of information.

There is of course a conceptual difficulty in identifying the Boltzmann entropy with the Clausius entropy. However, it has been shown in numerous examples that *changes* in one concept are equivalent to changes in the other, provided that the constant k is chosen as the Boltzmann constant, with the dimensions of energy over absolute temperature (K). Once one accepts the identification of the Clausius entropy with the Boltzmann entropy, then the interpretation of the entropy as uncertainty, information or MI is inevitable.

It is now about 100 years since Gibbs and Boltzmann developed statistical mechanics *based* on probability. It is over fifty years since Shannon laid down the foundation of information theory *based on* probability. Jaynes used the measure of information as defined by Shannon to develop statistical mechanics from the principle of maximum entropy.

I have already mentioned von Neumann's "*disservice*" to science in suggesting the term entropy.[17] My reason for agreeing with this statement is different to the ones expressed by Denbigh. I would simply say that I shall go back to Clausius' choice of the term, and suggest that he should have not used the term entropy in the first place. This term was coined at the time of the pre-atomistic view of thermodynamics. However, once the foundations of statistical mechanics were firmly established on probability and once information theory was established, von Neumann should have suggested to the scientific community *replacing* entropy by information (or by uncertainty or unlikelihood; see Chapter 3). In my view, the suggestion of the term entropy, is not only a disservice, but a corruption of a well-defined, well-interpreted, intuitively appealing and extremely useful and general concept of information with a term that means *nothing*.[18]

---

[17]See preface, page xviii
[18]See preface, page xvi

In an article entitled "How subjective is entropy?", Denbigh discussed the extent of subjectivity of information and the objectivity of entropy.[19]

Indeed there is a valid question regarding the two types of probabilities, subjective and objective. However, once we restrict ourselves to discussing only scientific probabilities, we regard these as objective quantities in the sense that everyone given the same information, i.e., the same conditions, will necessarily reach the same conclusion regarding the probabilities. See more on this in Chapter 2.

Information is defined in terms of probability distribution. Once we agree to deal only with scientific and objective probabilities, the corresponding measure of information becomes objective too.

The best example one can use to define Shannon's measure of information is the game of hiding a coin in M boxes (see Chapter 3). I hide a coin in one of the M boxes and you have to ask binary questions to be able to locate the coin. The missing information is: "Where is the coin?" Clearly, since I know where the coin is while you do not know where it is, I have *more information* on the location of the coin than you have. This information is clearly subjective — your ignorance is larger than mine. However, that kind of information is not the subject matter of information theory. To define the missing information, we have to formulate the problem as follows: "Given that a coin is hidden in one of the M boxes, how many questions does one need to ask to be able to find out where the coin is?" In this formulation, the MI is *built-in* in the problem. It is as objective as the *given* number of boxes, and it is indifferent to the person who hid the coin in the box.

Denbigh and Denbigh (1985), who thoroughly discussed the question of the subjectivity of information, asked:

> "*Whose information or uncertainty is being referred to? Where does the entropy reside?*"

---

[19]Denbigh, K. (1981), "How subjective is entropy," *Chemistry in Britain* **17**, 168–185.

And their conclusion was:

> *"The problem about whose information is measured by H remains obscure."*

In my view, the question of "whose information is being referred to" is not relevant to the quantity H, as defined by Shannon. Regarding the second question: Wherever the *number of boxes* (or the number of states of a thermodynamic system) "resides," there resides also the quantity H. Again, this question of the residency of H is irrelevant to the meaning of H.

Before ending this section on entropy and information, I should mention a nagging problem that has hindered the acceptance of the interpretation of entropy as information. We recall that entropy was *defined* as a quantity of heat divided by temperature. As such, it bears the units of energy divided by the absolute temperature (K) (i.e., Joules over K or J/K, K being the unit of the absolute temperature on the Kelvin scale). These are two tangible, measurable and well-defined concepts. How come "information," which is a dimensionless quantity, a number that has nothing to do with either energy or temperature, could be associated with entropy, a quantity that has been *defined* in terms of energy and temperature? I believe that this is a very valid point of concern which deserves some further pondering. In fact, even Shannon himself recognized that his measure of information becomes identical with entropy only when it is multiplied by the constant k (now known as the Boltzmann constant), which has the units of energy divided by temperature. This, in itself, does not help much in identifying the two apparently very different concepts. I believe there is a deeper reason for the difficulty of identifying entropy with information.

First, note that in the process depicted in Figure 1.8, the change in entropy does involve some quantity of heat (energy) transferred and temperature. But this is only one example of a spontaneous process. Consider the expansion of an ideal gas in Figure 1.3 or the mixing of two ideal gases in Figure 6.6. In both cases, entropy increases. However, in both cases, there is no change in energy, no heat transfer, and no dependence on temperature. If we carry out these two processes for an ideal gas as in an isolated system, then the entropy change will be fixed, independent of the temperature of the system, and obviously no heat is transferred from one body to another. These examples are only indicative that changes in entropy do not *necessarily* involve units of energy and temperature.

Second, the units of entropy (J/K) are not only unnecessary for entropy, but they *should not* be used to express entropy at all. The involvement of energy and temperature in the original definition of the entropy is a historical accident, a relic of the pre-atomistic era of thermodynamics.

Recall that temperature was defined earlier than entropy and earlier than the kinetic theory of heat. Kelvin introduced the absolute scale of temperature in 1854. Maxwell published the molecular distribution of velocities in 1859. This has led to the *identification* of temperature with the mean kinetic energy of the atoms or molecules in the gas. Once the identification of the temperature as a measure of the average kinetic energy of the atoms had been confirmed and accepted,[20] there was no reason to keep the old units of K. One should redefine a new absolute temperature, denote it tentatively as $\overline{T}$, to replace kT. The new temperature $\overline{T}$ would have the units of energy and there should be no need for the Boltzmann constant. The equation for the entropy would be simply $S = \ln W$,[21] and entropy would be rendered dimensionless!

Had the kinetic theory of gases preceded Carnot, Clausius and Kelvin, Clausius would have defined changes of entropy as energy divided by temperature. But now this ratio would have been dimensionless. This will not only simplify Boltzmann's formula for the entropy, but will also facilitate the *identification* of the *thermodynamic* entropy with Shannon's measure of information.

Thus, without entering into the controversy about the question of the subjectivity or objectivity of information, whatever it is, I believe that the entropy is *identical*, both conceptually *and* formally, with Shannon's measure of information. This identification is rendered possible by redefining temperature in units of

---

[20] see footnote 4 on page 7.

[21] see cover design.

energy.[22] This will automatically expunge the Boltzmann constant k_B from the vocabulary of physics. It will simplify the Boltzmann formula for the entropy, and it will remove the stumbling block that has hindered the acceptance of entropy as information for over a hundred years. It is also time to change not only the units of entropy to make it dimensionless,[23] but the term "entropy" altogether. Entropy, as it is now recognized does not convey "transformation," nor "change," nor "turn." It does convey some measure of *information*. Why not replace the term that means "nothing" as Cooper noted, and does not even convey the meaning it was meant to convey when selected by Clausius? Why not replace it with a simple, familiar, meaningful, and precisely defined term?" This will not only remove much of the mystery associated with the unfamiliar word entropy, but will also ease the acceptance of John Wheeler's view of regarding "*the physical world as made of information, with energy and matter as incidentals.*"[24]

Finally, it should be said that even when we identify entropy with information, there is one very important difference between the thermodynamic information (entropy) and Shannon's information, which is used in communications or in any other branch of science. There is a huge difference of the order of magnitudes between the two. The concept of information is, however, the same in whatever field it is being used. We shall discuss in Chapter 4 the passage from a system with a small number of degrees of freedom to thermodynamic systems having a huge number of degrees of freedom.

---

[22] As is effectively done in many fields of physics.

[23] Note that the entropy would still be an extensive quantity, i.e., the entropy of the system would be proportional to the size of the system.

[24] Quoted by Jacob Bekenstein (2003).

Chapter 2

Elements of Probability Theory

2.1 Introduction

Probability theory is a branch of mathematics. Its uses are in all fields of sciences, from physics and chemistry, to biology and sociology, to economics and psychology; in short, it is used everywhere and anytime in our lives.

It is an experimental fact that in many seemingly random events, high regularities are observed. For instance, in throwing a die once one cannot predict the occurrence of any single outcome. However, if one throws the die many times, then one can observe that the relative frequency of occurrence, of say, the outcome 3 is about $\frac{1}{6}$.

Probability theory is a relatively new branch of mathematics. It was developed in the 16th and 17th centuries. The theory emerged mainly from questions about games of chances addressed to mathematicians.

A typical question that is said to have been addressed to Galileo Galilei (1564–1642) was the following:

Suppose that we play with three dice and we are asked to bet on the *sum* of the outcomes of tossing the three dice simultaneously. Clearly, we feel that it would not be wise to bet our chances on the outcome 3, nor on 18; our feeling is correct (in a sense discussed below). The reason is that both 3 and 18 have only one way of occurring; 1:1:1 and 6:6:6 respectively, and we intuitively judge that these events are relatively rare. Clearly, choosing the sum 7 is better. Why? Because there are more *partitions* of the number 7 into three numbers (between 1 and 6), i.e., 7 can be obtained as a result of four possible partitions: 1:1:5, 1:2:4, 1:3:3, 2:2:3. We also *feel* that the larger the sum, the larger the number of partitions, up to a point, roughly in the center between the minimum of 3 to the maximum of 18. But how can we choose between 9 to 10? A simple count shows that both 9 and 10 have the same number of partitions, i.e., the same number of combinations of integers (between 1 and 6), the sum of which is 9 or 10. Here are all the possible partitions:

For 9: 1:2:6, 1:3:5, 1:4:4, 2:2:5, 2:3:4, 3:3:3

For 10: 1:3:6, 1:4:5, 2:2:6, 2:3:5, 2:4:4, 3:3:4

At first glance, we might conclude that since 9 and 10 have the same number of *partitions*, they must have the same chances of winning the game. That conclusion is wrong, however. The correct answer is that 10 has better chances of winning than 9. The reason is that, though the number of partitions is the same for 9 and 10, the total number of outcomes of the three dice that sum up to 9 is a little bit smaller than the total number of outcomes that sum up to 10. In other words, the number of *partitions* is the same, but each partition has different "*weight*," e.g., the outcome 1:4:4 can be realized in three different ways:

1:4:4, 4:1:4, 4:4:1

Since the dice are distinguishable, these three outcomes are different events. Therefore, they contribute the weight 3 to the partition 1:4:4.

This is easily understood if we use three dice having different colors, say blue, red and white, the three possibilities for 1:4:4 are:

blue 1, red 4 and white 4

blue 4, red 1 and white 4

blue 4, red 4 and white 1

When we count all possible partitions and all possible weights, we get the following results:

For 9:	1:2:6,	1:3:5,	1:4:4,	2:2:5,	2:3:4,	3:3:3	
Weights:	6	6	3	3	6	1	(total 25)

For 10:	1:3:6,	1:4:5,	2:2:6,	2:3:5,	2:4:4,	3:3:4,	
Weights:	6	6	3	6	3	3	(total 27)

Thus, the total distinguishable outcome is 25 for the sum of 9, and is 27 for the sum of 10. Therefore, the relative chances of winning with 9 and 10, is 25:27, i.e., favoring the choice of 10. Note that 10 and 11 have the same chances of winning.

But what does it mean that 10 is the best choice and that this is the "correct," winning number? Clearly, I could choose 10 and you could choose 3 and you might win the game. Does our calculation guarantee that if I choose 10, I will always win? Obviously not. So what does the ratio 25:27 mean?

The theory of probability gives us an answer. It does not predict the winning number, and it does not guarantee winning; it only says that if we play this game many times, the probability that the choice of 9 wins is 25/216, whereas the probability of the choice of 10 wins is slightly larger at 27/216 (216 being the total number of possible outcomes, $6^3 = 216$). How many times do we have to play in order to guarantee my winning? On this question, the theory is mute. It only says that in the limit of infinite number of games, the frequency of occurrence of 9 is 25/216, and the frequency of occurrence of 10 is 27/216. But an infinite number of games cannot be realized. So what is the meaning of these probabilities? At this moment, we cannot say anything more than that the ratio 27:25 reflects our degree of *belief* or our degree of confidence that the number 10 is more likely to win than the number 9.[1]

In the aforementioned discussion, we have used the term probability without defining it. In fact, there have been several attempts

[1]The usage of the word "belief" in the context of the definition of probability has led many to deem probability as a subjective concept. This is partially true if we use the term probability colloquially. However, it is not true when probability is used in the sciences. It should be stressed that the probability of an event does not depend on the number of trials; exactly as a mass of a piece of matter does not depend on the number of measurements caried out to determine it.

to *define* the term probability. As it turns out, each definition has its limitations. But more importantly, each definition uses the concept of probability in the very definition, i.e., all definitions are circular. Nowadays, the mathematical theory of probability is founded on an axiomatic basis, much as Eucledean geometry or any other branch of mathematics.

The concept of *probability* is considered to be a primitive concept that cannot be defined in terms of more primitive concepts, much like a point or a line in geometry are considered to be primitive concepts. Thus, even without having a proper definition of the term, probability is a fascinating and extremely useful concept.

Probability theory was developed mainly in the 17th century by Fermat (1601–1665), Pascal (1623–1705), Huyghens (1629–1695) and by J. Bernoulli (1654–1705). The main motivation for developing the mathematical theory of probability was to answer various questions regarding games of chance. A classical example is the problem of how to divide the stakes when the game of dice must be stopped (we shall discuss this problem in Section 2.6). In this chapter, we shall start with the axiomatic approach to probability theory. We shall discuss a few methods of calculating probabilities and a few theorems that are useful in statistical mechanics.

2.2 The Axiomatic Approach

The axiomatic approach to probability was developed mainly by Kolmogorov in the 1930s. It consists of the three elements denoted as $\{\Omega, F, P\}$, which together define the probability *space*. The three elements of the probability space are the following.

2.2.1 The sample space, denoted Ω

This is the set of all possible outcomes of a specific experiment (sometimes referred to as a trial).

Examples: The sample space of all possible outcomes of tossing a coin consists of two elements $\Omega = \{H, T\}$, where H stands for head and T stands for tail. The sample space of throwing a die consists of the six possible outcomes $\Omega = \{1, 2, 3, 4, 5, 6\}$. These are called simple events or "points" in the sample space. In most cases, simple events are equally probable, in which case they are referred to as *elementary events*.

Clearly, we cannot write down the sample space for every experiment. Some consist of an infinite number of elements (e.g., shooting an arrow on a circular board), some cannot even be described (e.g., how the world will look like next year). We shall be interested only in simple spaces where the counting of the outcomes referred to as elementary events is straightforward.

2.2.2 The field of events, denoted F

A compound event, or simply an *event*, is defined as a union, or a sum of elementary events. Examples of events are:

(a) The result of tossing a die is "even"; consists of the elementary events $\{2, 4, 6\}$, i.e., either 2 or 4 or 6 have occurred, or will occur in the experiment of tossing a die.

(b) The result of tossing a die is "larger than or equal to 5"; consists of the elementary events $\{5, 6\}$, i.e., either 5 or 6 have occurred.

In mathematical terms, F consists of all partial sets of the sample space Ω. Note that the event Ω itself belongs to F. Also, the empty event denoted ϕ also belongs to F.

We shall mainly discuss finite sample spaces. We shall also apply some of the results to infinite or even continuous spaces by using arguments of analogy. More rigorous treatment requires the tools of measure theory.

For finite sample space, each partial set of Ω is an event. If there are n elementary events in Ω, then the total number of events in F is 2^n.

This can be seen by straightforward counting[2]:

$\binom{n}{0}$, one event denoted by ϕ (the impossible or the empty event)

[2]The symbol $\binom{m}{n}$ means $\frac{m!}{(m-n)!n!}$. By definition $0! = 1$, hence $\binom{m}{0} = 1$.

$$
\begin{aligned}
n = \binom{n}{1} &\text{ simple (or elementary) events} \\
\binom{n}{2} &\text{ events consisting of two simple events} \\
\binom{n}{3} &\text{ events consisting of three simple events} \\
&\vdots \\
\binom{n}{n} &\text{ one event, consists of all the space } \Omega \text{ (the certain event).}
\end{aligned}
\tag{2.2.1}
$$

Altogether, we have 2^n events, i.e.,

$$\binom{n}{0} + \binom{n}{1} + \binom{n}{2} + \cdots + \binom{n}{n} = (1+1)^n = 2^n. \tag{2.2.2}$$

We have used the Newton Binomial theorem in performing the summation. A more general form of the theorem is (see Appendix A)

$$(x+y)^n = \sum_{i=0}^{n} \binom{n}{i} x^i y^{n-i}. \tag{2.2.3}$$

An alternative method of calculating the total number of events is as follows.

We denote the simple events of Ω by the numbers $(1, 2, 3, \ldots, n)$. This is a vector of n dimensions. Each compound event can also be described as a vector of n dimensions. For instance, the event that consists of the three elementary events (3,5,7) can be written as an n-dimensional vector of the form

$$(No, No, Yes, No, Yes, No, Yes, No, No, \ldots, No),$$

i.e., components one and two are not included, the component three is included, four is not included, five is included, and so on.

A compound event is a partial set of Ω. Therefore, we can describe the compound event simply by referring to the list of simple events that are included in that event. The components of the compound events are Yes or No according to whether or not a specific simple event is included or not. Thus, each event can be uniquely written as an n-dimensional vector consisting of Yes's and No's or "1" and "0." Clearly, since the length of the vector is n, and each component can be either "Yes" or "No," altogether we have 2^n such vectors corresponding to all the possible compound events. In this notation, the impossible event is written as

$$(No, No, \ldots, No)$$

and the certain event as

$$(Yes, Yes, \ldots, Yes).$$

At this stage, we introduce some notation regarding operations between events.

The event $A \cup B$ (or $A + B$) is called the union (or the sum) of the *two events*. This is the event: "either A or B has occurred."

The event $A \cap B$ (or $A \cdot B$) is called the intersection (or the product) of the *two events*. This is the event: " both A and B have occurred."

The complementary event, denoted $\bar{A}$ (or $\Omega - A$), is the event: "A did not occur."

The notation $A \subset B$ means: A is partial to B, or A is included in the event B, i.e., the occurrence of A implies the occurrence of B.

These relations between events are described in Figure 2.1.

2.2.3 The probability function, denoted P

For each event A belonging to F, we assign a number P called the probability of the event A. This number fulfills the following

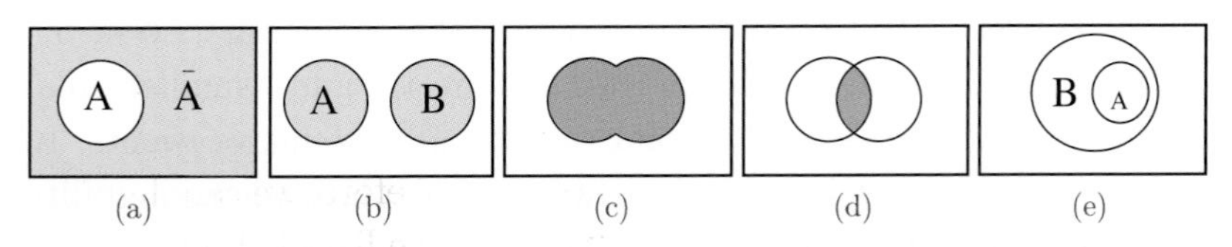

Figure 2.1. Some relations between events (sets): (a) event A and its complementary event $\bar{A}$, (b) disjoint events $A \cap B = \phi$, (c) union of overlapping events $A \cup B$, (d) intersection of overlapping events $A \cap B$, and (e) event A is included in the event B, $A \subset B$.

properties:

$$a: P(\Omega) = 1, \tag{2.2.4}$$

$$b: 0 \le P(A) \le 1, \tag{2.2.5}$$

c: If A and B are disjoint events (or mutually exclusive), then

$$P(A \cup B) = P(A) + P(B). \tag{2.2.6}$$

The first two conditions define the range of numbers for the probability function. The first condition simply means that the event Ω has the largest value of the probability. By definition, we assume that some result has occurred, or will occur, hence, the event Ω is also referred to as the certain event, and is assigned the value of one. The impossible event, denoted ϕ, is assigned the number zero, i.e., $P(\phi) = 0$.

The third condition is intuitively clear. Two events A and B are said to be *disjoint*, or mutually exclusive, when the occurrence of one event excludes the possibility of the occurrence of the other. In mathematical terms, we say that the intersection of the two events $(A \cap B)$ is empty, i.e., there is no simple event that is common to both A and B. As a simple example, consider the two events

$$A = \{\text{the outcome of throwing a die is } \textit{even}\},$$
$$B = \{\text{the outcome of throwing a die is } \textit{odd}\}.$$

Clearly, the events A and B are disjoint; the occurrence of one *excludes* the occurrence of the other. Now, define the event

$$C = \{\text{the outcome of throwing a die is larger than or equal to } 5\}.$$

Clearly, A and C, or B and C are not disjoint. A and C contain the elementary event 6. B and C contain the elementary event 5.

The events, "greater than or equal than 4," and "smaller than or equal to 2," are clearly disjoint or mutually exclusive events. In anticipating the discussion below, we can calculate the probability of the first event $\{4, 5, 6\}$ being $3/6$, and the probability of the second event $\{1, 2\}$ being $2/6$; hence, the combined (or the union) event $\{1, 2, 4, 5, 6\}$ has the probability $5/6$, which is the sum of $2/6$ and $3/6$.

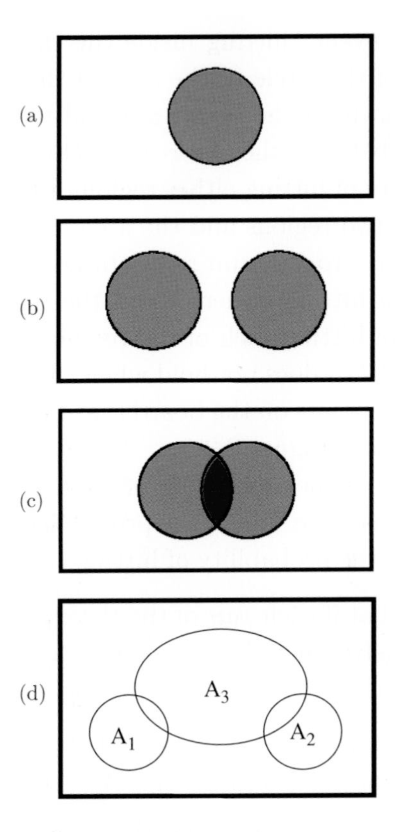

Figure 2.2. Venn diagrams.

If the two events are not disjoint, say "greater than or equal to 4" and "even," then the rule (c) should be modified [this can be proven from the properties listed in (2.2.4) and (2.2.5)]:

$$d: P(A \cup B) = P(A) + P(B) - P(A \cap B). \tag{2.2.6a}$$

A very useful way of visualizing the concept of probability and the sum rule is the Venn diagram (Figures 2.1 and 2.2).

Suppose we throw a dart on a rectangular board having a total area of Ω. We assume that the dart *must* hit some point within the area of the board. We now draw a circle within the board (Figure 2.2a), and ask what the probability of hitting the area within this circle is. We assume, by plain common sense, that the probability of the event "hitting inside the circle," is equal to the ratio of the area of the circle to the area of the entire board.[3]

Two regions drawn on the board are said to be disjoint if there is no overlapping between the two regions (Figure 2.2b). It is clear that the probability of hitting either region or the other is the ratio of the area of the two regions and the area of the whole board.

This leads directly to the sum rules stated in the axioms above. The probability of hitting either one of the regions is the sum of the probabilities of hitting each of the regions.

This sum rule (2.2.6) does not hold when the two regions overlap, i.e., when there are points on the board that belong to both regions, like the case shown in Figure 2.2c.

It is quite clear that the probability of hitting either of the regions A or B is, in this case, the sum of the probabilities of hitting each of the regions, minus the probability of hitting the overlapping region.

Exercise: Show that if each *pair* of the three events A_1, A_2 and A_3 are mutually exclusive, then the three events are mutually exclusive. Show an example that the reverse of this statement is not true.

Solution: Given that $A_1 \cdot A_2 = \phi$, $A_1 \cdot A_3 = \phi$ and $A_2 \cdot A_3 = \phi$, it follows directly from the definition that $A_1 \cdot A_2 \cdot A_3 = \phi$. If, however, it is known that $A_1 \cdot A_2 \cdot A_3 = \phi$, it does not necessarily follow that the events are mutually exclusive in pairs. An example is shown in Figure 2.2d.

On this relatively simple axiomatic foundation, the whole edifice of the mathematical theory of probability has been erected. It is not only extremely useful but is also an essential tool in all the sciences and beyond.

In the axiomatic structure of the theory of probability, the probabilities are said to be *assigned* to each event. These probabilities must subscribe to the three conditions a, b and c. The theory does not *define* probability, nor provide a method of calculating or measuring these probabilities. In fact, there is no way of calculating

[3]We exclude the possibility of hitting a specific point or a line — these have zero probability. We also assume that we throw the dart at *random*, not aiming at any particular area.

probabilities for any general event. It is still a quantity that measures our degree or extent of belief of the occurrence of certain events. As such, it is a highly subjective quantity. However, for some simple experiments, say tossing a coin or throwing a die, we have some very useful methods of calculating probabilities. They have their limitations and they apply to "ideal" cases, yet these probabilities turn out to be extremely useful. What is more important, since these are based on common sense reasoning, we should *all* agree that these are the "correct" probabilities, i.e., these probabilities turn from subjective quantities into objective quantities.[4] They "belong" to the events as much as mass belongs to a piece of matter. We shall describe two very useful "definitions" that have been suggested and commonly used for this concept.

2.3 The Classical Definition

This is sometimes referred to as the *a priori* definition.[5] Let $N(total)$ be the *total* number of outcomes of a specified experiment, e.g., for throwing a die $N(total)$ is six, i.e., the six outcomes (or six elementary events) of this experiment. We denote by $N(event)$, the number of outcomes (i.e., elementary events) that are included in the event in which we are interested. Thus, the probability of the "event," in which we are interested, is *defined* as the ratio

$$P(event) = \frac{N(event)}{N(total)}. \tag{2.3.7}$$

[4] There exists a voluminous literature discussing the definition and the meaning of the concept of probability. In this book we shall assume that the reader has an intuitive understanding of the meaning of probability. Denbigh and Denbigh (1985) make a distinction between a weak and a strong sense of objectivity. The former refers to objects that "can be publically agreed"; the latter refers to an object that has "a reality quite independent of man's presence in the world." Personally, I believe that even a weaker sense is sufficient for objectivity of probability. It suffices that all those who use the theory agree upon the values of probability assigned to the events.

[5] *A priori* here is only in the sense of contrasting the *a posteriori* sense of probability. The former refers to the case that no experiment should be carried out to determine the probability.

We have actually used this intuitively appealing definition when calculating the probability of the event "greater than or equal to 4." The total number of elementary events (outcomes) of tossing a die is $N(total) = 6$. The number of elementary events included in the event "greater than or equal to 4" is $N(event) = 3$, hence, the probability of this event is 3/6 or 1/2, which we all agree is the "correct" probability.

However, care must be used in applying (2.2.7) as a *definition* of probability. First, not every event can be "decomposed" into elementary events, e.g., the event "tomorrow it will start raining at ten o'clock." But more importantly, the above formula presumes that each elementary event has the same likelihood of occurrence. In other words, each elementary event is presumed to have the same *probability*; 1/6 in the case of a die. But how do we know that? We have given a formula for calculating the probability of an event based on the knowledge of the probabilities of each of the elementary events. This is the reason why the classical definition cannot be used as a bona fide *definition* of the concept of probability; it is a circular definition. In spite of this, the "definition" (or rather the method of calculating probabilities) is extremely useful. Clearly, it is based on our *belief* that each elementary event has equal probability: 1/6. Why do we believe in that assertion? The best we can do is to invoke the argument of symmetry. Since all faces of a die are presumed equivalent, their probabilities must be equal. This conclusion should be universally agreed upon, as much as the axiomatic assertion that two straight lines will intersect at, at most, a single point. Thus, though the probability of the event "it will rain tomorrow" is highly subjective, the probability that the outcome of the event "even" in throwing a die is 1/2, should be agreed on by anyone who intends to use probabilistic reasoning, as much as anyone who adopts the axioms of geometry, if he or she intends to use geometrical reasoning.

Feller (1957), in his classical text book on probability writes:

> "*Probabilities play for us the same role as masses in mechanics. The motion of the planetary system can be discussed without knowledge of the individual mass and without contemplating methods for their actual measurement.*"

Feller also refrained from discussing probabilities that one deemed to be a matter of "judgments," and limited himself to discussing what he calls "*physical*," or "*statistical probability*."

> "*In a rough way we may characterize this concept by saying that our probabilities do not refer to judgments, but to possible outcomes of a conceptual concept*" — "*There is no place in our system for speculations concerning the probability that the sun will rise tomorrow.*"

Similarly, Gnedenko (1962) asserts that probability expresses a certain objective property of the phenomenon that is studied, and adds: "It should be perfectly clear that a purely subjective definition of mathematical probability is quite untenable."

As in geometry, all of the probabilities as well as all theorems derived from the axioms strictly apply to ideal cases: a "fair" die, or a "fair" coin. There is no definition of what a fair die is. It is as "ideal" a concept as an ideal or Platonic circle or cube. All *real* dice, as all cubes or spheres, are only approximate replicas of ideal Platonic objects. In practice, if we do not have any reason to suspect that a die is not homogenous or symmetrical, we can assume that it is ideal.

In spite of this limitation, this procedure of calculating probabilities is very useful in applications. One of the basic postulates of statistical mechanics is that each of the microstates constituting a macrostate of a thermodynamic system has the same probability. Again, one cannot prove that postulate much less than one can "prove" the assertion that each outcome of throwing a die is $1/6$. We shall further discuss this postulate in Chapter 5. This brings us to the second "definition," or better, the second procedure of calculating probabilities.

2.4 The Relative Frequency Definition

This definition is referred to as the "*a posteriori*" or the "experimental" definition, since it is based on actual counting of the relative frequency of occurrence of events.

The simplest example would be tossing a coin. There are two possible outcomes; head (H) or tail (T). We exclude the rare events that the coin will fall exactly perpendicular to the floor, or break into pieces during the experiment, or even disappear from sight so that the outcome is undeterminable.

We proceed to toss N times. The frequency of occurrence of heads is recorded. This is a well-defined and feasible experiment. If $n(H)$ is the number of heads that occurred in $N(total)$ trials, then the frequency of occurrence of a head is $n(H)/N(total)$. The *probability* of occurrence of the event "H" is *defined* as the limit of the frequency $n(H)/N(total)$ when $N(total)$ tends to infinity. The frequency definition is thus

$$\Pr(H) = \lim_{N(total)\to\infty} \frac{n(H)}{N(total)}. \tag{2.2.8}$$

This limit may be interpreted in two different ways. Either one performs a *sequence* of experiments in time, and measures the limit of the relative frequency when the number of experiments is infinity, or throws infinite coins at once and counts the fraction of coins which turned out to be H.

Clearly, such a definition is not practical. First, because we cannot perform infinite trials. Second, even if we could, who could guarantee that such a limit exists at all? Hence, we only have to imagine what this limit will be. We believe that such a limit exists, and that it is unique, but, in fact, we cannot prove it.

In practice, we use this definition for very large N. Why? Because we believe that if N is large enough and if the coin is fair, then there is a *high probability* that the relative frequency of occurrence of head will be 1/2. We see that we have used again the *concept* of probability in the very definition of probability.

This method could be used to "prove" that the probability of each outcome of throwing a die is 1/6. Simply repeat the experiment many times and count how many times the outcome 4 (or any other outcome) has occurred. The relative frequency can serve as "proof" of the probability of that event. This is not a mathematical proof. As in any other branch of science, this reasoning rests on the our experience that if N is large enough, we should get the frequency of one out of six experiments. We cannot tell how many experiments are sufficient to determine the probability of an event. So, how do we estimate the probability of any event? The only answer we can give is that we believe in our *common sense*. We use common sense to judge that because of the symmetry of the die (i.e., all faces are equivalent), there must be equal probability for the outcome of any specific face. Likewise, when we record the number of times an H (or a T) occurs in many throws of a coin and see that the frequency of occurrence tends to some constant, we conclude that that constant is the probability of the event H.

It should be noted, however, that the actual decision of which events are the elementary events is not always simple or possible. We present one famous example to demonstrate this. Suppose we have N particles (say electrons) and M boxes (say energy levels). There are different ways of distributing the N particles in the M boxes. If we have no other information, we might assume that all possible configurations have equal likelihood. Figure 2.3 shows all the configurations for two particles ($N = 2$) and four boxes ($M = 4$).

We can assign equal probabilities to all of these 16 configurations. This is referred to as the "classical statistics" — not to be confused with the "classical" definition of probability. This would be true for coins or dice distributed in boxes. It would not work for molecular particles distributed in energy levels.

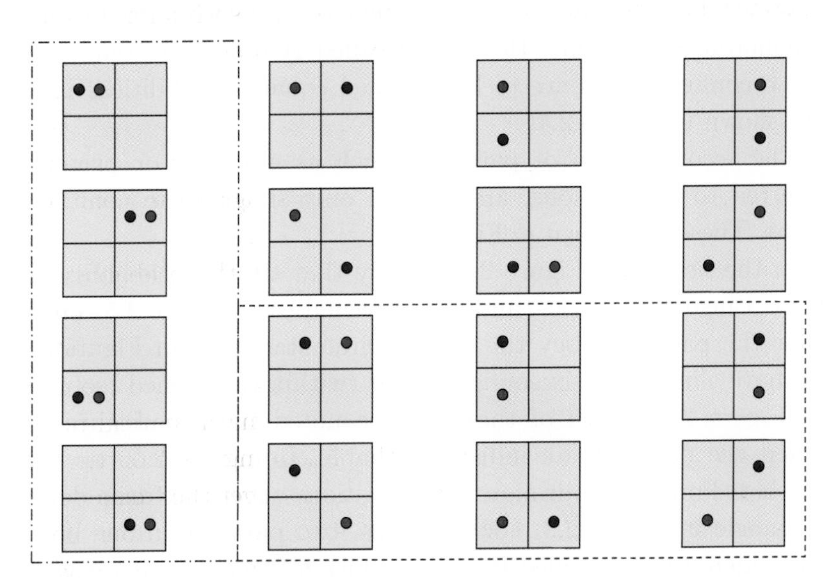

Figure 2.3. All possible configurations for two different particles in four boxes.

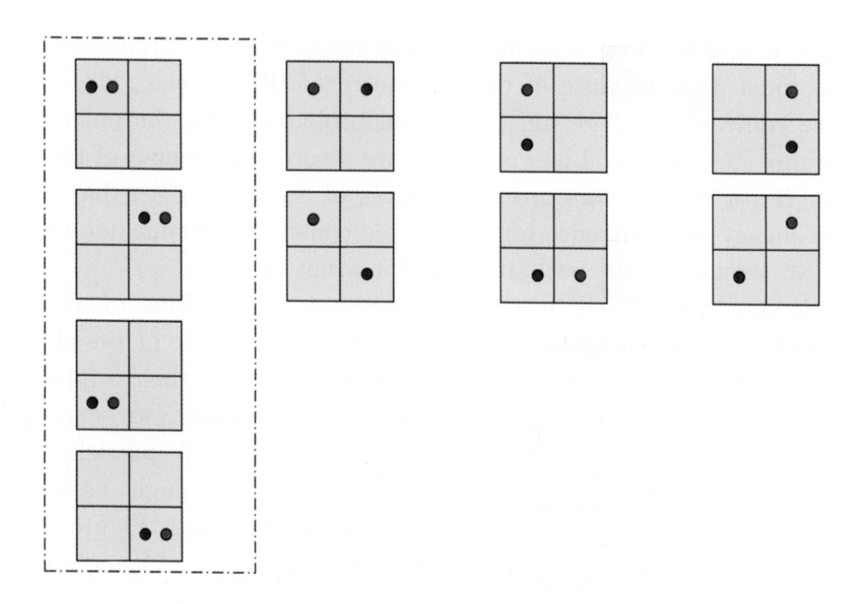

Figure 2.4. Bose–Einstein configurations.

It turns out that Nature imposes some restrictions as to which configurations are to be counted as elementary events. Nature also tells us that there are two ways of counting the elementary events depending on the type of particles. For one type of particles (such as photons or helium, ^{4}He atoms) called Bosons, only ten out of these configurations are to be assigned equal probabilities. These are shown in Figure 2.4.

The second group of particles (such as electrons or protons), referred to as Fermions, are allowed only six of these configurations. These are shown in Figure 2.5.

In the first case, Figure 2.4, we say that the particles obey the Bose–Einstein statistics, and in the second case, Figure 2.5, we say that the particles obey the Fermi–Dirac statistics. In Figure 2.4, we have eliminated six configurations (within the dashed rectangle in Figure 2.3). Each of these were counted twice in Figure 2.3, when the particles are indistinguishable. In Figure 2.5, we have further eliminated four more configurations within the dash-dotted rectangle in Figure 2.3. For Fermions, two particles in one box is forbidden. (This is called Pauli's Exclusion Principle). It turns out

event A given that an event B has occurred. We write this as $\Pr\{A/B\}$ (read: probability of A given the occurrence of B),[7] and define it by

$$\Pr(A/B) = \Pr(A \cdot B)/\Pr(B). \tag{2.5.2}$$

Clearly, if the two events are independent, then the occurrence of B has no effect on the probability of the occurrence of A. Hence, from (2.5.1) and (2.5.2), we get

$$\Pr(A/B) = \Pr(A). \tag{2.5.3}$$

We can define the correlation between the two events as[8]

$$g(A, B) = \frac{\Pr(A \cdot B)}{\Pr(A)\Pr(B)}. \tag{2.5.4}$$

We say that the two events are positively correlated when $g(A, B) > 1$, i.e., the occurrence of one event enhances or increases the probability of the second event. We say that the two events are negatively correlated when $g(A, B) < 1$, and that they are uncorrelated (sometimes referred to as indifferent) when $g(A, B) = 1$.

As an example, consider the following events:

$A = \{$The outcome of throwing a die is "4"$\}$,

$B = \{$The outcome of throwing a die is "even"$\}$ (i.e., it is one of the following: 2, 4, 6),

$C = \{$The outcome of throwing a die is "odd"$\}$ (i.e., it is one of the following: 1, 3, 5). (2.5.5)

We can calculate the following two conditional probabilities:

$$\Pr\{\mathit{of}\ A/\ \mathit{given}\ B\} = \frac{1}{3} > \Pr\{\mathit{of}\ A\} = \frac{1}{6}, \tag{2.5.6}$$

$$\Pr\{\mathit{of}\ A/\ \mathit{given}\ C\} = 0 < \Pr\{\mathit{of}\ A\} = \frac{1}{6}. \tag{2.5.7}$$

---

[7] This definition is valid for $\Pr(B) \neq 0$. Note also that $\Pr(B)$ is sometimes interpreted as the probability that the *proposition* B is true, and $\Pr(A/B)$ as the probability that the A is true given the evidence B.

[8] In the theory of probability, correlation is normally defined for a *random variable* (and its values range between -1 and $+1$). Here, we define the correlation differently; the values of g range from zero to infinity (see Section 2.7).

In the first example, the knowledge that B has occurred *increases* the probability of the occurrence of A. Without that knowledge, the probability of A is 1/6 (one out of six possibilities). Given the occurrence of B, the probability of A becomes larger: 1/3 (one of the three possibilities). But *given* that C has occurred, the probability of A becomes zero, i.e., *smaller* than the probability of A without that knowledge.

Exercise: It is known that Mrs. A has two children. You meet her in the street walking with a boy. She introduces the boy as her child. What is the probability that she has two boys?

Solution: The sample space contains four elementary events

$$BB,\ BG,\ GB,\ GG,$$

B for boy and G for girl (the order is important). These events are equally probable. We are given that there is at *least* one boy (we do not know if this is the first or the second). Therefore, the event GG is excluded. We are left with three possibilities BB, BG and GB. The probability of having two boys is one of these three equally likely possibilities, hence the probability is 1/3. In terms of conditional probability

$$\begin{aligned}\Pr(BB/\ \textit{at least one boy}) &= \frac{\Pr(BB \cdot \textit{at least one boy})}{\Pr(\textit{at least one boy})}\\ &= \frac{\Pr[BB \cap (BB \cup BG \cup GB)]}{\Pr[BB \cup BG \cup GB]} = \frac{1/4}{3/4} = \frac{1}{3}.\end{aligned}$$

Exercise: It is known that Mrs. A has three children. She is seen with one of her boys. Calculate the probabilities:

(1) that she has three boys,
(2) that she has one boy and two girls,
(3) that she has two boys and one girl.

Answers: (1) 1/7, (2) 3/7, (3) 3/7.

It is important to distinguish between *disjoint* (i.e., mutually exclusive) and *independent* events. Disjoint events are events that are mutually exclusive; the occurrence of one excludes the occurrence of the second. The events "even" and "5" are disjoint. In terms of Venn diagrams, two regions that are non-overlapping are

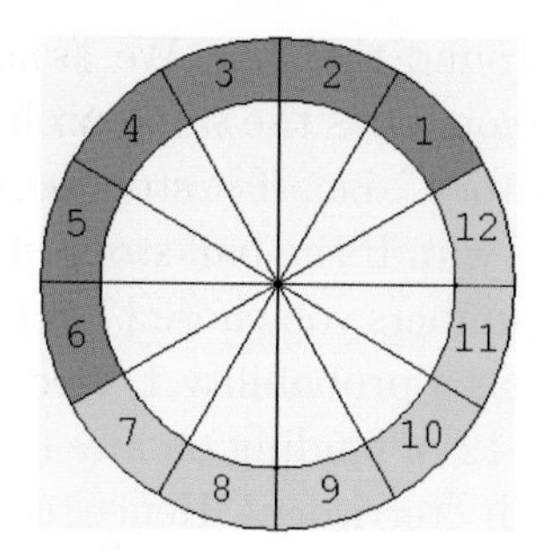

Figure 2.6. A roulette with 12 outcomes, with two regions A (blue) and B (yellow); see Section 2.5.

disjoint. If the dart hits one region, say A, in Figure 2.2b, then we know for sure that it did not hit the second region B.

Disjoint events are properties of the events themselves (i.e., the two events have no common elementary event). Independence between events is not defined in terms of the elementary events that are contained in the two events but is *defined* in terms of their probabilities. If the two events are disjoint, then they are strongly *dependent*. In the above example, we say that the two events are *negatively* correlated. In other words, the conditional probability of hitting one circle, given that the dart hit the other circle, is zero (which is smaller than the "unconditional" probability of hitting one circle).

If the two regions A and B overlap (i.e., they are not disjoint), then the two events could be either dependent or independent. In fact, the two events could either be positively or negatively correlated. The following two examples illustrate the relation between dependence and the extent of overlapping. The reader is urged to do the exercise carefully. We first present a simple example with a finite number of outcomes.

Let us consider the following case. In a roulette, there are altogether 12 numbers $\{1, 2, 3, 4, 5, 6, 7, 8, 9, 10, 11, 12\}$.

Each of us chooses a sequence of six consecutive numbers; say I choose the sequence

$$A = \{1, 2, 3, 4, 5, 6\},$$

and you choose the sequence

$$B = \{7, 8, 9, 10, 11, 12\}.$$

The ball is rolled around the ring. We assume that the roulette is "fair," i.e., each outcome has the same probability of 1/12. If the ball stops in my "territory," i.e., if it stops at any of the numbers I chose $\{1, 2, 3, 4, 5, 6\}$, I win. If the ball stops in your territory, i.e., if it stops at any of the numbers you chose $\{7, 8, 9, 10, 11, 12\}$, you win.

Clearly, each of us has a probability 1/2 of winning. The ball has equal probability of 1/12 of landing on any number, and each of us has six numbers in each "territory." Hence, each of us has the same chance of winning.

Now, suppose we run this game and you are told that I won. What is the probability that you will win if you chose B? Clearly, $\Pr\{B/A\} = 0 < 1/2$, i.e., the conditional probability of B given A is zero, which is *smaller* than the unconditional probability; $\Pr\{B\} = 1/2$.

Exercise: Calculate the following conditional probabilities. In each example, my choice of the sequence $A = \{1, \ldots, 6\}$ is *fixed.* Calculate the conditional probabilities for the following *different* choices of your sequence:

$$\Pr\{7, 8, 9, 10, 11, 12/A\}, \Pr\{6, 7, 8, 9, 10, 11/A\},$$
$$\Pr\{5, 6, 7, 8, 9, 10/A\}, \Pr\{4, 5, 6, 7, 8, 9/A\},$$
$$\Pr\{3, 4, 5, 6, 7, 8/A\}, \Pr\{2, 3, 4, 5, 6, 7/A\}, \Pr\{1, 2, 3, 4, 5, 6/A\}.$$

Note that in this game, both of us can win, or lose.

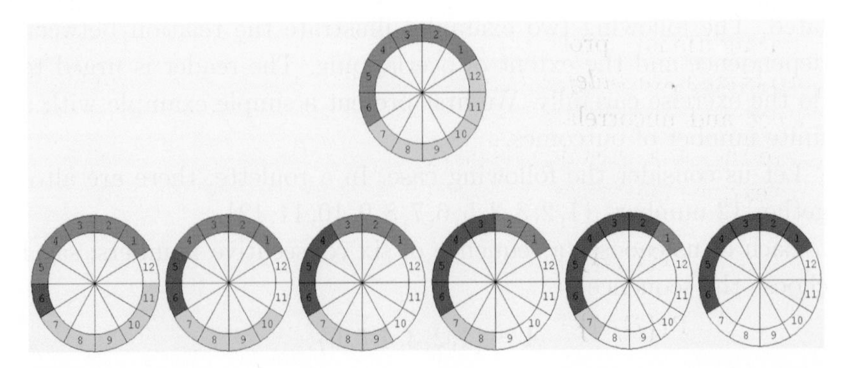

Figure 2.7. Different choices of the event B. The intersection of B and A is shown in green.

Solution: In all of these problems, my choice of a sequence is fixed: $\{1, 2, 3, 4, 5, 6\}$ and it has the probability of $1/2$ of winning. If you choose the *disjoint* event $\{7, 8, 9, 10, 11, 12\}$, the conditional probability is

$$\Pr\{B/A\} = \Pr\{7, 8, 9, 10, 11, 12/A\} = 0,$$

since knowing that A occurs excludes the occurrence of B. This is the case of extreme negative correlation. In the first example of choosing an overlapping sequence, we have

$$\Pr\{B/A\} = \Pr\{6, 7, 8, 9, 10, 11, /A\} = \frac{1}{6} < \frac{1}{2}.$$

Knowing that A occurred means that your winning is possible only if the ball landed on "6," and hence the conditional probability is $1/6$, which is smaller than $\Pr\{B\} = 1/2$, i.e., there is negative correlation.

Similarly,

$$\Pr\{B/A\} = \Pr\{5, 6, 7, 8, 9, 10/A\} = \frac{2}{6} < \frac{1}{2}.$$

Here, "given A," you win only if the ball lands on either "5" or "6," and hence the conditional probability is $2/6$, still smaller than $\Pr\{B\} = 1/2$, i.e., negative correlation.

In the third case,

$$\Pr\{B/A\} = \Pr\{4, 5, 6, 7, 8, 9/A\} = \frac{3}{6} = \frac{1}{2}.$$

Here, the conditional probability is $1/2$, exactly the same as the "unconditional" probability $\Pr(B) = 1/2$, which means that the two events are *independent*, or uncorrelated. (The terms independence and uncorrelated are different for random variables. They have the same meaning when applied to events; see Section 2.7.)

Next, we have three cases of positive correlations:

$$\Pr\{B/A\} = \Pr\{3, 4, 5, 6, 7, 8/A\} = \frac{4}{6} > \frac{1}{2},$$

$$\Pr\{B/A\} = \Pr\{2, 3, 4, 5, 6, 7/A\} = \frac{5}{6} > \frac{1}{2},$$

$$\Pr\{B/A\} = \Pr\{1, 2, 3, 4, 5, 6/A\} = \frac{6}{6} = 1 > \frac{1}{2}.$$

In the last example, knowing that A occurs makes the occurrence of B certain. In these examples, we have seen that overlapping events can be either positively correlated, negatively correlated, or non-correlated.

Note how the correlation changes from extreme negative ("given A" *certainly* excludes your winning in the first example), to extreme positive ("given A" assures your winning in the last example). At some intermediate stage, there is a choice of a sequence that is indifferent to the information "given A."

The second example of negative and positive correlations is shown in Figure 2.8 for a continuous sample space.

Consider the two rectangles A and B having the same area.

Since A and B have the same area, the probability of hitting A is equal to the probability of hitting B. Let us denote that by $\Pr\{A\} = \Pr\{B\} = p$, where p is the ratio of the area of A to the total area of the board, say, $p = 1/10$ in this illustration.

When the two rectangles are separated, i.e., there is no overlapping area, we have

$$\Pr\{A/B\} = 0.$$

This is the case of an extreme *negative* correlation. Given that the dart hit B, the probability that it hit A is zero (i.e., smaller than p).

The other extreme case is when A and B become congruent, i.e., the overlapping area is total. In this case, we have

$$\Pr\{A/B\} = 1.$$

This is the case of an extreme *positive* correlation. Given that the dart hit B, the probability that it hit A is one (i.e., larger than p).

When we move B towards A, the correlation changes continuously from maximal *negative* correlation (non-overlapping) to *maximal* positive correlation (total overlapping). In between, there is a point when the two events A and B are independent. Assume for simplicity that the total area of the board is unity, then the probability of hitting A is simply the area of A. The probability of hitting B is the area of B. Let $p = \Pr\{A\} = \Pr\{B\}$. The conditional probability is calculated by $\Pr\{A/B\} = \Pr\{A \cdot B\}/\Pr\{B\}$, where $\Pr\{A \cdot B\}$ is the

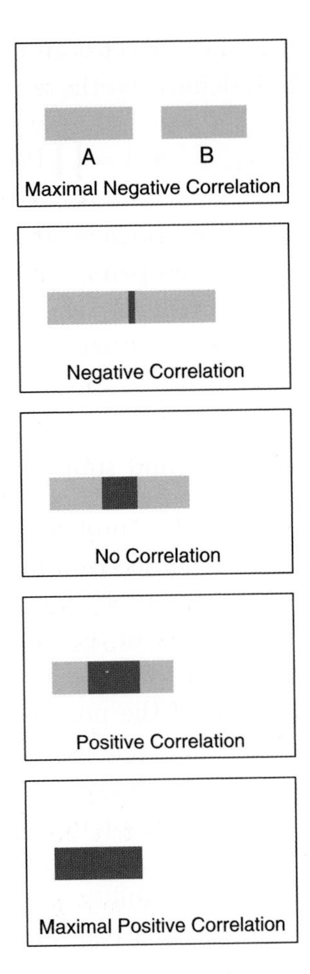

Figure 2.8. Various degrees of overlapping between two rectangular regions, and their corresponding extent of correlations.

probability of hitting both A and B. Denote this probability by x. The correlation between the two events, "hitting A," and "hitting B," is defined as $g = \Pr\{A \cdot B\} / \Pr\{A\} \Pr\{B\} = x/p^2$, when the overlapping area is $x = 0$, $g = 0$ (i.e., extreme negative correlation). When the overlapping area is $x = p$, we have $g = p/p^2 = 1/p$ or $\Pr\{A/B\} = p/p = 1$ (i.e., extreme positive correlation). When $x = p^2$, we have $g = p^2/p^2 = 1$ or $\Pr\{A/B\} = \Pr\{A\}$; in this case, there is no correlation between the two events, i.e., the two events are independent.

Finally, we note that the concept of independence between n events $A_1, A_2, \ldots, A_n$ is defined by the requirement that

$$\Pr(A_1, A_2, \ldots, A_n) = \prod_{i=1}^{n} \Pr(A_i).$$

It should be noted that independence between the n events does not imply independence between pairs, triplets, etc. For example, independence between three events A_1, A_2, A_3 does not imply independence between pairs and *vice versa*. An example is shown in Appendix C.

2.5.1 Conditional probability and subjective probability

There is a tendency to refer to "probability" as *objective*, and to *conditional* probability as subjective. These assertions are, in general, not true. First, note that probability is always *conditional.* When we say that the probability of the outcome "4" of throwing a die is 1/6, we actually mean that the *conditional probability* of the outcome 4 *given* that one of the possible outcomes 1,2,3,4,5,6 has occurred, or will occur,[9] and that the die is fair and that we threw it at random, and any other information that is relevant. We usually suppress this *given* information in our notation and refer to it as the unconditional probability. This unconditional probability is considered as an objective probability.

Now, let us consider the following two pairs of examples:

O_1: The conditional probability of an outcome "4," given that Jacob *knows* that the outcome is "even," is 1/3.

O_2: The conditional probability of an outcome "4," given that Abraham *knows* that the outcome is "odd," is zero.

S_1: The conditional probability that the "defendant is guilty," given that he was seen by the police at the scene of the crime is 0.9.

S_2: The conditional probability that the "defendant is guilty," given that he was seen by at least five persons in another city at the time of the crime's commission is nearly zero.

[9] When we ask for $\Pr(A)$, we actually mean $\Pr(A/\Omega)$, i.e., that one of the events in Ω has occurred.

In all of the aforementioned examples, there is a tendency (such statements are sometimes made in textbooks) to refer to *conditional probability* as a *subjective* probability. The reason is that, in all the abovementioned examples, we involved *personal* knowledge of the conditions. Therefore, we judge that it is highly subjective. However, that is not so. The two probabilities, denoted O_1 and O_2, are objective probabilities. The fact that we mention the names of the persons who are knowledgeable of the conditions does not make the conditional probability subjective. We could make the same statement as in O_1, but with Rachel instead of Jacob. The conditional probability of an outcome "4," given that Rachel knows that the outcome is even, is 1/3. The result is the same. The apparent subjectivity of this statement is a result of the involvement of the *name* of the person who "knows"[10] the condition. A better way of phrasing O_1 is: the conditional probability of an outcome "4," given that *we* know that the outcome is even, is 1/3, or even better; the conditional probability of an outcome "4," *given* that the outcome is "even," is 1/3.

In the last two statements, it is clear that the fact that Jacob, Rachel, or anyone of us *knows* the condition does not have any effect on the conditional probability. In the last statement, we made the condition completely impersonal. Thus, we can conclude that the *given condition* does not, in itself, convert an objective (unconditional) probability into a subjective probability.

To the best of my knowledge, the probabilities used in all cases in the sciences are objective. The reason is that the knowledge of the probabilities is usually explicitly or implicitly *given*. This is in accord with Laplace's statement: "When probability is unknown, all possible values of probability between zero to one should be considered equally likely."

There is a general agreement that there are essentially two distinct types of probabilities. One is referred to as the judgmental probability which is highly subjective, the two examples S_1 and

[10]This kind of argument is often used in connection with the use of information theory in the interpretation of entropy. "Information" is associated with "knowledge" and knowledge is deemed to be subjective. We shall discuss this in the next chapter.

S_2 fall into this category. The second is the physical or scientific and is considered as objective probability. Here, we shall use only the scientific probability. Scientists might disagree on the *meaning* of probabilities, but when probabilities are used, they all agree on their numerical values. The reason is that in all cases, the "correct" answer to a probability question is the answer you either already know, or at least you know a method of how to calculate it.

In using probability in all the sciences, we always presume that the probabilities are given either explicitly or implicitly by a given recipe on how to calculate these probabilities. Sometimes, these are very easy to calculate, sometimes extremely difficult, but you always assume that they are "there" in the event as much as mass is attached to any piece of matter.

Note also that at the most fundamental level, probability is based on our *belief*. Whether it is a belief that the six outcomes of a die are equally likely, or that the laws of quantum mechanics are correct and will be correct in the future, or that the tables of recorded statistics are reliable and that the same statistics will be maintained. If any of the conditions, or the *given* information, are changed, then the method of calculating probabilities must also be changed accordingly.

We have already quoted Callen (1983) on page 16. We repeat the beginning of that quotation here:

> "*The concept of probability has two distinct interpretations in common usage. 'Objective probability' refers to a* ***frequency****, or a* ***fractional occurrence****; the assertion that 'the probability of newborn infants being male is slightly less than one half' is a statement about census data. 'Subjective probability' is a measure of* ***expectation based on less than optimum information****.*"

As I have explained in Chapter 1 (see page 16), my views differ from Callen's in a fundamental sense. Both examples given by Callen could be subjective or objective depending on the *given* amount of information or on the given relevant knowledge.

An extra-terrestial visitor who has no information on the recorded gender of newborn infants would have no idea what the probabilities for a male or female are, and his assignment of probabilities would be totally subjective.[11]

On the other hand, *given* the *same* information and the same knowledge including the frequencies of boys and girls, including the reliability of all the statistical medical records, the assignment of probabilities will inevitably be *objective*.

Let us go back to the question regarding the gender of a new born, and reformulate the questions with an increasing amount of *given* information.

(1) What is the probability that the newborn is a boy (given only that a child was born, this condition is usually suppressed)?
(2) What is the probability that the newborn is a boy, given the census data on the ratio of boys and girls is about 1:1, and assuming that these statistics are reliable and will also be maintained in the future?
(3) What is the probability that the newborn is a boy given the information as in (2), and in addition, it is known that the ultrasound test shows a boy, and that this test is quite reliable, and it is known that in 9 out of 10 cases, the predictions are correct.
(4) What is the probability that the newborn is a boy given the information as in (2) and (3), and in addition, the level of some maternal hormones indicates that it is a boy, and that this is a very sensitive test, and that 10 out of 10 predictions based on this test are correct.

Clearly, the answer to question (1) can be anything. If nothing is known on the census data, we cannot give any "objective" estimate of the probability. This is a highly subjective probability though it is formulated as an unconditional probability (in the sense that the only information given is that the experiment has been performed, i.e., a child is born).

The answer to question (2) is $1/2$, to (3) is $9/10$ and to (4) is 1. These answers are objective in the sense that everyone who is given this information will necessarily give the same answer, again in

[11] The same is true for an extra-terrestial visitor who has never seen a die or a coin and is asked about the probability of a specific outcome of a die or a coin.

contrast to Callen's statement. The fact that the question is cast in the form of a conditional probability does not make the probability subjective. What makes a probability estimate subjective is when we have no sufficient information to determine the probabilities, or when different subjects have different information. This conclusion is far more general and also applies for dice or coins.

Thus, an extra-terrestial visitor who never saw or heard of tossing a coin would not know the answer to the ("unconditional") probability of outcome H. Any answer given is necessarily subjective. However, if he or she knows that the coin is fair, and knows that many experiments carried out by many gamblers show that the frequency ratio of H and T is nearly one, then the extra-terrestrial could correctly give the answer about the probability of H and T.

2.5.2 Conditional probability and cause and effect

The "condition" in the conditional probability of an event may or may not be the cause of the event. Consider the following two examples:

(1) The conditional probability that the patient will die of lung cancer, given that he or she has been smoking for many years is 0.9.
(2) The conditional probability that the patient is a heavy smoker given that he or she has lung cancer is 0.9.

Clearly, the information given in the first case is the *cause* (or the very probable cause) of the occurrence of lung cancer. In the second example, the information that is given — that the patient has cancer — certainly cannot be the *cause* for being a heavy smoker. The patient could have started to smoke at age 20, far earlier, in time than his cancer developed.

Although the two examples given above are clear, there are cases where conditional probability is confused with causality. We perceive causes as preceding the effect in the time axis. Similarly, condition in conditional probability is conceived as occurring earlier in the time axis.

Consider the following simple and very illustrative example that was studied in great detail (Falk 1979). You can view it as a simple exercise in calculating conditional probabilities. However, I believe this example has more to it. It demonstrates how we intuitively associate conditional probability with the arrow of time, confusing causality with conditional probabilistic argument.

The problem is very simple. An urn contains four balls, two whites and two blacks. The balls are well mixed and we draw one ball, blindfolded.

First we ask: What is the probability of the event "White ball on first draw"? The answer is immediate: 1/2. There are four equally probable outcomes; two of them are consistent with "white ball" event, and hence the probability of the event is $2/4 = 1/2$.

Next we ask: What is the conditional probability of drawing a white ball on a *second* draw, given that in the first draw we drew a white ball (the first ball is not returned to the urn). We write this conditional probability as $\Pr\{White_2/White_1\}$. The calculation is very simple. We know that a white ball was drawn on the first trial and was not returned. After the first draw, three balls are left; two blacks and one white. The probability of drawing a white ball is simply 1/3. We write this conditional probability as

$$\Pr\{White_2/White_1\} = \frac{1}{3}.$$

Now, the more tricky question: What is the probability that we *drew* a white ball in the *first* draw, given that the second draw was white? Symbolically, we ask for

$$\Pr\{White_1/White_2\} = ?.$$

This is a baffling question. How can an event in the "present" (white ball at the *second* draw), affect the event in the "past" (white drawn in the *first* trial)?

These questions were actually asked in a classroom. The students easily and effortlessly answered the question about $\Pr\{White_2/White_1\}$, arguing that drawing a white ball in the first draw has *caused* a change in the urn, and therefore has influenced the probability of drawing a second white ball.

However, asking about $\Pr\{White_1/White_2\}$ caused uproar in the class. Some claimed that this question is meaningless, arguing that an event in the present cannot affect the probability of an event in the past. Some argued that since the event in the present cannot affect the probability of the event in the past, the answer to the question is 1/2. They were wrong. The answer is 1/3. We shall revert to this problem and its solution after presenting Bayes theorem in Section 2.6.

The distinction between causality and conditional probability is important. Perhaps, we should add one characteristic property of causality that is not shared by conditional probability. Causality is transitive. This means that if A causes B, and B causes C, then A causes C, symbolically: if $A \Rightarrow B$ and $B \Rightarrow C$, then $A \Rightarrow C$. A simple example: if smoking causes cancer, and cancer causes death, then smoking causes death.

Conditional probability might or might not be transitive. We have already distinguished positive correlation (or supportive correlation) and negative correlation (counter or anti-supportive).[12]

If A supports B, i.e., the probability of the occurrence of B *given* A is larger than the probability of the occurrence of B, i.e., $\Pr\{B/A\} > \Pr\{B\}$; and if B supports C, (i.e., $\Pr\{C/B\} > \Pr\{C\}$ then in general, it does not follow that A supports C.

Here is an example where supportive conditional probability is not transitive. Consider the following three events in throwing a die:

$$A = \{1, 2, 3, 4\}, \quad B = \{2, 3, 4, 5\}, \quad C = \{3, 4, 5, 6\}.$$

Clearly, A supports B (i.e., $\Pr\{B/A\} = 3/4 > Pr\{B\} = 2/3$), B supports C (i.e., $\Pr\{C/B\} = 3/4 > Pr\{C\} = 2/3$), but A does not support C (i.e., $\Pr\{C/A\} = 1/2 < Pr\{C\} = 2/3$).

2.5.3 Conditional probability and probability of joint events

Suppose that you tossed a coin 1,000 times and all of the outcomes turned out to be heads $\{H\}$. What is the probability of the next outcome being H? Most people would answer that the chances

[12] More details in Falk (1979).

of having 1,001 heads are extremely small. That is correct. The chances are $(1/2)^{1001}$, extremely small indeed. But the question was not about the probability of having 1,001 heads but on the *conditional* probability of a result H given 1,000 heads in the last 1,000 tosses. This conditional probability is one half (assuming that all the events are independent).

The psychological reason for the confusion is that you know that the probability of H and T are half-half. Therefore, if you throw 1,000 times, it is most likely that you will get about 500 H and 500 T. The event "the first 1,000 throws result in heads," though very rare, is possible. You might feel that "it is time that the chances will turn in favor of T," and the sequence of outcome must behave properly. Therefore, you feel that the chances of a tail T, given that 1,000 heads have occurred, are now close to one. That is wrong, however. In fact, if a coin shows 1,000 outcomes in a row to be H, I might suspect that the coin is unbalanced, therefore I might conclude that the chances of the next H are larger than 1/2.

To conclude, if we are given a fair coin, and we toss it at random (which is equivalent to saying that the probability of H is 1/2), the probability of having 1,000 outcomes of H is very low $(1/2)^{1000}$. But the *conditional* probability of having the next H, given "1,000-heads-in-a-row" is still 1/2. This is of course true presuming that the events at each toss are independent.

2.6 Bayes' Theorem

Bayes' theorem is an extremely useful tool both for calculating probabilities and for making plausible reasoning. We shall start with a simple and intuitively clear theorem known as the theorem of total probability.

Consider n events that are pairwise disjoint and that their union (or sum) covers the entire space Ω (this is sometimes referred to as a partition). In mathematical terms, we assume

$$(1)\ A_i \cdot A_j = \phi \quad \text{for each pair of events } \underset{i \neq j}{i, j} = 1, 2, \dots, n, \tag{2.6.1}$$

$$(2)\ \Omega = \textstyle\sum_{i=1}^{n} A_i \quad (\text{or} \bigcup_{i=1}^{n} A_i) \tag{2.6.2}$$

Example: Consider the outcomes of throwing a die. Define the events

$$A_1 = \{\text{even outcome}\} = \{2, 4, 6\},$$
$$A_2 = \{\text{odd and larger than one}\} = \{3, 5\},$$
$$A_3 = \{1\}. \tag{2.6.3}$$

Clearly, these three events are disjoint and their union covers the entire range of possible outcomes.

Exercise: Show that for any n events $A_1, A_2, \ldots A_n$; if they are pairwise mutually exclusive (i.e., $A_i \cdot A_j = \phi$ for each pair of $i \neq j$), then it follows that each group of events are mutually exclusive. The inverse of this statement is not true. For instance, events A, B and C can be mutually exclusive (i.e., $A \cdot B \cdot C = \phi$) but not pairwise mutually exclusive.

For the second example, we use again the squared board of unit area and any division of the area into several mutually exclusive areas, as in Fig. 2.9a.

Clearly, by construction, each pair of events consists of mutually exclusive events, and the sum of the events is the certain event, i.e., hitting any place on the board (neglecting the borderlines) is unity.

Let B be any event. We can always write the following equalities:

$$B = B \cdot \Omega = B \cdot \sum_{i=1}^{n} A_i = \sum_{i=1}^{n} B \cdot A_i. \tag{2.6.4}$$

The first equality is evident (for any event B, its intersection with the total space of events Ω, gives B). The second equality follows from the assumption (2.6.2). The third equality follows from the distributive law of the product of events.[13]

From the assumption that all A_i are mutually exclusive, it also follows that all $B \cdot A_i$ are mutually exclusive.[14] Therefore, it follows that the probability of B is the sum of the probabilities of all the

[13]Show that $A \cdot (B + C) = A \cdot B + A \cdot C$ for any three events A, B, C and by generalization $B \cdot \sum A_i = \sum B \cdot A_i$.

[14]This follows from the commutative property of the product of events i.e., $B \cdot A_i \cdot B \cdot A_j = B^2 \cdot A_i \cdot A_j = B \cdot \phi = \phi$.

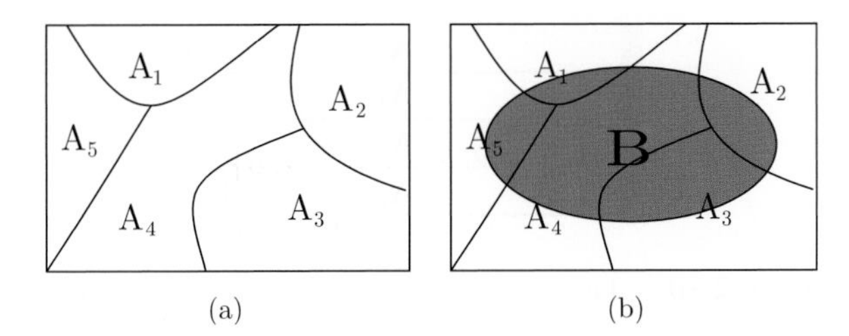

Figure 2.9. (a) Five mutually exclusive events, the union of which is the certain event. (b) A region B intersects the events A_i.

$B \cdot A_i$, (see Figure 2.9b):

$$P(B) = P\left(\sum_{i=1}^{n} B \cdot A_i\right) = \sum_{i=1}^{n} P(B \cdot A_i) = \sum_{1=1}^{n} P(A_i)P(B/A_i). \tag{2.6.5}$$

The last equality follows from the definition of conditional probabilities.

Theorem (2.6.5) is very simple and intuitively clear. We have started with n events that cover the entire space of events Ω. We took an arbitrary "cut" from Ω that we called B. The total probability theorem states that the area of B is equal to the sum of the areas that we have cut from each A_i as shown in the shaded B areas in Figure 2.9b.

From the definition of the conditional probability and theorem (2.6.5), it follows immediately that

$$P(A_i/B) = \frac{P(A_i \cdot B)}{P(B)} = \frac{P(A_i)P(B/A_i)}{\sum_{j=1}^{n} P(A_j)P(B/A_j)}. \tag{2.6.6}$$

This is one formulation of Bayes' theorem. It is a very simple theorem, but as we shall soon see, it is extremely useful for solving problems that seem intractable.

The events A_i are sometimes referred to as *a priori* events. Suppose we know the probabilities of all the events A_i and also all the conditional probabilities (B/A_i). Then (2.6.6) allows us to calculate the (*a posteriori*) probabilities $P(A_i/B)$. The assignments of the terms *a priori* and *a posteriori* should not be regarded as ordered in time. The reason will be clear from the examples that we shall work out.

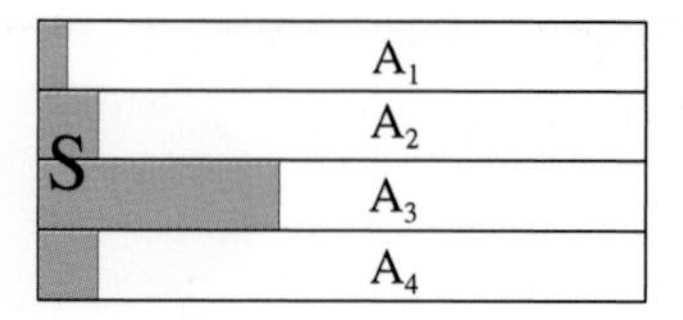

Figure 2.10. The four events described in the example.

Example: Suppose that our board of unit area is divided into four equal and mutually exclusive areas (see Figure 2.10). Clearly, the probability of hitting any one of these areas is

$$P(A_1) = P(A_2) = P(A_3) = P(A_4) = 1/4. \tag{2.6.7}$$

Now, suppose each of the areas A_i has a shaded part. Let us denote all the shaded areas in Figure 2.10 by S. (S does not need to be a connected region, as in Figure 2.10.)

We are given all the conditional probabilities

$$P(S/A_1) = \frac{1}{20}, \quad P(S/A_2) = \frac{1}{10}, \quad P(S/A_3) = \frac{2}{5},$$
$$P(S/A_4) = \frac{1}{10}, \tag{2.6.8}$$

i.e., $P(S/A_i)$ is the ratio of the shaded area to the total area of each A_i.

Bayes' theorem simply states that if we know all the areas A_i, and all the fractions of the shaded area in each of the A_i (2.6.8), then given that we hit the area S, we can calculate the probability that we have hit a specific area A_i. Clearly, time does not play any role in this theorem.

Solution of the urn problem discussed in Section 2.5.2.

We are asked to calculate the probability $\Pr(White_1/White_2)$, i.e., given that the outcome on the second draw is white, what is the probability that the outcome of the first draw is white?

Using Bayes' theorem, we write

$$\Pr(White_1/White_2) = \frac{\Pr(White_1 \cdot White_2)}{\Pr(White_2)}$$
$$= \frac{\Pr(White_2/White_1)\Pr(White_1)}{\Pr(White_2/White_1)\Pr(White_1) + \Pr(White_2/Black_1)\Pr(Black_1)}. \tag{2.6.9}$$

Note that on the right-hand side of (2.6.9), all the probabilities are known. Hence, we can calculate the required conditional probability

$$\Pr(White_1/White_2) = \frac{1/3 \times 1/2}{1/3 \times 1/2 + 2/3 \times 1/2} = 1/3. \quad (2.6.10)$$

This is the same as the conditional probability $\Pr(White_2/White_1)$, as in Section 2.5.2.

Exercise: Generalize the urn problem as in Section 2.5.2 for n white balls and m black balls. Calculate $\Pr(White_2/White_1)$ and $\Pr(White_1/White_2)$ in the general case.

We shall further discuss this problem from the information-theoretical point of view in Section 3.7.

An important application of the Bayes' theorem is the following. Suppose we have a test for a virus which gives a *positive* result (+) if a person is a carrier with probability 0.99, this means that for each hundred carriers (C) who are tested, 99% show positive results. The test also shows false positive results, i.e., a non-carrier tests positive, say with probability 10^{-4}. This sounds like a reliable and valuable test.

Denote by C the event "being a carrier of the virus." Suppose that $P(C)$ is the probability of carriers in a given country (i.e. the fraction of carriers in a given population). Denote by $P(+/C)$ the conditional probability that the test is positive given that the person is a carrier. We are given that $P(+/C) = 0.99$. On the other hand, it is known that a false positive result occurs in one out of 10,000, which means that given that a person is not a carrier $\bar{C}$, the probability of having a positive result is very low, i.e.,

$$P(+/\bar{C}) = 10^{-4}. \quad (2.6.11)$$

Clearly, the closer $P(+/C)$ is to unity, and the smaller the false positive results $P(+/\bar{C})$, the more reliable the test is. Yet, some unexpected and counterintuitive results can be obtained.

Suppose a person chosen at random from the population takes the test and the result is positive, what is the probability that the person is actually a carrier? The question is thus, what is $P(C/+)$?

According to Bayes' theorem

$$P(C/+) = \frac{P(+/C)P(C)}{P(C)P(+/C) + P(\bar{C})P(+/\bar{C})}. \quad (2.6.12)$$

This is a straightforward application of (2.6.6). In our example, we know $P(+/\bar{C}) = 10^{-4}$, and $P(+/C) = 0.99$ and we need to know

$$P(C/+) = \frac{0.99P(C)}{0.99P(C) + (1 - P(C))10^{-4}}. \quad (2.6.13)$$

As can be seen from this equation and from Figure 2.11, the probability that a person who tested *positive* is actually a carrier, depends on $P(C)$, i.e., on the percentage of the carriers in the population. Intuitively, we feel that since $P(+/\bar{C}) = 10^{-4}$ and $P(+/C) = 0.99$, the test is very sensitive and we should expect a high value for $P(C/+)$, and a small value of $P(\bar{C}/+)$. Indeed, if the population of carriers is larger than, say, $P(C) = 0.001$ then $P(C/+)$ is nearly one. However, if the population of carriers is very low, say, $P(C) = 0.0001$, then $P(C/+)$ can be about 0.5. This is understandable. If the population of carriers is extremely small, then testing positive becomes a rare event; in the case of $P(C) = 0.0001$, the denominator of (2.6.13) is

$$P(+) = P(+/C)P(C) + P(+/\bar{C})P(\bar{C}) = 0.00019899. \quad (2.6.14)$$

Hence, the ratio in (2.6.13) is

$$P(C/+) = \frac{0.99 \times 10^{-4}}{0.00019899} \approx 0.5. \quad (2.6.15)$$

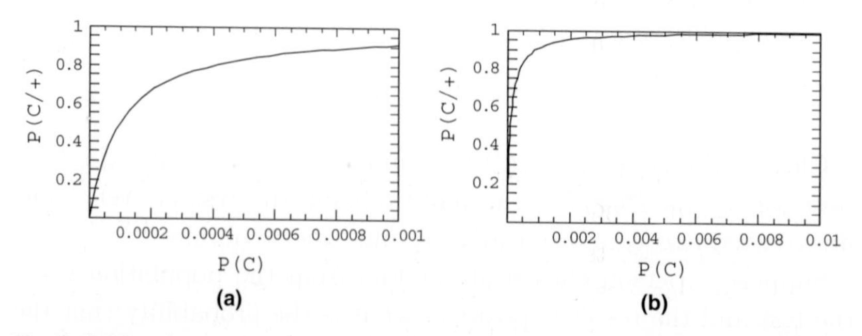

Figure 2.11. Plot of the function $P(C/+)$ as a function $P(C)$ for low and high fractions of the population $P(C)$.

As Figure 2.11 shows, the conditional probability $P(C/+)$ is a monotonic increasing function of $P(C)$. For large $P(C)$ (Figure 2.11b), the value of $P(C/+)$ is nearly one. However, for very small $P(C)$ (Figure 2.11a), the values of $P(C/+)$ could be very small, which means $P(\bar{C}/+)$ is very large, i.e., nearly one.

The reason that we are surprised to see such a relatively small value of $P(C/+)$ or a large $P(\bar{C}/+)$ is that intuitively we confuse the *given conditional probability*, $P(+/C)$, with the calculated conditional probability, $P(C/+)$, the latter of which can be any number between 0 to 1. In an extreme case when the population of carriers is extremely low, say $P(C) = 10^{-6}$, then $P(C/+) \approx 0.01$, or $P(\bar{C}/+) = 0.99$. An even more extreme case is when $P(C) = 0$, i.e., given that none in the population is a carrier. In this case, *given* a positive test, the probability of being a carrier must be zero, and the probability of being a non-carrier is one! We shall further analyze this problem from the informational point of view in Section 3.6.

A slightly different interpretation of (2.6.12) is the following. $P(C)$ is the fraction of the population which is known to be carriers. Therefore, prior to any test, we can judge that the probability that a person selected at random from the population is found to be a carrier is $P(C)$. This is sometimes called the *prior* probability. Now we perform the test and find that the result is positive. Equation (2.6.12) tells us how to modify our *prior* probability $P(C)$ to obtain the posterior probability $P(C/+)$, given the new information. Similarly, one can extend the same reasoning whenever more information is available (say additional tests) to assess and modify our probabilities given the new information. It should be stressed that the terms prior and posterior, or "before" and "after," as used in the aforementioned reasoning, do not necessarily imply events that are *ordered* in time.

Exercise: Consider the following problem. A metal detector in an airport goes off in 999 out of a 1,000 of the cases when a person carrying metal objects passes through (i.e., $P(+/C) = \frac{999}{1000}$). There are also false alarm cases, i.e., where the detector goes off even if the passenger is not carrying any metallic objects, i.e., $P(+/\bar{C}) = 10^{-6}$. Given that the detector went off, what is the probability that the passenger is a carrier of a metallic object?

2.6.1 A challenging problem

The following is a problem of significant historical value. It is considered to be one of the problems the solution of which has not only crystallized the concept of probability, but has also transformed the reasoning about chances taken in gambling saloons into mathematical reasoning occupying the minds of mathematicians. This problem was addressed to Blaise Pascal by his friend Chevalier de Méré in 1654.[15]

Two players bet $10 each. The rules of the game are very simple. Each one chooses a number between one and six. Suppose Dan chose 4 and Linda chose 6. They roll a single die and record the sequence of the outcomes. Every time an outcome "4" appears, Dan gets a point. When a "6" appears, Linda gets a point. The player who collects three points first wins the total sum of $20. For instance, a possible sequence could be

$$1, 4, 5, 6, 3, 2, 4, 6, 3, 4 \tag{2.6.16}$$

Once a "4" appears for the third time, Dan wins the entire sum of $20 and the game ends.

Now, suppose the game is started and at some moment the sequence of outcomes is the following:

$$1, 3, 4, 5, 2, 6, 2, 5, 1, 1, 5, 6, 2, 1, 5. \tag{2.6.17}$$

At this point, there is some emergency and the game must be stopped! The question is how to divide the sum of $20.

Note that the problem does not arise if the rules of the game explicitly instruct the player on how to divide the sum should the game be halted. But in the absence of such a rule, is it not clear how to divide the sum.

Clearly, one feels that since Dan has "collected" one point, and Linda "collected" two points, Linda should get the larger portion of the $20. But how much larger? The question is, what is the *fairest* way of splitting the sum, given that sequence of outcomes? But what does it mean, the *fairest splitting of the sum*? Should

[15]The historical account of these and other earlier probabilistic problems can be found in David (1962).

Linda get twice as much as Dan because she has gained twice as many points in her favor? Or perhaps, simply split the sum into two equal parts since the winner is undetermined? Or perhaps, let Linda collect the total sum because she is "closer" to winning than Dan.

A correspondence between Blaise Pascal and Pierre de Fermat ensued for several years. These were the seminal thoughts which led to the theory of probability. Recall that in the 17th century, the concept of probability was far from being crystallized yet. The difficulty was not only of finding the mathematical solution to the problem. It was not less difficult to clarify what the problem was, i.e., what does it mean to find a *fair* method of splitting the sum?

The answer to the last question is the following:

Having no specific rule on how to divide the sum in case of halting the game, the "fairest" way of splitting the sum is according to the ratio of the probabilities of the two players winning the game, had the game continued.

In stating the problem in terms of probabilities, one hurdle was overcome. We now have a well-formulated problem. But how do we calculate the *probabilities* of either player winning? We feel that Linda has a better chance of winning since she is "closer" to collecting three points than Dan.

We can easily calculate that the probability of Dan winning, on the *next* throw is *zero*. The probability of Linda winning on the next throw is $1/6$, and the probability of neither one of them winning on the next throw is $5/6$. One can calculate the probabilities of winning after two throws, three throws, etc. The calculations become very complicated for the fourth, fifth and larger number of throws. Try calculating the probability of each player winning on the next two throws, and for the next three throws and see just how messy it gets. There is a simple solution based on the theorem of total probability that involves a one-unknown equation.

The solution to the problem is this. Denote by X the probability of Linda winning. Clearly, in the next throw after the game is halted, there are three mutually exclusive possibilities:

$$\begin{aligned}
&\text{(I) outcome } \{6\} \text{ with probability } 1/6,\\
&\text{(II) outcome } \{4\} \text{ with probability } 1/6, \qquad (2.6.18)\\
&\text{(III) outcome } \{1,2,3,5\} \text{ with probability } 4/6.
\end{aligned}$$

Let us denote the event "Linda wins" by LW. Using the theorem of total probability, the following equation holds:

$$
\begin{aligned}
X = \Pr(LW) &= \Pr(I)\Pr(LW/I) + \Pr(II)\Pr(LW/II) \\
&\quad + \Pr(III)\Pr(LW/III) \\
&= 1/6 \times 1 + 1/6 \times 1/2 + 4/6 \times X. \qquad (2.6.19)
\end{aligned}
$$

This is an equation with one unknown $6X = 3/2 + 4X$. The solution is $X = 3/4$. Note that the events (I), (II), and (III) refer to the possible outcomes on the *next* throw. However, the event "LW" refers to *Linda wins*, regardless of the number of subsequent throws. Equation (2.6.19) means that the probability of Linda winning is the sum of the three probabilities of the three mutually exclusive events. If (I) occurs, then she wins with probability one. If (II) occurs, then she has a probability 1/2 of winning (since in this case both will have two points). If (III) occurs, then the probability of her winning is X, the same as at the moment of halting the game. Therefore, the two players should divide the total sum in such a way that Linda gets 3/4 and Dan gets 1/4.

Exercise: Solve the same problem for the following two cases:

(1) Dan has collected two points and Linda zero points.
(2) Dan has collected one point and Linda zero points.

2.6.2 A more challenging problem: The three prisoners' problem

This problem is important, interesting, challenging and enjoyable. Going through the solution should be a rewarding experience for the following reasons:

(1) It is a simple problem having a counter-intuitive solution.
(2) It illustrates the power of Bayes' theorem in solving probabilistic problems.
(3) It involves the concepts of "information" in making a "probabilistic decision" but in a very different sense from the way this term is used in information theory.

This problem could be justifiably referred to as the *prisoner dilemma*. As we shall present it, it does involve a very serious dilemma. However, the term "prisoner dilemma" is already used in game theory for quite a different problem. Therefore, we have chosen the title of: "The three prisoner problem." This problem is equivalent to the problem sometimes referred to as the Monty Hall problem, which arises in connection with the TV show "Let's Make a Deal."[16] We shall describe here the more dramatic version of the problem involving a matter of life and death.

The problem

Three prisoners on death row, named A, B, and C await their execution scheduled for the first of January. A few days before the said date, the king decides to free one of the prisoners. The king does not know any of these prisoners, so he decides to throw a die, the result of which will determine who shall be set free. On the faces of this die are inscribed the letters A,A,B,B,C,C. The die is fair. The probability of each of the letters appearing is 1/3. The king then instructs the warden to free the prisoner bearing the name that came up on the face of the die. He warns the warden not to divulge to the prisoners who the lucky one is until the actual day of the execution.

The facts that the king has decided to free one prisoner, that the prisoner was chosen at random with a 1/3 probability, and that the warden has full knowledge of the results but is not allowed to divulge to anyone, are known to the prisoners. It is also common knowledge that the warden is indifferent to the prisoners. Whenever he has to make a choice, he would toss a coin. A day before the execution, prisoner A approaches the warden and asks him: "I know you are not allowed to tell me or anyone else, who shall be freed tomorrow, however, I know that two of the three prisoners must be executed tomorrow. Therefore, *either* B or C must be executed. Please tell me which one of these two will be executed.

Clearly, by revealing to prisoner A the identity of one prisoner who will be executed, the warden does not reveal the name of the one who will be freed. Therefore, he can answer prisoner A's query

[16] See Falk (1993).

without defying the king's order. The warden tells prisoner A that B is going to be executed the next day.

The problem is this. Originally, prisoner A knew that the three prisoners have 1/3 probability of getting freed. Now, he got some more *information*. He knows that B is doomed to die. Suppose A is given the opportunity to exchange names with C (note that the king decided to free the prisoner, who on the day of the execution carries the *name* that appeared on the face of the die). Prisoner A's dilemma is this: Should he switch names with C (to render it more dramatic, we could ask: should A offer C, say \$100 to swap their names?) In probability terms, the question is the following: Initially, prior to asking the warden, A knows that the probability of his survival is 1/3. He also knows that the warden says the truth, and that the warden is not biased (i.e., if he has to make a choice between two answers, he will choose between the two with equal probability). After receiving the information, the question is: "What is the conditional probability of A's survival given the information supplied by the warden?"

Intuitively, we feel that since initially A *and* C had the same probability of being freed, there should be the same probability for A *and* C to be freed after the warden informed A that B will be executed.

This problem can be easily solved using Bayes' theorem. The solution is counter-intuitive. The reader is urged to solve the problem before reading the solution in Appendix P.

2.7 Random Variables, Average, Variance and Correlation

A random variable (rv)[17] is a real-valued function defined on a sample space. If $\Omega = \{w_1, \ldots, w_n\}$, then for any $w_i \in \Omega$, the quantity $X(w_i)$ is a real number.[18] Note that the outcomes of the

[17] The term "random variable" is somewhat misleading. A rv is actually a function over the domain Ω, i.e., the variables w are the outcomes of the experiment and X has a real value for each $w \in \Omega$.

[18] We shall use capital letters X, Y for the random variables, but use $X(w)$ or $Y(w)$ when referring to the components of these functions, or the value of the function X at the point w.

experiment, i.e., the elements of the sample space, are not necessarily numbers. The outcomes can be colors, different objects, different figures, etc. For instance, the outcomes of tossing a coin are $\{H, T\}$. Another example: suppose we have a die, the six faces of which have different colors, say white, red, blue, yellow, green and black. In such a case, we cannot plot the function $X(w)$, but we can write the function explicitly as a table. For example,

$$\begin{aligned} X(w = \textit{white}) &= 2, \\ X(w = \textit{red}) &= 2, \\ X(w = \textit{blue}) &= 2, \\ X(w = \textit{yellow}) &= 1, \\ X(w = \textit{green}) &= 0, \\ X(w = \textit{black}) &= 1. \end{aligned} \tag{2.7.1}$$

Note that we always have the equality

$$P(\Omega) = \sum_{w \in \Omega} P(w) = \sum_i P[X(w) = x_i] = 1. \tag{2.7.2}$$

The first sum is over all the elements $\omega \in \Omega$. The second sum is over all values of rv, x_i.

In general, even when w_i are numbers, the values of $X(w_i)$ are not necessarily equal to w_i. For instance, the outcomes of an ordinary die are the numbers $1, 2, 3, 4, 5, 6$. An rv X can be defined as

$$X(w_i) = w_i^2 \quad \text{or} \quad Y(w_i) = \exp(w_i). \tag{2.7.3}$$

Clearly, in these cases $X(w_i)$ and $Y(w_i)$ differ from w_i.

In the general case when the outcomes w_i are numbers and their corresponding probabilities are p_i, we can define the average outcome of the experiment as usual. For instance, for the ordinary die, we have

$$\sum_{i=1}^{6} w_i p_i = \frac{1}{6} \sum_{i=1}^{6} w_i = 3.5. \tag{2.7.4}$$

However, it is meaningless to talk of an average of the outcomes of an experiment when the outcomes are not numbers. For instance, for the six-colored die described above, there is no "average" outcome that can be defined for this experiment. However, if an *rv* is defined on the sample space, then one can always *define* the average (or the mean value) of the *rv* as[19]

$$\bar{X} = \langle X \rangle = E(X) = \sum_x xP\{w : X(w) = x\}$$
$$= \sum_w P(w)X(w). \tag{2.7.5}$$

The first sum in (2.7.5) is over all possible values of x; the second sum is over all possible outcomes $\omega \in \Omega$.

The term $\{w : X(w) = x\}$ is an *event*,[20] consisting of all the outcomes w, such that the value of the *rv*, $X(w)$ is x. In the example (2.7.1), these events are

$$\{w : X(w) = 0\} = \{green\},$$
$$\{w : X(w) = 1\} = \{yellow, black\},$$
$$\{w : X(w) = 2\} = \{white, red, blue\}. \tag{2.7.6}$$

The corresponding probabilities are (assuming the die is "fair")

$$P\{X(w) = 0\} = \frac{1}{6},$$
$$P\{X(w) = 1\} = \frac{2}{6},$$
$$P\{X(w) = 2\} = \frac{3}{6}. \tag{2.7.7}$$

[19]We shall use capital letters X, Y, etc., for the *rv* and lower case letters x, y, etc., for the values of the *rv*. We shall either use $\overline{X}$ or $\langle X \rangle$ for an average. In the mathematical theory of probability, the average is denoted by $E(X)$ and referred to as the "expected value." This is a somewhat unfortunate term for an average. The value of the average is, in general, not an *expected* outcome. For instance, in (2.7.4), the average value is 3.5; this is certainly not an *expected* outcome.

[20]In the mathematical theory of probability, the distribution of an *rv* is defined by

$$F(x) = P(X = x)$$

and for a continuous *rv*, the density of the distribution is defined such that

$$f(x)\,dx = P(x \le X \le x + dx).$$

Sometimes we shall use the short hand notation for $\Pr\{w : X(w) = x\}$ as $P(X = x)$ or even $P(x)$. If we have several random variables, we shall use the notation $P_X(x)$ and $P_Y(y)$ for the distribution of the *rv* X and Y, respectively.

Hence, the average of the *rv* is

$$\bar{X} = 0 \times \frac{1}{6} + 1 \times \frac{2}{6} + 2 \times \frac{3}{6} = \frac{4}{3}. \qquad (2.7.8)$$

There are many other examples of outcomes that are not numbers for which an average quantity is not defined. Consider the following example. Suppose that two drugs A and B when administered to a pregnant woman have the following probabilities for the gender of the child at birth:

	Boy	Girl
Drug A	1/2	1/2
Drug B	2/3	1/3

Which drug is better then? There is no answer to this question because the outcomes are not numbers. All we can say is that if we used drug A, we shall have on average 50% boys and 50% girls. But there is no "average" outcome. On the other hand, if we assign numbers to the outcomes, say

$$X(boy) = 10,$$
$$X(girl) = 3,$$

then we can say that the average of the *rv* X for each of the two drugs is

$$\frac{1}{2} \times 10 + \frac{1}{2} \times 3 = \frac{13}{2} = 6.5 \qquad \text{(for drug A)},$$
$$\frac{2}{3} \times 10 + \frac{1}{3} \times 3 = \frac{23}{3} = 7\frac{2}{3} \qquad \text{(for drug B)}.$$

Thus, in a society where infant boys are more welcome, there is a higher value placed on the outcome "boy" rather than on the outcome "girl." In this case, we may say that drug B is "better" than drug A — better, only in the sense that the average of the *rv* X has a higher value for A than for B.

In a different society, there might be different values assigned to boys and girls, and the corresponding average values of the *rv* would be different. For instance, if one defines a different *rv* on the same sample space $\{boy, girl\}$

$$Y(boy) = 1,$$
$$Y(girl) = 1,$$

then the average value of this *rv* would be

$$1/2 \times 1 + 1/2 \times 1 = 1 \quad \text{(for drug A)},$$
$$2/3 \times 1 + 1/3 \times 1 = 1 \quad \text{(for drug B)}.$$

In this case, the two drugs are considered equal.

Exercise: Suppose that drugs A and B administered to pregnant women produce the following number of offspring with probabilities:

Number of offspring	0	1	2	3	4	5
Probabilities for Drug A	0.5	0.1	0.2	0.1	0.1	0.0
Probabilities for Drug B	0.0	0.2	0.1	0.2	0.2	0.3

Which drug is better then? Clearly, the answer to this question depends on the *value* we assign to the *rv*. Here, the outcomes are the number of offspring. We can calculate the average number of offspring expected for each drug. If more offspring are considered "better" than fewer, then the second drug, B, is "better." However, if we consider a larger number of offspring to be a burden, then we shall judge drug A as the better one.

A special, very useful random variable is the "indicator function," sometimes referred to as the characteristic function. This is defined for any event A as follows:

$$I_A(w) = \begin{cases} 1 & \text{if } \omega \in A, \\ 0 & \text{if } \omega \in A. \end{cases} \tag{2.7.9}$$

The distribution of this *rv* is: $P(I_A = 1) = P(A), P(I_A = 0) = 1 - P(A) = P(\bar{A})$.

As a *rv*, $I_A(w)$ is defined for any point $\omega \in \Omega$. An important property of this *rv* is

$$\langle I_A \rangle = E(I_A) = 0 \times P(I_A = 0) + 1 \times P(I_A = 1) = P(A), \tag{2.7.10}$$

i.e., the average, or the expected, value of I_A is simply the probability of the event A.

The extension of the concept of *rv* to two or more *rv* is quite straightforward. Let X and Y be the two *rv*. They can be defined

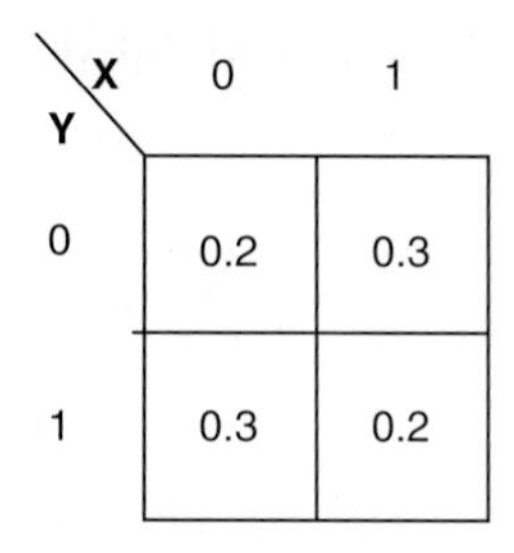

Figure 2.12. Joint probabilities; see Section 2.7.[21]

on the same or on different sample spaces. The joint distribution of X and Y is defined as the probability that X attains the value x and Y attains the value y, and is denoted $P(X = x, Y = y)$. The distribution

$$P(X = x) = \sum_y P(X = x, Y = y), \qquad (2.7.11)$$

is referred to as the marginal distribution of X.[21] Similarly, the marginal distribution of Y is defined as

$$P(Y = y) = \sum_x P(X = x, Y = y), \qquad (2.7.12)$$

where the summation is over all y values in (2.7.11) and over all x values in (2.7.12). Similarly, for a continuous rv, we replace the distributions by the densities of the distribution and instead of (2.7.11) and (2.7.12), we have

$$f_X(x) = \int f_{X,Y}(x, y)\, dy, \qquad (2.7.13)$$

$$f_Y(y) = \int f_{X,Y}(x, y)\, dx. \qquad (2.7.14)$$

Suppose we toss two connected coins, the outcomes of the two coins are dependent events. Define $X(H) = 1, X(T) = 0$, and similarly for Y. We write the distribution $P(X = x, Y = y)$ in a table form, Figure 2.12. In this example, the marginal distributions are $(0.5, 0.5)$ for both X and Y.

The average value could be different for different rv defined on the same sample space. For instance, for the outcomes of an ordinary,

[21]The marginal distributions are written on the *margins* of the table. Hence, the term "marginal."

fair die, the averages of X and Y defined in (2.7.3) on the sample space $\Omega = \{1, 2, 3, 4, 5, 6\}$ are

$$\bar{X} = \sum_{i=1}^{6} w_i^2 \frac{1}{6} = \frac{1}{6} \sum_{i=1}^{6} w_i^2 = 15.167, \tag{2.7.15}$$

$$\bar{Y} = \sum_{i=1}^{6} \exp(w_i) \frac{1}{6} = \frac{1}{6} \sum_{i=1}^{6} \exp(w_i) = 106.1. \tag{2.7.16}$$

For the continuous case, the probability density of the rv X is defined as

$$f_X(x)dx = P\{w : x \leq X(w) \leq x + dx\}. \tag{2.7.17}$$

This is the probability of the event in the curly brackets, i.e., the set of all w such that $X(w)$ falls between x and $x + dx$. The corresponding average is now defined by

$$\bar{X} = \int_{-\infty}^{\infty} x f_X(x) dx. \tag{2.7.18}$$

Some properties of the average are:

(i) For any constant number c

$$E(X + c) = E(X) + c \tag{2.7.19}$$

(ii) For two rv X and Y, the sum $X + Y$ is also an rv and

$$E(X + Y) = E(X) + E(Y) \tag{2.7.20}$$

(iii) For any real number c, we have

$$E(cX) = cE(X) \tag{2.7.21}$$

(iv) In general, for any number of random variables $X_1, \ldots, X_n$, and real numbers $a_1, \ldots, a_n$, we have

$$E\left(\sum_{i=1}^{n} a_i X_i\right) = \sum a_i E(X_i). \tag{2.7.22}$$

This is called the linear property of the expectation functional or of the average. All the above properties are easily derived from the definition of the average quantity. A special average of a random variable is the *variance* denoted by σ^2, which measures how much the distribution is spread out or dispersed. This is defined as

$$\begin{aligned}\sigma^2 = Var(X) \equiv \overline{(X - \bar{X})^2} = E(X - E(X))^2 &= \overline{X^2} - 2\bar{X}\bar{X} + \bar{X}^2 \\ &= \overline{X^2} - \bar{X}^2 \geq 0.\end{aligned} \tag{2.7.23}$$

For the continuous case, we have

$$\sigma^2 = \int_{-\infty}^{\infty} (x - \bar{X})^2 f_X(x) dx \geq 0. \tag{2.7.24}$$

The positive square root of σ^2 is referred to as standard deviation. Note that from its definition, it follows that σ^2 is always non-negative.[22] For two random variables X and Y, one defines the *covariance* of X and Y as

$$\begin{aligned}&Cov(X, Y) = E[(X - E(X))(Y - E(Y))] \\ &= E(X, Y) - E(X)E(Y) = \overline{(X - \bar{X})(Y - \bar{Y})} = \overline{XY} - \bar{X}\bar{Y}\end{aligned} \tag{2.7.25}$$

and the *correlation coefficient* of X and Y as

$$R(X, Y) = Cor(X, Y) = \frac{Cov(X, Y)}{\sqrt{Var(X)\,Var(Y)}}, \tag{2.7.26}$$

where the denominator is introduced to render the range of values of $Cor(X, Y)$ between -1 and $+1$.

Exercise: Prove that $|Cor(X, Y)| \leq 1$. See Appendix D.

Two discrete random variables are said to be independent if and only if for any value of x and y

$$P(X = x, Y = y) = P(X = x)P(Y = y). \tag{2.7.27}$$

A similar definition for the continuous case in terms of the densities is

$$f_{X,Y}(x, y) = f_X(x) f_Y(y). \tag{2.7.28}$$

[22] Important variances in statistical mechanics are the fluctuations in energy, in volume, and in the number of particles.

When two *rv* are independent, then $Cov(X, Y) = 0$. This follows from the definition of the average

$$
\begin{aligned}
E(X \cdot Y) = \langle X \cdot Y \rangle &= \sum_{x,y} P(X = x, Y = y)xy \\
&= \sum_{x} P(X = x)x \sum_{y} P(Y = y)y = E(X)E(Y) \\
&= \langle X \rangle \langle Y \rangle.
\end{aligned}
\tag{2.7.29}
$$

Two *rv* for which $E(X \cdot Y) = E(X)E(Y)$ are called uncorrelated *rv*.

Note, however, that if $Cov(X, Y)$ is zero, it does not necessarily follow that X and Y are independent.[23] The reason is that independence of X and Y applies for *any* value of x and y, but the non-correlation applies only to the *average* value of the *rv*, $X \cdot Y$.

Clearly, independence of *rv* implies that the *rv* are uncorrelated, but the converse of this statement is, in general, not true.

For two uncorrelated random variables X and Y, we have

$$
Var(X + Y) = Var(X) + Var(Y). \tag{2.7.30}
$$

It is clear that this relationship holds for two independent *rv*. Here, we show that (2.7.30) is valid under weaker conditions, i.e., that the two *rv* are uncorrelated. To show this, we start from the definition of the variance:

$$
\begin{aligned}
Var(X + Y) &= E\{[(X + Y) - E(X + Y)]^2\} \\
&= E[(X - E(X))^2] + E[(Y - E(Y))^2] \\
&\quad + 2E[(X - E(X))(Y - E(Y))] \\
&= Var(X) + Var(Y).
\end{aligned}
\tag{2.7.31}
$$

In the last equality, we used the condition (2.7.29) that the two *rv* are uncorrelated.

The variance is not a linear functional of the *rv*. The following two properties are easily derived from the definition of the variance.

[23] In the theory of probability, correlation is normally defined for random variables. For random variables, "independent" and "uncorrelated events" are different concepts. For single events, the two concepts are identical. See Appendix D.

For any real number c, we have

$$Var(cX) = c^2\, Var(X), \tag{2.7.32}$$

$$Var(X+c) = Var(X) + Var(c) = Var(X). \tag{2.7.33}$$

For n random variables that are mutually uncorrelated in pairs having the same average $E(X_i) = \mu$ and the same variance $Var(X_i) = \sigma^2$, we have for the arithmetic mean of the *rv*'s:

$$\bar{X} \equiv \left(\frac{\sum_{i=1}^n X_i}{n}\right), \quad Var(\bar{X}) = Var\left(\frac{\sum_{i=1}^n X_i}{n}\right)$$

$$= \frac{\sum_i Var(X_i)}{n^2} = \frac{\sigma^2}{n}, \tag{2.7.34}$$

or equivalently

$$\sigma(\bar{X}) = \frac{\sigma}{\sqrt{n}}. \tag{2.7.35}$$

Exercise: X is the *rv* defined on the sample space of the outcomes of a fair die, i.e., $P(X=i) = \frac{1}{6}$ for $i = 1, 2, \ldots, 6$. X_1 and X_2 are two *rv* defined on the same sample space, but for two independent fair dice thrown together. We define the new *rv*, $Y = |X_2 - X_1|$. The possible outcomes of Y are $\{0, 1, 2, 3, 4, 5\}$. Write the joint probabilities of $X = X_1$, and Y, the averages $E(X), E(Y)$ and $E(XY)$ and the covariance of X and Y.

Solution: The joint probabilities are given in Table 2.1. Each entry is $P(X = i, Y = j)$.

Table 2.1.

$X \backslash Y$	0	1	2	3	4	5	$P_X(i)$
1	1/36	1/36	1/36	1/36	1/36	1/36	1/6
2	1/36	2/36	1/36	1/36	1/36	0	1/6
3	1/36	2/36	2/36	1/36	0	0	1/6
4	1/36	2/36	2/36	1/36	0	0	1/6
5	1/36	2/36	1/36	1/36	1/36	0	1/6
6	1/36	1/36	1/36	1/36	1/36	1/36	1/6
$P_Y(j)$	6/36	10/36	8/36	6/36	4/36	2/36	1

The marginal distributions P_X and P_Y are shown at the right-hand column and the bottom row, respectively. The averages of X and Y are:

$$E(X) = \sum_{i=1}^{6} iP_X(i) = 3.5, \tag{2.7.36}$$

$$E(Y) = \sum_{j=0}^{5} jP_Y(j) = \frac{35}{18}, \tag{2.7.37}$$

$$Var(X) = \frac{35}{12}, \quad Var(Y) = \frac{665}{324}, \tag{2.7.38}$$

$$E(XY) = \sum_{i,j} ijP_{XY}(i,j) = \frac{245}{36}, \tag{2.7.39}$$

$$Cov(X,Y) = 0. \tag{2.7.40}$$

Note: This is an example of two *rv* that are *uncorrelated* [see (2.7.40)] but *dependent*. To show dependence, it is sufficient to show that in at least one case, the joint probability of the two events is unequal to the product of the probabilities of the two events, e.g., $P_{XY}(2,5) = 0$, but $P_X(2)P_Y(5) = \frac{1}{6} \times \frac{2}{36} \neq 0$. Hence, the two *rv* are dependent.

2.8 Some Specific Distributions

We shall discuss here very briefly three important distributions that are frequently used in statistical thermodynamics.

2.8.1 The binomial distribution

Suppose we have a series of experiments, each with only two outcomes, say H or T in tossing a coin, or a particle being in the left or the right compartment, or a dipole moment pointing "up" or "down." Let p be the probability of the occurrence of one of the outcomes, say H, and $q = 1 - p$ is the probability of the occurrence of other, say T. If the series of trials are independent, then the probability of any *specific* sequence of outcomes, say

$$H\ T\ H\ H\ T\ T\ H \tag{2.8.1}$$

is simply the product of the probabilities of each of the outcomes, i.e., if $P(H) = p$ and $P(T) = 1 - p = q$, we have for the specific sequence (2.8.1) of heads and tails, the probability

$$pqppqqp = p^4 q^3. \tag{2.8.2}$$

Similarly, if we have seven particles distributed in two compartments, say the right (R) and the left (L), the probability of finding exactly particles (1), (2), (4) and (7) in R and particles (3), (5), and (6) in L is also p^4q^3. [In this case $p = P(R)$ and $q = P(L)$].

In most cases, in statistical mechanics, we are not interested in the specific configuration, i.e., which particle is in which compartment, but only in the *number* of particles in each compartment.

Clearly, we have many possible *specific* sequences of outcomes for which the number of particles in L is constant. A specific configuration is written as a sequence, say $RLLR$, which means first particle in R, second particle in L, third particle in L and fourth particle in R. For four particles in two compartments, we list all possible specific configurations in the left-hand column of Table 2.2. There

Table 2.2.

Specific Configuration	Number of Particles in L	Number of Specific Events in Each Group
$RRRR$	0	1
$LRRR$	1	4
$RLRR$	1	
$RRLR$	1	
$RRRL$	1	
$LLRR$	2	6
$LRLR$	2	
$LRRL$	2	
$RLLR$	2	
$RLRL$	2	
$RRLL$	2	
$LLLR$	3	4
$LLRL$	3	
$LRLL$	3	
$RLLL$	3	
LLLL	4	1

is only one configuration for which all particles are in R, four configurations for which one particle is in L, and three are in R, etc.

Clearly, any two *specific* configurations are disjoint, or mutually exclusive events. Therefore, the probability of occurrence of all the four particles in R is p^4. The probability of occurrence of one particle in L (regardless of which particle is in L) is simply the sum of the probabilities of the four events listed in the table, i.e.,

$$P(\text{one particle in } L) = 4p^3q. \tag{2.8.3}$$

Similarly,

$$P(\text{two particles in } L) = 6p^2q^2, \tag{2.8.4}$$

$$P(\text{three particles in } L) = 4pq^3, \tag{2.8.5}$$

$$P(\text{four particles in } L) = q^4. \tag{2.8.6}$$

In general, for N particles (or N coins), the probability of occurrence of the event "n particles in L" (or n coins showing T), is

$$P_N(n) = \binom{N}{n} p^n q^{N-n}. \tag{2.8.7}$$

Note that in constructing the expression (2.8.7), we used the product rule for the probabilities of the independent events "R" and " L," and the sum rule for the disjoint events "n specific particles in L" and $N-n$ specific particles in R. The number of specific disjoint events is simply the number of ways of selecting a group of n particles out of N identical particles. The distribution (2.8.7) is referred to as the binomial distribution. The coefficients $\binom{N}{n}$ are the coefficients in the binomial expansion. For any a and b, the binomial expansion is[24]

$$(a+b)^N = \sum_{n=0}^{N} \binom{N}{n} a^{N-n} b^n. \tag{2.8.8}$$

Sometimes, the distribution $P_N(n)$ is also referred to as the Bernoulli distribution (particularly in the case $p = q = 1/2$).

Figure 2.13 shows the binomial distribution $P_N(n)$ defined in (2.8.7) for the case $p = q = 1/2$ for different values of N. Note that

[24]The normalization of $P_N(n)$ follows from (2.8.8), i.e., $\sum_{n=0}^{N} P_N(n) = 1$.

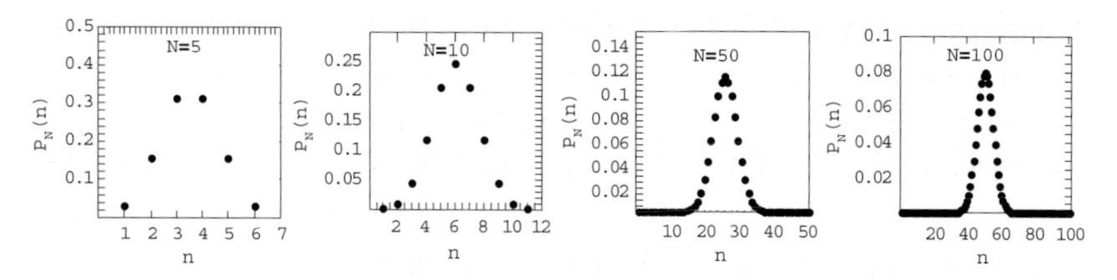

Figure 2.13. The binomial distribution $P_N(n)$ in (2.87) for different values of N.

as N increases, the form of the curve becomes more and more similar to a bell-shaped curve, or the normal distribution (see below).

Exercise: Show that the average and the variance of the binomial distribution are:

$$E(X) = \langle X \rangle = \sum_{n=0}^{N} nP_N(n) = Np, \tag{2.8.9}$$

$$\sigma^2 = E(X^2) - [E(X)]^2 = Np(1-p). \tag{2.8.10}$$

Solution:

$$E(X) = \bar{n} = \sum_{n=0}^{N} nP_N(n) = \sum_n n\binom{N}{n} p^n q^{N-n}, \tag{2.8.11}$$

where $q = 1-p$. We now formally view p and q as two independent variables, and write the identity

$$p\frac{\partial}{\partial p}\left[\sum_n \binom{N}{n} p^n q^{N-n}\right] = p\sum_n \binom{N}{n} np^{n-1}q^{N-n}$$
$$= \sum_n \binom{N}{n} np^n q^{N-n}. \tag{2.8.12}$$

Using this identity in (2.8.11) and (2.8.8), we get

$$\bar{n} = \sum_n nP_N(n) = p\frac{\partial}{\partial p}\left[\sum_n \binom{N}{n} p^n q^{N-n}\right] = p\frac{\partial}{\partial p}(p+q)^N$$
$$= pN(p+q)^{N-1} = pN. \tag{2.8.13}$$

Note that the identity (2.8.12) is valid for any p and q, whereas in (2.8.13), we used a particular pair of p and q such that $p+q=1$.

For the variance, we can use the same trick as above to obtain

$$\sigma^2 = \sum_{n=0}^{N} (n - \bar{n})^2 P_N(n) = \overline{(n^2)} - (\bar{n})^2, \tag{2.8.14}$$

$$\begin{aligned}\overline{(n^2)} &= \sum_n n^2 P_N(n) = \sum_n n^2 \binom{N}{n} p^n q^{N-n} \\ &= p\frac{\partial}{\partial p} p\frac{\partial}{\partial p}\left[\sum_n \binom{N}{n} p^n q^{N-n}\right] = p\frac{\partial}{\partial p} p\frac{\partial}{\partial p}[(p+q)^N] \\ &= p^2 N(N-1)(p+q)^{(N-2)} + pN(p+q)^{(N-1)}.\end{aligned} \tag{2.8.15}$$

For the particular case $p + q = 1$, from (2.8.15), we get

$$\overline{(n^2)} = N(N-1)p^2 + Np. \tag{2.8.16}$$

Hence,

$$\sigma^2 = \overline{(n^2)} - (\bar{n})^2 = N(N-1)p^2 + Np - (Np)^2 = Npq. \tag{2.8.17}$$

2.8.2 The normal distribution

As we have seen in Figure 2.13, when N is very large, the form of the distribution function becomes very similar to the normal, or the Gaussian distribution. We shall now show that for large N, we can get the normal distribution as a limiting form of the binomial distribution.

We start with the binomial distribution (2.8.7) and treat n as a continuous variable.

The average and the variance are

$$\bar{n} = \langle n \rangle = \sum_{n=0}^{N} n P_N(n) = pN, \tag{2.8.18}$$

$$\sigma^2 = \langle n^2 \rangle - \langle n \rangle^2 = Npq. \tag{2.8.19}$$

It is easy to show that the distribution $P_N(n)$ has a sharp maximum at $\langle n \rangle$ (Figure 2.13). We expand $\ln P_N(n)$ about the average

$\bar{n} = \langle n \rangle$, and take the first few terms

$$\ln P_N(n) = \ln P_N(\bar{n}) + \left.\frac{\partial \ln P_N(n)}{\partial n}\right|_{n=\bar{n}} (n - \bar{n}) + \left.\frac{1}{2}\frac{\partial^2 \ln P_N(n)}{\partial n^2}\right|_{n=\bar{n}} (n - \bar{n})^2 \cdots, \tag{2.8.20}$$

where the derivatives are evaluated at the point $n = \bar{n}$.

At the maximum, the first derivative is zero. Therefore, we need to consider the expansion (2.8.20) from the second term. Taking the second derivative of $\ln P_N(n)$ and using the Stirling approximation (Appendix E), we get

$$\left.\frac{\partial^2 \ln P_N(n)}{\partial n^2}\right|_{n=\bar{n}} = -\frac{1}{Npq}. \tag{2.8.21}$$

Note that the second derivative is always negative, which means that $P_N(n)$ [or $\ln P_N(n)$] has a maximum at $n = \bar{n}$.

We next rewrite Equation (2.8.20) (neglecting all higher order terms in the expansion that can be shown to be negligible for large N) as

$$P_N(n) = C \exp\left[-\frac{(n - \bar{n})^2}{2Npq}\right]. \tag{2.8.22}$$

Since we have introduced an approximation in (2.8.20) and (2.8.22), we need to normalize this function, i.e., we require that

$$\int_{-\infty}^{\infty} P_N(n) dn = 1. \tag{2.8.23}$$

This integral is well-known. Thus, from (2.8.22) and (2.8.23), we get the normalization constant

$$C = (2\pi Npq)^{-1/2}. \tag{2.8.24}$$

Substituting (2.8.24) into (2.8.22), we get

$$P_N(n) = \frac{1}{\sqrt{2\pi Npq}} \exp\left[-\frac{(n - \bar{n})^2}{2Npq}\right]. \tag{2.8.25}$$

This is the normal, or the Gaussian distribution.

Denoting

$$\sigma^2 = Npq, \tag{2.8.26}$$

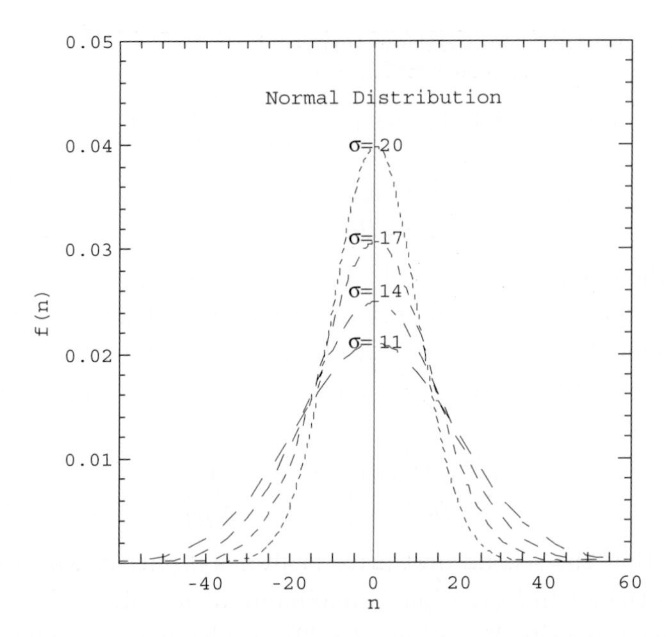

Figure 2.14. The normal distribution (2.8.27) with different values of σ.

we can rewrite this function as

$$f(n) = \frac{1}{\sqrt{2\pi\sigma^2}} \exp\left[-\frac{(n-\bar{n})^2}{2\sigma^2}\right]. \tag{2.8.27}$$

Figure 2.14 shows the function $f(n)$ for various values of σ and $\bar{n} = 0$. It is easy to see by direct integration, that $\bar{n}$ and σ^2 are the average and the variance of the normal distribution.

A very useful approximation for calculating sums of the form

$$\Pr(n_1, n_2) = \sum_{n=n_1}^{n=n_2} \binom{N}{n} p^N (1-p)^{N-n} \tag{2.8.28}$$

is given by the DeMoivre–Laplace theorem. For very large N, we have from (2.8.27) and (2.8.28)

$$\Pr(n_1, n_2) \xrightarrow{\text{Large } N} \frac{1}{\sqrt{2\pi Npq}} \int_{n_1}^{n_2} \exp\left[\frac{-(n-Np)^2}{2Npq}\right] dn. \tag{2.8.29}$$

Exercise: A coin is thrown 1,000 times. What is the probability of finding the outcome H between 450 to 550 of the times? Calculate the exact and the approximate probabilities.

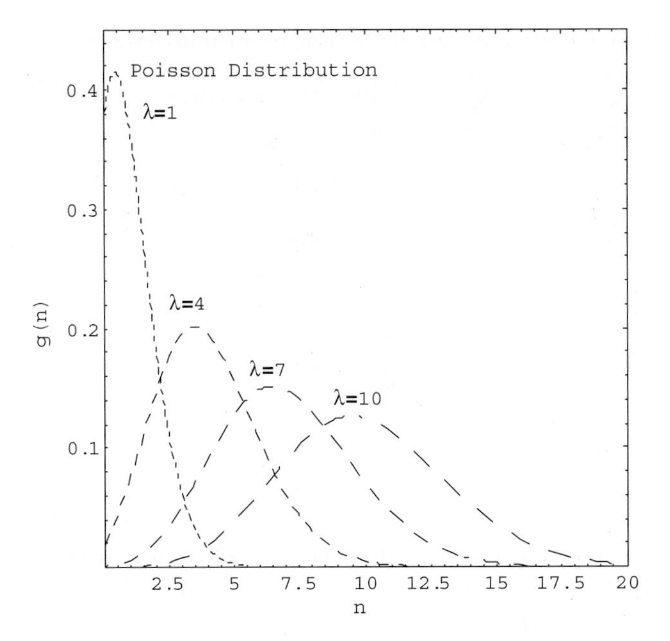

Figure 2.15. The poisson distribution for different values of the average λ.

2.8.3 The Poisson distribution

A second important limit of the binomial distribution is the Poisson distribution. This occurs when $p \to 0$ and $N \to \infty$, but the average $\lambda = \langle n \rangle = pN$ is constant. In this case, we have

$$\frac{N!}{(N-n)!} = N(N-1)\cdots(N-n+1) \xrightarrow{N\to\infty} N^n \tag{2.8.30}$$

and

$$q^{N-n} = (1-p)^{N-n} = \left(1 - \frac{\lambda}{N}\right)^N \left(1 - \frac{\lambda}{N}\right)^{-n} \xrightarrow{N\to\infty} \exp(-\lambda). \tag{2.8.31}$$

Hence, we can write (2.8.7) in this limit as

$$P_N(n) \to g(n) = \frac{\lambda^n \exp(-\lambda)}{n!}, \tag{2.8.32}$$

which is the Poisson distribution. Figure 2.15 shows the function $g(n)$ for several values of λ.

It is easy to see that this function is normalized, i.e.,

$$\sum_{n=0}^{\infty} g(n) = \sum_{n=0}^{\infty} \frac{\lambda^n \exp(-\lambda)}{n!} = e^{-\lambda} e^{\lambda} = 1. \tag{2.8.33}$$

The average and the variance of the Poisson distribution are

$$\begin{aligned} \bar{n} &= \sum_{n=0}^{\infty} n g(n) = \sum_{n=0}^{\infty} n \frac{\lambda^n \exp(-\lambda)}{n!} \\ &= \lambda \exp(-\lambda) \sum_{n=1}^{\infty} \frac{\lambda^{n-1}}{(n-1)!} = \lambda, \end{aligned} \tag{2.8.34}$$

$$\sigma^2 = \overline{(n-\bar{n})^2} = \lambda. \tag{2.8.35}$$

The proof of (2.8.35) is left as an exercise.

As an example, suppose we have a gas in a volume V, and density $\rho = \frac{N}{V}$. At equilibrium, the density at each point in the system is constant. The probability of finding a specific particle in a small region v is simply

$$p = \frac{v}{V}. \tag{2.8.36}$$

The average number of particles in v is

$$\lambda = Np = \frac{Nv}{V} = \rho v. \tag{2.8.37}$$

If $v \ll V$, the probability p is very small. The probability of finding n particles in v follows the Poisson distribution, i.e.,

$$g(n) = \frac{\lambda^n \exp(-\lambda)}{n!} = \frac{(\rho v)^n \exp(-\rho v)}{n!}. \tag{2.8.38}$$

Exercise: The density of molecules in a gas is $\rho = 10^{20}$ particles per cm^3. Calculate the probability of finding no more than ten particles in a region of volume $0.001\,\text{cm}^3$.

2.9 Generating Functions

Generating functions (GFs) are a kind of bridge between discrete mathematics and continuous analysis. In probability theory, generating functions are extremely powerful tools for solving discrete problems. Sometimes they offer a simple and elegant solution to a seemingly intractable problem.

We shall discuss here only random variables that attain integral values, e.g., $k = 1, 2, 3, 4, 5, 6$ in throwing a die.

Definition: Let $a_0, a_1, a_2, \ldots,$ be a sequence of real numbers. We define the generating function of this set as the function

$$A(t) = a_0 + a_1 t + a_2 t^2 + \cdots . \tag{2.9.1}$$

If the sequence $\{a_i\}$ is finite, then $A(t)$ is simply a polynomial in the variable t. If $\{a_i\}$ is infinite, then $A(t)$ is a power series in t, in which case, it might be convergent in some range of values, say $t_1 \leq t \leq t_2$. The function $A(t)$ is referred to as *generating* function.[25] Whenever $A(t)$ is known in some closed form, one can *generate*, or calculate the coefficients in (2.9.1), simply by taking the derivatives of $A(t)$, i.e.,

$$A(0) = a_0, \tag{2.9.2}$$

$$\left.\frac{\partial A}{\partial t}\right|_{t=0} = a_1, \tag{2.9.3}$$

$$\left.\frac{\partial^2 A}{\partial t^2}\right|_{t=0} = 2a_2, \tag{2.9.4}$$

and in general

$$a_i = \frac{1}{i!}\left.\frac{\partial^i A}{\partial t^i}\right|_{t=0}. \tag{2.9.5}$$

Examples:

(i) The GF of the infinite sequence $a_i = 1$, for $i = 0, 1, 2, \ldots$ is

$$A(t) = \sum_{i=0}^{\infty} 1 \times t^i. \tag{2.9.6}$$

This sum is convergent for $|t| < 1$ and the sum is

$$A(t) = \frac{1}{1-t}. \tag{2.9.7}$$

---

[25]This particular generating function is also referred to as *ordinary*, or *probability* generating function.

(ii) The GF of the finite sequence (for fixed n)

$$a_i = \binom{n}{i}$$

is

$$A(t) = \sum_{i=0}^{n} \binom{n}{i} t^i = (1+t)^n. \tag{2.9.8}$$

(iii) The GF of the infinite sequence

$$a_i = \frac{1}{i!}$$

is

$$A(t) = \sum_{i=0}^{\infty} \frac{t^i}{i!} = \exp(t). \tag{2.9.9}$$

In the above examples, we have defined the GF for a few series of numbers. In application to probability theory, we shall apply the concept of GF to a random variable $(rv)X$ that attains only integral numbers. As a shorthand notation, we denote

$$p_i = P\{w : X(w) = i\}, \tag{2.9.10}$$

i.e., p_i is the probability of the event $\{w : X(w) = i\}$ consisting of all w for which $X(w) = i$, where i is an integer.

The GF associated with the *rv* X is defined by

$$P_X(t) = \sum_{i=0}^{\infty} p_i t^i. \tag{2.9.11}$$

If the sequence is finite, then the sum is reduced to a polynomial. If it is infinite, then since all $p_i \leq 1$, the series converges at least in the range $-1 < t < 1$.

From the GF we can easily get the average and the variance of the *rv* X. The first derivative is

$$P'_X(t) = \sum_{i=0}^{\infty} p_i i t^{i-1}, \tag{2.9.12}$$

and hence, and evaluating at $t = 1$, we get

$$P'_X(t=1) = \sum_{i=0}^{\infty} i p_i = \langle x \rangle, \tag{2.9.13}$$

which is the average of the *rv* X [note that this average does not necessarily exist if the sum (2.9.13) does not converge].

Similarly, the variance may be easily obtained from

$$\begin{aligned}\sigma^2 &= Var(X) = P''_X(t=1) + P'_X(t=1) - (P'_X(t=1))^2 \\ &= \langle X^2\rangle - \langle X\rangle^2. \end{aligned} \tag{2.9.14}$$

The most important and useful theorem which renders many difficult problems soluble is the *convolution theorem.*

Let X and Y be the two independent random variables that attain non-negative integral numbers. We assume that the distributions of the two $r\nu$ are given:

$$P(X=i) = a_i, \tag{2.9.15}$$

$$P(Y=i) = b_i. \tag{2.9.16}$$

We define the sum of the two $r\nu$:

$$S = X + Y, \tag{2.9.17}$$

which is itself an $r\nu$. We are interested in the distribution of the sum S. Let us denote it by

$$P(X+Y=r) = P(S=r) = c_r. \tag{2.9.18}$$

Clearly, we can write

$$\begin{aligned}c_r &\equiv P(X+Y=r) \\ &= \sum_i P(X=i, Y=r-i) = \sum_i a_i b_{r-i}. \end{aligned} \tag{2.9.19}$$

In the second equality, we sum over all i. We can let i change over the range of all integers. Those integers for which the probability is zero will not contribute to the sum. If $X=i$, then necessarily $Y=r-i$, and since we have assumed independence of the two $r\nu$, X and Y, we get the last equality in (2.9.19).

As an example, suppose that X and Y are the $r\nu$ of throwing independent dice. Clearly, the new $r\nu$ $S=X+Y$ has the range of values $r=2,3,\ldots,12$. For instance,

$$\begin{aligned}c_7 &= P(S=7) = \sum_i P(X=i, Y=7-i) \\ &= a_1b_6 + a_2b_5 + \cdots + a_6b_1. \end{aligned} \tag{2.9.20}$$

As we see, for two $r\nu$ like these, it is relatively easy to calculate the distribution of the sum. It is very difficult, sometimes impossible, to do the same for more complicated cases, e.g., for three or more dice. In these cases, the convolution theorem comes to our aid.

Let $A(t)$ and $B(t)$ be the GF of X and Y, respectively, i.e.,

$$A(t) = \sum_i a_i t^i, \tag{2.9.21}$$

$$B(t) = \sum_j b_j t^j. \tag{2.9.22}$$

The convolution theorem states that the GF of the $r\nu$ $S = X + Y$ is given by the product of $A(t)$ and $B(t)$, i.e.,

$$C(t) = A(t)B(t). \tag{2.9.23}$$

This is easy to prove by starting from the result (2.9.23) and rewriting it as

$$\begin{aligned} A(t)B(t) &= \sum_{i=0}^{\infty} a_i t^i \sum_{j=0}^{\infty} b_j t^j \\ &= \sum_{i=0}^{\infty}\sum_{j=0}^{\infty} a_i b_j t^{i+j}. \end{aligned} \tag{2.9.24}$$

We can now regroup the terms in (2.9.24). Instead of the summation on columns and rows, as in (2.9.24), we can sum over the elements of the diagonals, i.e., elements for which the sum $i + j = r$ is constant (Figure 2.16).

Hence, we write

$$\begin{aligned} \sum_{i=0}^{\infty}\sum_{j=0}^{\infty} a_i b_j t^{i+j} &= \sum_{r=0}^{\infty} t^r \sum_{k=0}^{r} a_k b_{r-k} \\ &= \sum_{r=0}^{\infty} t^r c_r = C(t). \end{aligned} \tag{2.9.25}$$

Thus, we have started with the product of $A(t)$ and $B(t)$ in (2.9.24) and obtained $C(t)$, which is the required result (2.9.23).

The generalization of the convolution theorem to any number of independent $r\nu$ is quite straightforward. Let X_k with $k = 1, 2, \ldots, n$ be a sequence of $r\nu$.

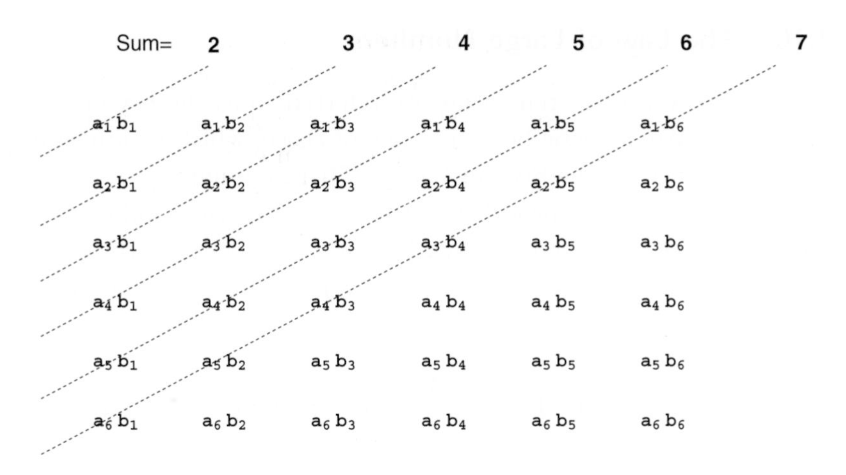

Figure 2.16. Summation along the diagonal lines.

We denote by

$$p_i^{(k)} = P(X_k = i) \tag{2.9.26}$$

the distribution of the kth $r\nu$, i.e., $p_i^{(k)}$ is the probability of the event $\{X_k = i\}$. Define the sum of these random variables:

$$S = X_1 + X_2 + \cdots + X_n. \tag{2.9.27}$$

The generating function of the sum S is the product of the generating functions of all the X_k, i.e.,

$$P_S(t) = P_{X_1}(t)P_{X_2}(t)\cdots P_{X_n}(t). \tag{2.9.28}$$

A special case is when all of the X_k have the same distribution, i.e.,

$$p_i^{(k)} = P(X_k = i) = p_i, \tag{2.9.29}$$

where p_i is independent of k. For this case, (2.9.28) reduces to

$$P_S(t) = \prod_{k=1}^{n} P_{X_k}(t) = [P_X(t)]^n, \tag{2.9.30}$$

where $P_X(t)$ is the GF of one of the random variables X_k.

The generalized convolution theorem can be easily proven by mathematical induction.

2.10 The Law of Large Numbers

There are several theorems that are referred to as the law of large numbers.[26] We describe here only one of these, which is important in connection with the Second Law of Thermodynamics.

In the axiomatic approach to probability, the probability of an event A is *given*, and denoted by p. The theory does not provide the value of p, nor a method of calculating p. However, the following theorem holds. For any $\varepsilon > 0$

$$\lim_{N_T \to \infty} \Pr\left\{\left|\frac{n(A)}{N_T} - p\right| < \varepsilon\right\} = 1, \tag{2.10.1}$$

where $n(A)$ is the number of experiments (or trials) in which the event A occurred, and N_T is the total number of experiments.

The theorem states that if we make a very large number of experiments and record the number of outcomes A, then for any $\varepsilon > 0$ the probability of finding the *distance* between $n(A)/N_T$ and p, smaller than ε, is, in the limit of $N_T \to \infty$, unity. In other words the frequency $n(A)/N_T$ tends to the probability p at very large numbers of experiments. This is what we expect intuitively.

We have already seen the DeMoivre–Laplace theorem (Section 2.8), which is also a limiting form of the Bernoulli distribution:

$$\binom{N}{n} p^n q^{N-n} \to \frac{1}{\sqrt{2\pi Npq}} \exp[-(n - pN)^2/2Npq], \tag{2.10.2}$$

where $q = 1 - p$.

We now transform the inequality $|\frac{n(A)}{N_T} - p| < \varepsilon$ as follows. We first rewrite it as

$$-\varepsilon < \frac{n(A)}{N_T} - p < \varepsilon \tag{2.10.3}$$

or equivalently

$$-\varepsilon N_T < n(A) - pN_T < p < \varepsilon N_T. \tag{2.10.4}$$

[26] See, for example, Feller (1957) and Papoulis (1990).

We now define n_1 and n_2:

$$n_1 \equiv -\varepsilon N_T + pN_T < n(A) < \varepsilon N_T + pN_T \equiv n_2 \tag{2.10.5}$$

Therefore, we can rewrite the left-hand side of (2.10.1) as

$$\Pr\left\{\left|\frac{n(A)}{N_T} - p\right| < \varepsilon\right\}$$
$$= \Pr\{-\varepsilon N_T + pN_T < n(A) < \varepsilon N_T + pN_T\}$$
$$\xrightarrow{(2.10.2)} \frac{1}{\sqrt{2\pi N_T pq}} \int_{-\varepsilon N_T + pN_T}^{\varepsilon N_T + pN_T} \exp[-(n - pN_T)^2/2N_T pq]dn, \tag{2.10.6}$$

where $q = 1 - p$.

It is convenient to define the following function

$$erf(x) = \frac{1}{\sqrt{2\pi}} \int_0^x \exp[-y^2/2]dy. \tag{2.10.7}$$

This function is referred to as the error function. This function is applied in many branches of mathematics, not only in error analysis.

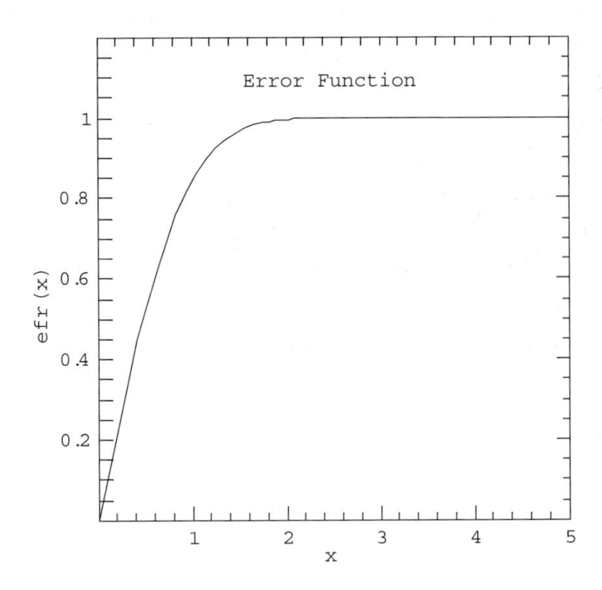

Figure 2.17. The form of the error function (2.10.7).

Hence, (2.10.6) can be written as

$$\Pr\left\{\left|\frac{n(A)}{N_T} - p\right| < \varepsilon\right\}$$
$$\to erf\left[\frac{n_2 - pN_T}{\sqrt{N_T pq}}\right] - erf\left[\frac{n_1 - pN_T}{\sqrt{N_T pq}}\right]$$
$$= erf\left[\frac{\varepsilon N_T}{\sqrt{N_T pq}}\right] - erf\left[\frac{\varepsilon N_T}{\sqrt{N_T pq}}\right] = 2erf\left[\varepsilon\sqrt{\frac{N_T}{pq}}\right]. \quad (2.10.8)$$

The last equality follows from the property of the error function

$$erf(-x) = -erf(x). \quad (2.10.9)$$

Figure 2.7 shows the form of the function $erf(x)$.

As $N_T \to \infty$, we have

$$\lim_{N_T\to\infty} \Pr\left\{\left|\frac{n(A)}{N_T} - p\right| < \varepsilon\right\} = \lim_{N_T\to\infty} 2erf\left[\varepsilon\sqrt{\frac{N_T}{pq}}\right] = 1. \quad (2.10.10)$$

Thus, for any fixed p (and $q = 1-p$) and any $\varepsilon > 0$, this probability tends to one. We shall use this result in connection with the Second Law of Thermodynamics (Section 6.12) for the special case $p = q = 1/2$:

$$\sum_{n=\frac{1}{2}N-\varepsilon N}^{\frac{1}{2}N+\varepsilon N} \left(\frac{1}{2}\right)^N \binom{N}{n} \approx 2erf[2\varepsilon\sqrt{N}] \xrightarrow{N\to\infty} 1. \quad (2.10.11)$$

Hence, the probability of finding the system in a small neighborhood of the maximum ($n = 1/2N$), i.e., ($\frac{1}{2}N - \varepsilon N < n < \frac{1}{2}N + \varepsilon N$) tends to one when N is large. In applying the error function for a thermodynamic system, N is of the order of 10^{23}, therefore $\varepsilon > 0$ can be chosen to be very small in such a way that the value of the error function is almost one!

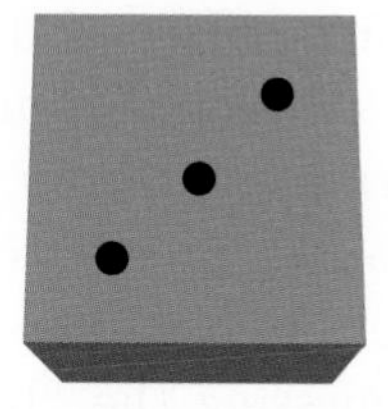

Chapter 3

Elements of Information Theory

Information theory was officially born in 1948.[1] The theory was developed in connection with the problem of transmission of information along communication channels. The main concerns of the theory were to search for a method of transmitting information through a noisy channel with high efficiency and reliability, i.e., at a reasonably high rate but with minimal errors.[2] Earlier measures of information, notably Hartley's measure[3] were applied in the theory of communication. However, it was Shannon's measure that was found useful in many branches of science in general, and in statistical mechanics, in particular. As noted in the introduction (Section 1.3), the *concept* of information was mentioned in connection with the thermodynamic entropy and the Second Law of Thermodynamics long before the development of information theory.

"Information," like probability, is a very general, qualitative, imprecise and very subjective concept. The same "information"

---

[1] Shannon (1948).

[2] Shannon has shown that even in noisy channels (i.e., channels that introduce errors), it is possible to transmit information at a non-zero rate with any arbitrary small probability of errors.

[3] Hartley (1927). It is interesting to note that Hartley developed "a quantitative measure of 'information' which is based on physical as contrasted with psychological consideration." Here, the "physical" and the "psychological" refers to what we call objective and subjective. Hartley defined as a "practical measure of information, the logarithm of the number of possible symbol sequences." As we shall see, Shannon's measure is an extension of Hartley's measure of information.

might have different meanings, effects and values to different people. Yet, as probability theory developed from a subjective and imprecise concept, so did information theory, which was distilled and developed into a quantitative, precise, objective and very useful theory. In the present book, *information* is a central concept. It is used both as a tool for guessing the "best" probability distributions, as well as for interpretation of fundamental concepts in the statistical mechanical theory of matter. In using the term *information* in information theory, one must keep in mind that it is not the *information* itself, nor the amount of information carried by a message, that we are concerned with, but the amount, or the size, of the message that carries the information. We shall further discuss this aspect of information in Section 3.3.

In this chapter, we present the basic definition, properties and meanings of the measure of information as introduced by Shannon. Today, this measure is used in many diverse fields of research from economics to biology, psychology, linguistics and many others.

We shall start with a qualitative idea of the measure of information. We then proceed to define the Shannon measure of information and list some of its outstanding properties. We shall devote Section 3.3 to discussing various interpretations of the quantity $-\sum p_i \ln p_i$. Finally, we discuss in Section 3.4 the method of "guessing" the best distribution that is consistent with all the known relevant information on the system.

3.1 A Qualitative Introduction to Information Theory

Let us start with a familiar game. I choose a person, and you have to find out who the person I chose is by asking binary questions, i.e., questions which are only answerable by "Yes" or "No." Suppose I have chosen Einstein. Here are two possible "strategies" for asking questions:

Dumb "Strategy"	Smart "Strategy"
1) Is it Nixon?	1) Is the person a male?
2) Is it Gandhi?	2) Is he alive?
3) Is it me?	3) Is he in politics?

4) Is it Marilyn Monroe?	4) Is he a scientist?
5) Is it you?	5) Is he very well-known?
6) Is it Mozart?	6) Is he Einstein?
7) Is it Bohr?	
8) ...	

In listing the two strategies above, I have qualified the two strategies as "dumb" and "smart." The reason is that, if you use the first "strategy," you might, if you are lucky, hit upon the right answer on the first question, while with the smart "strategy," one cannot possibly win after one question. However, in the dumb "strategy," hitting upon the right answer on the first guess is highly improbable. It is more likely that you will keep asking for a long time specific questions like those in the list, and in principle, never find the right answer. The reason for preferring the second "strategy" is that for each question you ask, you get more *information* (see below for a more precise definition), i.e., after each answer you get, you exclude a large number of possibilities (ideally, half of the possibilities; see below). In the smart "strategy," if the answer to the first question is "Yes," then you have excluded a huge number of possibilities — all females. If the answer to the second question is "No," then you have excluded all living persons. In each of the further answers you get, you narrow down further the *range* of possibilities, each time excluding a large group. In the dumb "strategy" however, assuming you are not so lucky to hit upon the right answer on the first few questions, at each point you get an answer, you exclude only *one* possibility, and practically, you almost have not changed the range of unknown possibilities. Even though we have not yet *defined* the term information, it is intuitively clear that in the smart "strategy," you gain more *information* from each answer than in the dumb "strategy." It feels correct to invest patience in choosing the smart "strategy," than to rush impatiently, and try to get the right answer quickly.

All that was said above is very qualitative. The term "information" used here is very imprecise. Shortly, we shall make the game more precise, the term "information" more quantitative, and the term "strategy" clearer so as to justify the reference to the two sets of questions as "dumb" and "smart." To achieve that, we have to "reduce" this type of game in such a way that it becomes more amenable to a quantitative treatment, and devoid of any traces of subjectivity.

The new game, described below is essentially the same game as before, but in its distilled form, it is much simpler and more amenable to a precise, quantitative and objective treatment.

Suppose we have eight equal boxes (Figure 3.1). I hide a coin in one of the boxes and you have to find where I hid it. All you know is that the coin *must* be in one of the boxes, and that I have no "favored" box. The box was chosen at random, or equivalently, there is a chance of 1/8 of finding the coin in any specific box.

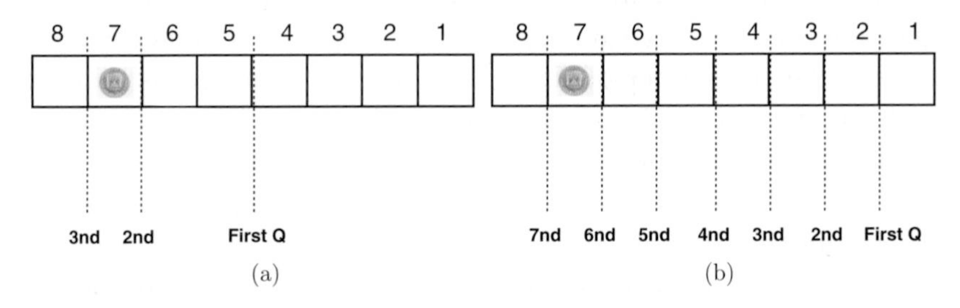

Figure 3.1. A coin is hidden in one of eight boxes: (a) and (b) are two strategies (smart and dumb) of asking binary questions, as described in Section 3.1.

Note that in this game, we have completely removed any traces of subjectivity – the *information* we need is "where the coin is." The fact that I, or you, or anyone else does not know where the coin is hidden does not make this information a subjective one. The information is a "built in" concept in the game and is not dependent on the person who plays or does not play the game.

Clearly, what one needs to acquire by asking questions is the *information* as to "where the coin is." To acquire this information, you are allowed to ask only binary questions. Thus, instead of an indefinite number of persons in the previous game, we have only eight possibilities.

Again, there are many strategies for asking questions. Here are two extreme and well-defined strategies.

The Dumbest Strategy	The Smartest Strategy
1) Is the coin in box 1?	1) Is the coin in the right half (of the eight)?
2) Is the coin in box 2?	2) Is the coin in the right half (of the remaining four)?
3) Is the coin in box 3?	3) Is the coin in the right half (of the remaining two)?
4) Is the coin in box 4?	4) I know the answer!
5) ...	

Although we did not *define* the two strategies in the new game, it is clear that these two correspond to the two strategies of the previous game. Now, however, it is clear that we can define the smartest strategy by asking questions to locate the coin in which *half* of the possibilities at each point. In the previous game, it was not clear what *all* the possibilities are, and even less clear if division by half is possible.

Note also that here I used the adjectives "dumbest" and "smartest." (I could not do that in the previous game, so I just wrote "dumb" and "smart.") The reason is that one can prove mathematically that if you choose the smartest strategy and if you play the game many, many times, you will out beat any other possible strategy, including the worst one denoted the "dumbest." Before turning to mathematics, let us examine qualitatively why the "smartest" strategy is far better than the "dumbest" one.

Qualitatively, if you choose the "dumbest" strategy, you might hit and guess the right box on the first question. But this happens with a probability of 1/8 and you miss with a probability of 7/8. Presuming you missed on the first question (which is more likely and far more likely with a larger number of boxes), you will have a chance of a hit with a probability of 1/7 and a miss with a probability of 6/7, on the second question, and so on. If you miss in six questions, after the seventh question you will *know* the answer, i.e., you will have the information as to where the coin is. If on the other hand, you choose the "smartest" strategy, you will certainly fail to get the required information on the first question. You will also fail on the second question, but you are *guaranteed* to obtained the required information on the third question.

Thus, although you might choose the dumbest strategy, and with low probability, hit upon the *first* question and win, you are better off choosing the second. The qualitative reason is the same as in the previous game: with the dumbest strategy, you can win on the first question but with very low probability. If you fail after the first question, you have eliminated only the first box and decreased slightly the number of remaining possibilities; from 8 to 7. On the other hand, in the smartest strategy, the first question eliminates *half* of the possibilities, leaving only four possibilities. The second question eliminates another half, leaving only two, and in the third question, you get the required information.

In information theory, the amount of missing information (MI), i.e., the amount of information one needs to acquire by asking questions, is *defined* in terms of the distribution of probabilities. In this example, the distribution is: $\{1/8, 1/8, 1/8, 1/8, 1/8, 1/8, 1/8, 1/8\}$. In asking the smartest question, one gains from each answer the maximum possible information (this is referred to as one bit of information). We shall soon see that maximum information is obtained in each question when you divide the space of all possible outcomes in two *equally probable* parts. In this particular case, we have divided at each step, all of the possibilities into two halves which have probability 1/2. In more general cases, it might not be possible to divide into two equally probable halves, say when the boxes are not equally probable, or when the number of boxes is odd.

Note also that the *amount* of information that is required is the same, no matter what strategy you choose. The choice of strategy allows you to get the same amount of information by different numbers of questions. The smartest strategy guarantees that you will get it, on average, by the minimum number of questions.

The important point to be noted at this stage is that no matter what strategy you choose, the larger the number of boxes, the larger the amount of information you need in order to find the coin; hence, the larger the number of questions needed to gain that information. This is clear intuitively. The amount of information is *determined* by the distribution [which in our case is $\{1/M, \ldots, 1/M\}$ for

M equally probable boxes]. The *same* information is collected on *average* with the minimal number of questions if you choose the smartest strategy.

Clearly, the information itself might have different meanings and values. What is relevant in information theory is only the *size* of the message, which we can measure as the number of questions one needs to ask in order to obtain the missing information (MI). Clearly, in our simple game of a coin hidden in M boxes, the number of boxes, M (or $\log M$) is an objective quantity "belonging" to the game.

The fact that one person knows where the coin is hidden and another person is ignorant of the location of the coin does not make the quantity of the MI a subjective quantity.

As we shall see in the next section, the quantity H defined by Shannon increases with the number of boxes. However, the information (in the general sense of the word) about "where the coin is?" is clearly independent of the number of boxes. The answer to the question "where the coin is?" is simply "the coin is in box x." Clearly, this is almost independent of M, (depending on how you express x; if x is expressed in words then there is a minor difference in the size of the message when we say, in the "second" or in the "tenth" box). What depends on M is the average number of questions one needs to ask to get that information (in communication theory the analog of M would be the size of the alphabet in the language).

To make the game completely impersonal, thereby removing any traces of "subjectivity," you can think of playing against a computer.[4] The computer chooses a box and you ask the computer the binary questions. Suppose you pay a cent for each answer you get for your binary questions. Certainly, you would like to get the required information (where the coin is hidden) by paying the least amount of money. By choosing the smartest strategy, you will get full value for your money.

Even without mathematical analysis, it is clear that in this particular example (where all boxes are equally probable), the larger

[4]Or better yet, two computers playing the game. One chooses a box and the second asks binary questions to find the chosen box.

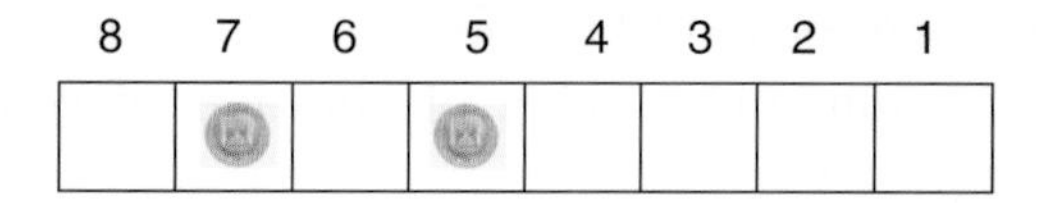

Figure 3.2. Two coins hidden in eight boxes.

the number of boxes, the larger the number of questions that have to be asked. If we use the smartest strategy, doubling the number of boxes requires only adding one more question. However, in the dumbest strategy, you would need to ask many more questions.

Let us go one step further. We are told that *two* different coins were hidden in N boxes (Figure 3.2). Let us assume that the two coins were placed in two *different* boxes (i.e., no two coins in the same box). Again, we know that the boxes were chosen at random. For the first coin, we have one out of N boxes; for the second, we have one out of $(N-1)$ boxes, etc. In the next chapter, we shall discuss several cases of such systems where n particles are distributed over N boxes.

3.2 Definition of Shannon's Information and Its Properties

In the mathematical theory of information as developed by Shannon, one starts by considering a random variable X, or an experiment (or a game). The probability distribution of X, $p_1, p_2, \ldots, p_n$, is presumed to be given.[5] The question posed by Shannon is the following: "*Can we find a measure of how much 'choice' is involved in the selection of the event, or of how much uncertain we are of the outcome?*" Shannon then assumed that if such a function, denoted $H(p_1, \ldots, p_n)$, exists, it is reasonable to expect that it will have the following properties.

(i) H should be continuous in all the p_i.

[5]Here we are *given* the distribution and we are asked to construct a function that measures information. In the application of information theory to statistical thermodynamics, we shall use the quantity referred to as information to obtain the "best" distribution. See Section 3.4.

(ii) If all the p_i are equal, i.e., $p_i = 1/n$, then H should have a maximum value and this maximum value should be a monotonic increasing function of n.

(iii) If a choice is broken down into successive choices, the quantity H should be the weighted sum of the individual values of H.

These are not only desirable properties of H, but also reasonable properties that we expect from such a quantity. The first assumption is reasonable in the sense that if we make an arbitrarily small change in the probabilities, then we expect that the change in the uncertainty should also be small. The second requirement is also plausible. For a fixed n, if the distribution is uniform, we have the least information on the outcome of the experiment, or the maximum uncertainty as to the outcome of the experiment. Clearly, the larger the number n is, the larger the required information. The third requirement is sometimes referred to as *the independence on the grouping of the events*. This requirement is equivalent to the statement that the missing information should depend only on the distribution $p_1, \ldots, p_n$ and not on the specific way we acquire this information, for instance, by asking binary questions using different strategies. We shall see ample examples in the following sections. With these plausible requirements of the expected function, Shannon then proved that the only function that satisfies these three requirements is[6]

$$H(p_1, \ldots, p_n) = -K \sum_{i=1}^{n} p_i \log p_i, \tag{3.2.1}$$

where K is some positive constant. In this book, we shall always take $K = 1$. The proof is given in Appendix F.

In this chapter, we shall define the quantity H, as originally defined by Shannon in (3.2.1) and examine its properties, and its

[6]Sometimes, we shall use the notation $H(X)$ instead of $H(p_1, \ldots, p_n)$. But one should be careful not to interpret $H(X)$ as implying that H is a function of X. H is a function of the distribution $p_1, \ldots, p_n$ pertaining to the *rv* X. The shorthand notation $H(X)$ should be read as the quantity H associated with the *rv* X. Note also that if $p_i = 0$, then $\log p_i = -\infty$, but $p_i \log p_i = 0$.

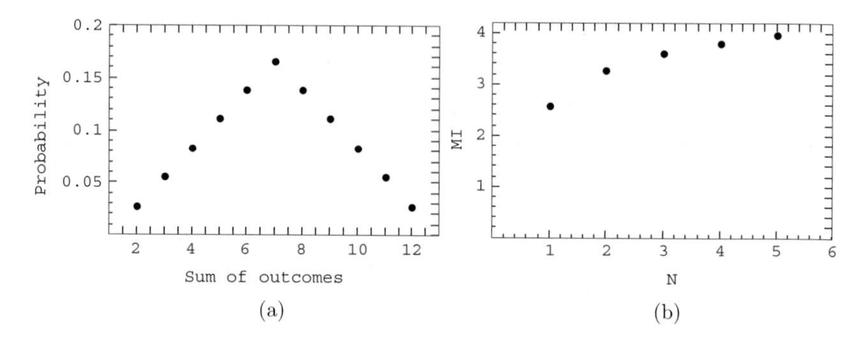

Figure 3.3. (a) Probability distribution for the sum of two dice. (b) The MI, defined in (3.2.1), for different numbers of dice, N (exercise, Section 3.2).

applications. We shall use log to mean logarithm in any base, and *ln* for logarithm to the base e.

Exercise: Two fair, identical and independent dice are thrown. Calculate the probability distribution of the sum of the outcomes and the associated quantity H.

Solution: Let X and Y be the *rv*, the outcomes of which are $\{1, 2, 3, 4, 5, 6\}$. We are interested in the sum $S = X + Y$ whose outcomes are $\{2, 3, 4, \ldots, 12\}$. The corresponding probabilities are plotted in Figure 3.3a. The entropy associated with this distribution is $H_2 = -\sum_{i=2}^{12} p_i \log_2 p_i = 3.274$.

Exercise: Suppose that we have N dice and define the sum of the outcomes of N dice, $S_N = X_1 + X_2 + \cdots + X_N$. Calculate the quantity H_N associated with the distribution S_N as a function of N.

In Figure 3.3a, we plot the probabilities of finding the sum of outcomes for two dice. In Figure 3.3b, we show the dependence of the MI on N.

Results: $H_1 = 2.58, H_2 = 3.27, H_3 = 3.59, H_4 = 3.81, H_5 = 3.98, H_6 = 4.11$.

3.2.1 Properties of the function H for the simplest case of two outcomes

Before discussing the general case of n outcomes, it is instructive to study the properties of H for some simple cases.

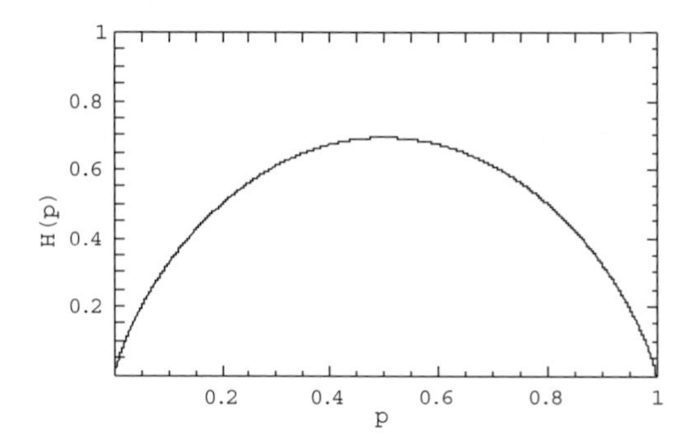

Figure 3.4. The function H for two outcomes; (3.2.2).

If there is only one possible outcome of an experiment, then $\Omega = A$, and $P(A) = 1$. In this case, we are *certain* of the outcome and no information is required, hence $H = 0$. The more interesting case is when we have only two possible outcomes, say A_1 and $A_2 = \bar{A}_1$, with probabilities p_1 and p_2, respectively. In this case, we write $p_1 = p$ and $p_2 = 1 - p$ and the H function is

$$H = -\sum p_i \log_2 p_i = -p \log_2 p - (1-p) \log_2(1-p). \qquad (3.2.2)$$

Figure 3.4 shows the function $H(p)$.[7] The function is positive everywhere, it is concave (or concave downward), it has a maximum at $p = 1/2$, and it is zero at both $p = 0$ and $p = 1$. It is easy to show that the function H tends to zero either at $p = 0$ or at $p = 1$.[8] The reason for this limit is that x tends to zero "faster," than $\ln x$ tends to minus infinity, therefore the product tends to zero as $x \to 0$. This property is consistent with what we intuitively expect from a quantity that measures the amount of information. If $p = 1$, then we know for certain that A_1 occurred. If $p = 0$, then we know for certain that A_2 occurred. In both cases, we do not need any further information. In other words, if we know that A_1

[7] Actually, we should write $H(p, q)$, where $q = 1 - p$. But in this case, H is a function of one variable p.

[8] Note that the function $x \ln x$ tends to zero as x tends to zero. This is easy to prove by L'Hôpital's theorem:

$$\lim_{x \to 0} [x \ln x] = \lim_{x \to 0} \frac{\ln x}{1/x} = \lim_{x \to 0} \frac{\frac{d}{dx}[\ln x]}{\frac{d}{dx}\left[\frac{1}{x}\right]} = \lim_{x \to 0} \frac{\frac{1}{x}}{\frac{-1}{x^2}} = 0.$$

(or A_2) has occurred, then there is no need to ask any questions to obtain this information. On the other hand, when $p = 1/2$, the two outcomes have equal probabilities. Hence, our uncertainty as to which event has occurred is maximal. If we use base 2 for the logarithm, for this case we get

$$H = -\frac{1}{2}\log_2\frac{1}{2} - \frac{1}{2}\log_2\frac{1}{2} = \log_2 2 = 1. \qquad (3.2.3)$$

The numerical *value* of the missing information in this case is one. This is also called one bit (binary digit) of information.

It is clear that for any value of $0 < p < 1/2$, we have more information than in the case $p = 1/2$. For instance, if we play with an uneven (or unfair) coin, and we know that it is more likely to fall with head upward, say $P(head) = 3/4$, we clearly have more information (or less MI) than when the coin is fair (but less if we know that it *always* falls with head upward). We can use this information to bet on the outcome "head," and on average, we can beat anyone who does not have that information and chooses randomly between head and tail.[9]

3.2.2 Properties of H for the general case of n outcomes

Let X be a random variable, with probability distribution given by $P_X(i) = P\{X(w) = i\} = p_i$. The subscript X is usually dropped when we know which random variable is considered.[10] Thus, we write

$$H = -\sum_{i=1}^{n} p_i \log p_i \qquad (3.2.4)$$

---

[9] When we say "choose randomly," we mean choose randomly with uniform distribution. In this section, we have used the personal language of "I know," "you know," or "we know," this or that information. The measure of information, however, is an objective quantity independent of the person possessing that information.

[10] We shall define H for any random variable, the probability distribution of which is given. Alternative terms that are used are: an *experiment* [Yaglom and Yaglom (1983)] or a scheme [Khinchin (1957)]. Here, we use log without specifying the base of the logarithm. In most of the applications in statistical thermodynamics, we shall use the natural logarithm denoted "ln".

with the normalization condition

$$\sum_{i=1}^{n} p_i = 1. \tag{3.2.5}$$

The most important property of H is that it has a maximum when all the p_i are equal. To prove that, we apply the method of Lagrange undetermined multipliers of finding a maximum of a function subject to a constrained condition (Appendix G).

We define the auxiliary function

$$F = H(p_1, \ldots, p_n) + \lambda \sum_i p_i. \tag{3.2.6}$$

Taking the partial derivatives of F with respect to each of the p_i, we have[11]

$$\left(\frac{\partial F}{\partial p_i}\right)_{p_i'} = -\log p_i - 1 + \lambda = 0 \tag{3.2.7}$$

or

$$p_i = \exp(\lambda - 1). \tag{3.2.8}$$

Substituting (3.2.8) into (3.2.5), we obtain

$$1 = \sum_{i=1}^{n} p_i = \exp(\lambda - 1) \sum_{i=1}^{n} 1 = n \exp(\lambda - 1).$$

Hence, from (3.2.8), we obtain

$$p_i = \frac{1}{n}. \tag{3.2.9}$$

This is an important result. It says that the maximum value of H, subject only to the condition (3.2.5), is obtained when the distribution is uniform. This is a generalization of the result we have seen in Figure 3.4.

The value of H at the maximum is

$$H_{\max} = -\sum_{i=1}^{n} p_i \log p_i = -\sum_{i=1}^{n} \frac{1}{n} \log \frac{1}{n} = \log n. \tag{3.2.10}$$

[11]The symbol p_i' stands for the vector $(p_1, p_2, \ldots, p_{i-1}, p_{i+1}, \ldots, p_n)$, i.e., all the components except p_i. Note also that here log stands for natural logarithm unless otherwise stated.

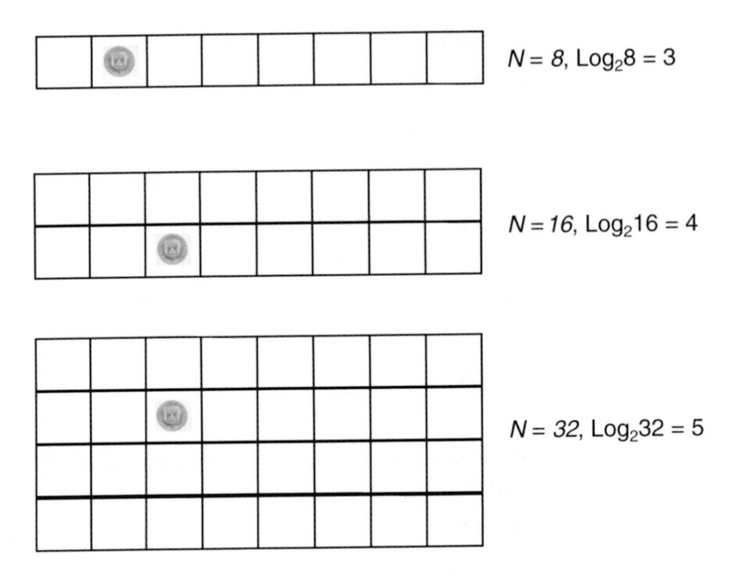

Figure 3.5. The MI for different numbers of boxes.

Clearly, when there are n equally likely events, the amount of missing information is larger, the larger the number of possible outcomes. It is instructive to consider again the game of hiding a coin in n boxes. For simplicity, let us assume that we have $n = 8, 16, 32$ boxes (Figure 3.5). It is intuitively clear that the larger the number of boxes, the larger the information we shall need to locate the hidden coin. For the three cases in Figure 3.5, the numbers of questions we need to ask using the smartest strategy are $3, 4, 5$, respectively. Incidentally, if we choose base 2 for the logarithm in (3.2.10), for these cases we get[12]

$$H(8) = \log_2 8 = 3, \quad H(16) = 4, \quad H(32) = 5,$$

i.e., the larger the information we need, the larger the number of binary questions we need to ask. By choosing the smartest strategy, we obtain the maximum information from each question, which is one bit of information. Hence, the amount of missing information is *equal* to the number of questions we need to ask to obtain the required information. Using any other strategy of asking questions

[12]Note again that $H(8)$ is a shorthand notation for $H(p_1, p_2, \ldots, p_8)$ when $p_i = 1/8$.

will be less "efficient," in the sense that each answer provides less than one bit of information, and hence we need to ask more questions to acquire the same information. Thus, if you pay, say, one cent for each answer, then on average, you will pay the least to get the same information when choosing the smartest strategy. It is clear now that the smartest strategy is the one in which we divide all possible events into two groups of equal probability.[13] As shown in the previous section, binary questions about two events of equal probabilities result in maximum information.

It should be stressed that the *amount* of missing information is fixed once the game is fully described in terms of the distribution. It does not depend on the way we ask questions, or on whether we ask questions at all. The missing information is *there* in the very description of the game, or more generally in the specification of the random variable. The number of questions and the amount of information obtained by each question can vary, but the total amount of information is fixed.

Exercise: In throwing two fair dice, it is known that the sum of the outcomes is an even number. What is the missing information H?

Solution: There are altogether 36 possible outcomes, 18 of which have a sum that is an even number. Therefore, $H = \log_2 18 = 4.17$.

It is clear that when the distribution is not uniform, the missing information is smaller than the maximal value of H. This in turn means that, on average, fewer questions need to be asked. An extreme case of a non-uniform distribution is the following. The quantity H is zero if and only if one event is certain (i.e., has probability one) and all others have probability zero. Clearly, if say $p_1 = 1$ and $p_i = 0$ for all $i = 2, \ldots, n$, it follows that

$$H = -\sum p_i \log p_i = 0. \tag{3.2.11}$$

This is a straightforward generalization of the result we have seen in Section 3.2.1. Conversely, if $H = 0$, then

$$-\sum p_i \log p_i = 0. \tag{3.2.12}$$

[13]In the examples given in this chapter, we have discussed only cases where M is a power of 2. However, the relation between H and the average number of questions is more general. See also Section 3.5.

But since each term is non-negative, it follows that each term in (3.2.12) must be zero individually. Hence,

$$p_i \log p_i = 0 \quad \text{for } i = 1, \ldots, n. \tag{3.2.13}$$

From (3.2.13), it follows that either $p_i = 0$ or $p_i = 1$. Since $\sum p_i = 1$, we must have all $p_i = 0$, except one, say p_1 which is one. The intuitive meaning of this result is clear. The amount of missing information is zero if we know that one specific event has occurred.

Suppose we have two random variables, X and Y with distributions $P_X(i) = P\{X = x_i\}$ and $P_Y(j) = P\{Y = y_j\}, i = 1, 2, \ldots, n$ and $j = 1, 2, \ldots, m$. Let $P(i, j)$ be the joint probability of occurrence of the events $\{X = x_i\}$ and $\{Y = y_j\}$. The H function defined on the probability distribution $P(i, j)$ is[14]

$$H(X, Y) = -\sum_{i,j} P(i, j) \log P(i, j). \tag{3.2.14}$$

The marginal probabilities are

$$p_i = \sum_{j=1}^{m} P(i, j) = P_X(i) \tag{3.2.15}$$

and

$$q_j = \sum_{i=1}^{n} P(i, j) = P_Y(j). \tag{3.2.16}$$

The information associated with the *rv* X and Y are

$$H(X) = -\sum_{i=1}^{n} P_X(i) \log P_X(i) = -\sum_{ij} P(i, j) \log \sum_{j=1}^{m} P(i, j), \tag{3.2.17}$$

$$H(Y) = -\sum_{j=1}^{m} P_Y(j) \log P_Y(j) = -\sum_{ij} P(i, j) \log \sum_{i=1}^{n} P(i, j). \tag{3.2.18}$$

[14]We shall use the shorthand notations: $p_i = P_X(i) = P\{X = x_i\}, q_j = P_Y(j) = P\{Y = y_j\}$ and $P_{ij} = P_{XY}(i, j) = P(i, j) = P\{X = x_i, Y = y_j\}$.

It is easy to show that for any two distributions $\{p_i\}$ and $\{q_i\}$ such that $\sum_{i=1}^{n} q_i = 1$ and $\sum_{i=1}^{n} p_i = 1$, the following inequality holds[15] (see Appendix H):

$$H(q_1, \ldots, q_n) = -\sum_{i=1}^{n} q_i \log q_i \leq -\sum_{i=1}^{n} q_i \log p_i. \tag{3.2.19}$$

From (3.2.14), (3.2.17) and (3.2.18), we obtain

$$\begin{aligned} &H(X) + H(Y) \\ &\quad = -\sum_i P_X(i) \log P_X(i) - \sum_j P_Y(j) \log P_Y(j) \\ &\quad = -\sum_{i,j} P(i,j) \log P_X(i) - \sum_{i,j} P(i,j) \log P_Y(j) \\ &\quad = -\sum_{i,j} P(i,j) \log \sum_{j=1} P(i,j) - \sum_{i,j} P(i,j) \log \sum_{i=1} P(i,j). \end{aligned} \tag{3.2.20}$$

Applying the inequality (3.2.19) to the two distributions $P(i,j)$ and $P_X(i)P_Y(j)$, we get

$$H(X) + H(Y) \geq -\sum_{i,j} P(i,j) \log P(i,j) = H(X,Y). \tag{3.2.21}$$

The equality holds when the two *rv* are independent, i.e., for

$$P(i,j) = P_X(i)P_Y(j), \quad \text{for each } (i,j), \tag{3.2.22}$$

we have

$$H(X,Y) = H(X) + H(Y). \tag{3.2.23}$$

The last two results simply mean that if we have two experiments (or two games), the outcomes of which are independent, then the missing information on the outcome of the two experiments is the sum of the missing information on the outcomes of each one of the experiments. On the other hand, if there is a dependence between the two sets of outcomes, then the missing information on

[15]This can be proven directly from the elementary inequality $\ln x \leq x - 1$ (see H.18 in Appendix H). Choosing $x = \frac{p_i}{q_i}$, we get $\ln \frac{p_i}{q_i} \leq \frac{p_i}{q_i} - 1$. Multiplying by q_i and summing over i, we get $\sum q_i \ln \frac{p_i}{q_i} \leq \sum p_i - \sum q_i = 0$, from which (3.2.19) follows.

the compound experiment (X, Y) will always be smaller than the missing information on the two experiments separately.

For dependent experiments, we use the conditional probabilities

$$P(y_j/x_i) = \frac{P(x_i \cdot y_j)}{P(x_i)} \tag{3.2.24}$$

to define the corresponding conditional quantity as[16]

$$\begin{aligned} H(Y/X) &= \sum_i P(x_i) H(Y/x_i) \\ &= -\sum_i P(x_i) \sum_j P(y_j/x_i) \log P(y_j/x_i) \\ &= -\sum_{i,j} P(x_i \cdot y_j) \log P(y_j/x_i) \\ &= -\sum_{i,j} P(x_i \cdot y_j) \log P(x_i \cdot y_j) + \sum_{i,j} P(x_i \cdot y_j) \log P(x_i) \\ &= H(X, Y) - H(X). \end{aligned} \tag{3.2.25}$$

Thus, $H(Y/X)$ measures the difference between the missing information on X and Y and the missing information on X. This can be rewritten as

$$\begin{aligned} H(X, Y) &= H(X) + H(Y/X) \\ &= H(Y) + H(X/Y), \end{aligned} \tag{3.2.26}$$

which means that the missing information in the two experiments is the sum of the missing information in one experiment plus the missing information when the outcome of the second experiment is known.

From (3.2.21) and (3.2.25), we also obtain the inequality

$$H(Y/X) \leq H(Y), \tag{3.2.27}$$

which means that the missing information on Y can never increase by knowing X. Alternatively, $H(Y/X)$ is the average uncertainty

[16] The quantity $H(Y/X)$ is sometimes denoted by $H_X(Y)$. $P(x_i \cdot y_j)$ is shorthand notation for the probability of the joint event $\{X = x_i, Y = y_j\}$, i.e., X attains x_i *and* Y attains y_j. The term "mutual information" is potentially misleading. It is better to refer to this quantity as a measure of the dependence between the two random variables.

that remains about Y when X is known. This uncertainty is always smaller than or equal to the uncertainty about Y. If X and Y are independent, then the equality sign in (3.2.27) holds. This is another reasonable property that we can expect from a quantity that measures MI or uncertainty. The quantity $H(Y/X)$ has been referred to as conditional information, or equivocation (Shannon 1948).

Another useful quantity is the *mutual* information defined by[17]

$$I(X;Y) \equiv H(X) + H(Y) - H(X,Y). \tag{3.2.28}$$

From the definition (3.2.28) and from (3.2.15) and (3.2.16), we have

$$\begin{aligned} I(X;Y) &= -\sum_{i=1}^{n} P_X(x_i)\log P_X(x_i) - \sum_{j=1}^{m} P_Y(y_j)\log P_Y(y_j) \\ &\quad + \sum_{i,j} P(i,j)\log P(i,j) \\ &= \sum_{i,j} P(i,j)\log\left[\frac{P(i,j)}{P_X(x_i)P_Y(y_j)}\right] \\ &= \sum_{i,j} P(i,j)\log g(i,j) \geq 0, \end{aligned} \tag{3.2.29}$$

where $g(i,j)$ is the correlation between the two events, $\{X = x_i\}$ and $\{Y = y_j\}$.

Note that $I(X;Y)$ is defined symmetrically with respect to X and Y. Sometimes, $I(X;Y)$ is referred to as the average amount of information conveyed by one *rv* X on the other *rv* Y, and *vice versa*. Thus, we see that $I(X;Y)$ is a measure of an average correlation [actually, it is the average of $\log g(i,j)$], i.e., it is a measure of the extent of dependence between X and Y. The inequality (3.2.29) is a result of the general inequality (3.2.21). The equality holds for independent X and Y.[18]

[17]Note that sometimes instead of $I(X;Y)$, the notation $H(X;Y)$ is used instead. This is potentially a confusing notation because of the similarities between $H(X,Y)$ and $H(X;Y)$.

[18]Note that the correlations $g(i,j)$ can be either larger than one or smaller than one, i.e., $\log g(i,j)$ can either be positive or negative. However, the average of $\log g(i,j)$ is always non-negative.

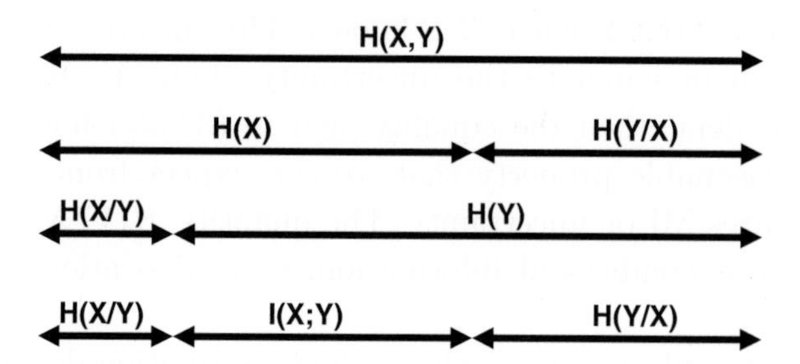

Figure 3.6. The relationships between the quantities $H(X), H(Y), H(X,Y)$ and $I(X;Y)$.

From (3.2.28) and (3.2.26), we can also write

$$I(X;Y) = H(Y) - H(Y/X) \geq 0$$

or

$$I(X;Y) = H(X) - H(X/Y) \geq 0. \tag{3.2.30}$$

The mutual information is always non-negative. It is zero when the two experiments are independent. Thus, $I(X;Y)$ measures the average reduction in MI about X resulting from knowing Y, and *vice versa*. The relations between $H(X), H(Y), H(X,Y)$ and $I(X;Y)$ are shown in Figure 3.6.[19]

Clearly, from the definition of $I(X;Y)$, it follows that

$$I(X;Y) = I(Y;X)$$

and

$$I(X;X) = H(X). \tag{3.2.31}$$

All the definitions of the quantities $H(X), I(X;Y)$, etc., can be extended straightforwardly to any number of random variables. For instance, if $X_1, \ldots, X_N$ are N random variables, the corresponding

[19] Some authors use Venn-like diagrams to show the relations between the various quantities $H(X), H(Y), H(X,Y)$ and $I(X;Y)$. This might be potentially confusing. In probabilities, two non-overlapping regions represent *mutually exclusive* events. On the other hand, $H(X)$ and $H(Y)$ are represented by non-overlapping regions when the random variables X and Y are *independent*.

function H is defined as

$$H(X_1, X_2, \ldots, X_N) = -\sum_{x_1, x_2, \ldots, x_N} P(x_1, x_2, \ldots, x_N) \times \log P(x_1, \ldots, x_N), \qquad (3.2.32)$$

where $P(x_1, x_2, \ldots, x_N)$ is the probability distribution of the joint event, $\{X_i = x_i\}$ (for $i = 1, 2, \ldots, N$). It is also easy to show that

$$H(X_1, \ldots, X_N) \leq H(X_1) + \cdots + H(X_N) \qquad (3.2.33)$$

and that the equality holds if and only if all the random variables are independent, in the sense that

$$I(X_1; \ldots; X_N) = \prod_{i=1}^{N} P(x_i). \qquad (3.2.34)$$

Actually, complete independence requires that this factorization of the joint probability will hold for any group of *rv*.

For dependent random variables, one can extend the definition of the mutual information as follows

$$I(X_1; X_2, \ldots, X_N) = \sum_{i=1}^{n} H(X_i) - H(X_1, \ldots, X_N) \geq 0. \quad (3.2.35)$$

We shall use this quantity in connection with the indistinguishability of the particles in Chapters 4 and 6.

Similarly, one can define the conditional amount of missing information for any set of random variables. For instance

$$H(X_{k+1}, \ldots, X_N / X_1, \ldots, X_k) = H(X_1, \ldots, X_k, X_{k+1}, \ldots, X_N) - H(X_1, \ldots, X_k) \leq H(X_{k+1}, \ldots, X_N), \qquad (3.2.36)$$

where the equality sign holds if and only if the random variables $(X_1, \ldots, X_k)$ are independent of the random variables $(X_{k+1}, \ldots, X_N)$.

Finally, we show that if we start with any distribution $\{p_1, \ldots, p_n\}$ and make a small change in the distribution toward uniformity, then H will always increase. This property follows from the existence of a single maximal value of H at the uniform distribution.

However, it is instructive to see the "mechanism" of how H changes upon the variations of the distribution.

Consider any arbitrary initial distribution $\{p_1, \ldots, p_n\}$. We take any two of these probabilities, say p_1 and p_2, such that $p_1 \neq p_2$, and make a small change, say from p_1 to $p_1 + dp$ and from p_2 to $p_2 - dp$, keeping all the other probabilities $p_3, \ldots, p_n$ fixed. The change in H is

$$\begin{aligned} dH &= \frac{\partial H}{\partial p_1} dp_1 + \frac{\partial H}{\partial p_2} dp_2 \\ &= (-\log p_1 - 1) dp_1 + (-\log p_2 - 1) dp_2. \end{aligned} \tag{3.2.37}$$

Since we need to conserve the sum $\sum p_i = 1, dp_1 = -dp_2 = dp$, from (3.2.37), we get

$$dH = -\log\left(\frac{p_1}{p_2}\right) dp. \tag{3.2.38}$$

If initially $p_1 < p_2$, then we must increase p_1 towards p_2, i.e., we must take $dp > 0$, which makes $dH > 0$. Similarly, if $p_1 > p_2$, we need to increase p_2 towards p_1, i.e., we must take $dp < 0$, and again we get $dH > 0$. Clearly, we can repeat this process of "equalization" in a pairwise fashion until we reach the uniform distribution, for which the quantity H attains its maximum value.

We can now summarize the main relationships between the concept of probability and information. Each random variable X defines a probability distribution $p_1, \ldots, p_n$. On this distribution, we define the MI (or the uncertainty, see Section 3.3), which we write as $H(p_1, \ldots, p_n)$, or $H(X)$ for short. Conditional information corresponds to a conditional probability (though we have to make a distinction between information on X given an event y_j, or the information on X given Y). For two independent random variables, the information on X and Y is the *sum* of the information on X and the information on Y. The correlation between the two *rv* is a measure of the extent of dependence between the two *rv*. The corresponding concept is the mutual information, which is also a measure of the average correlation between the two random variables.

3.2.3 The consistency property of the missing information (MI)

The third requirement as stated in Section 3.2 is the condition of consistency. This condition essentially states that the amount of information in a given distribution $(p_1, \ldots, p_n)$ is independent of the path, or of the number of steps we choose to acquire this information. In other words, the same amount of information is obtained regardless of the way or the number of steps one uses to acquire this information. In its most general form, the statement is formulated as follows.

Suppose we have n outcomes $A_1, \ldots, A_n$ of a given experiment and the corresponding probabilities are $p_1, \ldots, p_n$. We regroup all of the outcomes as, for instance,

$$\begin{aligned}
&\{A_1, A_2, A_3, A_4, A_5, A_6, A_7, \ldots, A_n\}\\
&\{A_1, A_2, A_3\}, \{A_4, A_5, A_6, A_7\} \cdots \{A_{n-2}, A_{n-1}, A_n\}\\
&A'_1, A'_2, \ldots, A'_r.
\end{aligned}$$

A'_1 is a new event consisting of the original events A_1, A_2, A_3. The corresponding probabilities are

$$p'_1, p'_2, \ldots, p'_r. \tag{3.2.39}$$

Thus, from the initial set of n outcomes, we constructed a new set of r new events $\{A'_1, A'_2, \ldots, A'_r\}$. Since all A_i are mutually exclusive, the probabilities of the new events are

$$p'_1 = \sum_{i=1}^{m_1} p_i, \quad p'_2 = \sum_{i=m_1+1}^{m_1+m_2} p_i, \ldots. \tag{3.2.40}$$

Thus, the event A'_1 consists of m_1 of the original events with probability p'_1, A'_2 consists of m_2 of the original events, with probability p'_2, and so on. Altogether, we have split the n events into r groups, each containing $m_k (k = 1, \ldots, r)$ of the original events so that

$$\sum_{k=1}^{r} m_k = n. \tag{3.2.41}$$

The consistency requirement is written as

$$H(p_1, \ldots, p_n) = H(p'_1, \ldots, p'_r) + \sum_{k=1}^{r} p'_k H\left(\frac{p_{l_{k-1}+1}}{p'_k}, \ldots, \frac{p_{l_k}}{p'_k}\right), \tag{3.2.42}$$

where we denoted $l_k = \sum_{i=1}^{k} m_i$, $l_0 = 0, l_r = n$.

The meaning of this equation is quite simple, though it looks cumbersome. The missing information in the original system $H(p_1, \ldots, p_n)$ is equal to the missing information of the set of the new r events $H(p'_1, \ldots, p'_n)$, plus the average missing information associated with each of the groups.

We shall defer the proof of the uniqueness of the function H to Appendix F. Here, we shall discuss a few simple examples of this condition.

Consider the case of four boxes, in one of which a coin is hidden. We assume that the probabilities of finding the coin in any one of the boxes are equal, 1/4. The amount of missing information is

$$H\left(\frac{1}{4}, \frac{1}{4}, \frac{1}{4}, \frac{1}{4}\right) = \log_2 4 = 2. \tag{3.2.43}$$

That is, we need two bits of information to locate the coin. We can obtain this information along different routes. The consistency requirement means that the amount of information obtained should be independent of the route, or of the "strategy" of questioning.

The first route is to divide the total number of boxes into two halves, each half having probability 1/2 (Figure 3.7a). For this case,

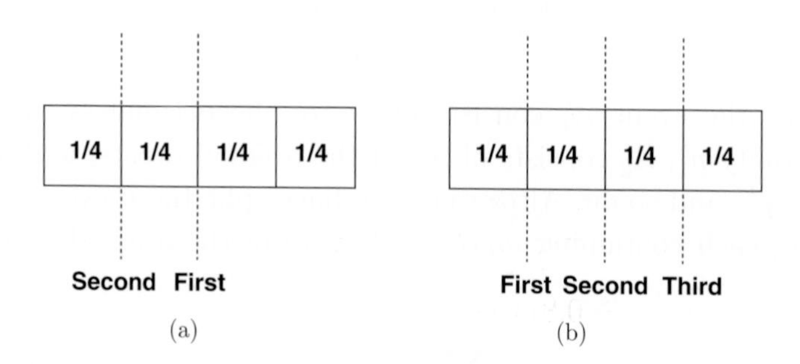

Figure 3.7. The two strategies of asking binary questions for four boxes.

the general equality (3.2.42) reduces to

$$H\left(\frac{1}{4},\frac{1}{4},\frac{1}{4},\frac{1}{4}\right) = H\left(\frac{1}{2},\frac{1}{2}\right) + \left[\frac{1}{2}H\left(\frac{1}{2},\frac{1}{2}\right) + \frac{1}{2}H\left(\frac{1}{2},\frac{1}{2}\right)\right] \tag{3.2.44}$$

which simply means that the MI in the original set of events is the sum of the MI associated with the two groups (left and right halves) and the average MI *within* the newly formed groups. After the first question, we do not know where the coin is. The reduction of the MI at this stage is

$$H\left(\frac{1}{2},\frac{1}{2}\right) = 1\,\text{bit}. \tag{3.2.45}$$

At the first step, we only know in which half of the boxes the coin is. In the next step, the average reduction in MI is

$$\left[\frac{1}{2}H\left(\frac{1}{2},\frac{1}{2}\right) + \frac{1}{2}H\left(\frac{1}{2},\frac{1}{2}\right)\right] = 1\,\text{bit}. \tag{3.2.46}$$

Thus, the sum of (3.2.45) and (3.2.46) is the same as in (3.2.43), i.e., the total missing information is 2 bits.

Now suppose we take a different route (Figure 3.7b). Instead of dividing into two halves, we divide into two groups; one box and three boxes. In this case, by applying (3.2.42), we have[20]

$$\begin{aligned}
H\left(\frac{1}{4},\frac{1}{4},\frac{1}{4},\frac{1}{4}\right) &= H\left(\frac{1}{4},\frac{3}{4}\right) + \left[\frac{1}{4}H(1) + \left(\frac{3}{4}\right)H\left(\frac{1}{3},\frac{1}{3},\frac{1}{3}\right)\right] \\
&= \left[H\left(\frac{1}{4},\frac{3}{4}\right)\right] + \left[\frac{3}{4}H\left(\frac{1}{3},\frac{2}{3}\right)\right] \\
&\quad + \frac{3}{4}\left[\frac{1}{3}H(1) + \frac{2}{3}H\left(\frac{1}{2},\frac{1}{2}\right)\right] \\
&= -\left[\frac{1}{4}\log_2\frac{1}{4} + \frac{3}{4}\log_2\frac{3}{4}\right] \\
&\quad - \frac{3}{4}\left[\frac{1}{3}\log_2\frac{1}{3} + \frac{2}{3}\log_2\frac{2}{3}\right] + \frac{1}{2} \\
&\cong 0.8113 + 0.6887 + 0.5 = 2.
\end{aligned} \tag{3.2.47}$$

[20]Note that $H(1) = 0$.

Note that by the first route, we gain one bit of information at each step. Therefore, we need exactly two steps to gain the required MI = 2. On the other hand, on the second route, the average amount of information gained at the first step is only 0.8113 bits. In the second step, 0.6887 bits, and in the third step 0.5 bits. Clearly, the total MI is again 2 bits. However, by this route, we shall need, on average, more steps or more questions to gain the same amount of information.

We shall return to this problem and similar ones in Section 3.5.

Exercise: Suppose you are given 26 letters (27 if we include the space between words) from the English alphabet. It is known that the frequency of occurrence of these letters is as in Table 3.1. I pick a letter at random (e.g., pick up a book and choose a page, and within the page point at a letter while the eyes are blindfolded). What is the best strategy for you to ask binary questions in order to find out what the letter is? How many binary questions do you need to ask on average if you use the best strategy? How much MI is there in this game (calculate with logarithm to the base 2)? Why is there a difference between the MI and the number of questions?

Table 3.1. The frequencies of the letters in the English language.

i	Letter	p_i	$-\log_2 p_i$	i	Letter	p_i	$-\log_2 p_i$
1	a	0.0575	4.1	15	o	0.0689	3.9
2	b	0.0128	6.3	16	p	0.0192	5.7
3	c	0.0263	5.2	17	q	0.0008	10.3
4	d	0.0285	5.1	18	r	0.0508	4.3
5	e	0.0913	3.5	19	s	0.0567	4.1
6	f	0.0173	5.9	20	t	0.0706	3.8
7	g	0.0133	6.2	21	u	0.0334	4.9
8	h	0.0313	5.0	22	v	0.0069	7.2
9	i	0.0599	4.1	23	w	0.0119	6.4
10	j	0.0006	10.7	24	x	0.0073	7.1
11	k	0.0084	6.9	25	y	0.0164	5.9
12	l	0.0335	4.9	26	z	0.0007	10.4
13	m	0.0235	5.4	27	–	0.1928	2.4
14	n	0.0596	4.1				

Solution: The total MI for this distribution is

$$H = -\sum_{i=1}^{27} p_i \log_2 p_i = 4.138\,\text{bits}. \qquad (3.2.48)$$

This means that in the "best strategy," we need to ask, on average, 4.138 questions. However, since we cannot divide the events into two groups with exactly equal probabilities at each step, the best we can do is divide into two groups of about equal probability. For example, in the first step choose the first 12 letters (with probability 0.504, and the remaining letters with probability 0.496), and so on in the next steps. In this way, you can get the answer in five questions. Clearly, this is much better than asking: Is the letter A? Or B? And so on.

Information theory was originally developed to deal with information transmitted along communication lines. We shall use this example to clarify some of the concepts we use in this book.

Consider the following message:

A: "Each of the houses in this street costs one million dollars" (3.2.49)

Clearly, the message A conveys *information* on the cost of houses in a certain street. This kind of information is not dealt with in information theory. If this message was never sent to me, I would obviously *miss* this information, but this is not the missing information we are dealing with.

Information theory deals only with a measure of the size of the message — not with the information carried by the message. The best way to describe the "size of the message" is to view each letter in the alphabet as a coin hidden in one of the 26 boxes (or 27 if we add the space between words). The average number of binary questions we need to ask to find out which letter was sent is between 4 and 5. Clearly, the more uniform the frequency of occurrence of the letters, the more questions we need to ask. This is the value of H per letter. Now multiply by the number of letters in the message and you get a feeling for the kind of the *size* of the message dealt with by information theory.

It is true that the concepts of "information," "missing information" or "amount of missing information," that feature in this book may or may not be objective quantities. However, the size of the message is objective, as objective as the number of letters in the alphabet, or the number of letters in the message itself.

Thus, the *same* information expressed in different languages or by different words can have different *sizes* in the above sense. On the other hand, totally different information might have the same *size*.

The quantity

$$R = \frac{\log_2 N - H(p_1, \ldots, p_N)}{\log_2 N}$$

is called the *redundancy* of the alphabet with N letters (or symbols). Since $\log N \geq H$ (for any *rv* with N outcomes), R varies between zero and one. It is zero when all the letters have equal probability. The other extreme case is when one letter has probability nearly one, and all others nearly zero. In this case $H = 0$, and the redundancy is one. In general, the redundancy is somewhere between zero and one. For the English language, $\log_2 N = \log_2 27 = 4.755$ and $R = 0.13$.

3.2.4 The case of an infinite number of outcomes

The case of discrete infinite possibilities is straightforward. First, we note that for a finite and uniform distribution, we have

$$H = \log n, \tag{3.2.50}$$

where n is the number of possibilities. Taking the limit of $n \to \infty$, we get

$$H = \lim_{n\to\infty} \log n = \infty, \tag{3.2.51}$$

which means that the missing information tends to infinity. Note, however, that the probabilities $1/n$ tend to zero.

For a non-uniform distribution, the quantity H might or might not exist, depending on whether the quantity

$$H = -\sum_{i=1}^{\infty} p_i \log p_i \tag{3.2.52}$$

converges or diverges.

The case of a continuous distribution is problematic.[21] If we start from the discrete case and proceed to the continuous limit, we get into some difficulties.[22] We shall discuss this problem in Appendix I. Here, we shall follow Shannon's treatment for a continuous distribution for which a density function $f(x)$ exists. By analogy with the definition of the H function for the discrete probability distribution, we define the quantity H for a continuous distribution. Let $f(x)$ be the density distribution, i.e., $f(x)dx$ is the probability of finding the random variable having values between x and $x + dx$.

We defined the H function as

$$H = -\int_{-\infty}^{\infty} f(x) \log f(x) dx. \tag{3.2.53}$$

A similar definition applies to an n-dimensional distribution function $f(x_1, \ldots, x_n)$. In (3.2.53), H is defined as the expectation value of the random variable $-\log f(X)$, where X is a random variable having distribution density $f(x)$.

In the following, we shall discuss three important cases that are relevant to statistical thermodynamics: the uniform distribution of the *location* of a particle in a one-dimensional box of length L, the normal distribution of the velocities in one direction, say v_x, and the Boltzmann distribution.

It should be noted that there is a fundamental difference between the first two distributions. Regarding the locational distribution, particles are always confined to some finite-size box, in any n-dimensional space.[23] On the other hand, there are in principle

[21] In Appendix I, we shall further discuss the difference between the MI as defined for discrete and continuous cases. Here, we note that H for the continuous case as defined in (3.2.53) may be negative, whereas in the discrete case, it is always positive.

[22] Shannon himself was apparently not worried about the mathematical difficulties involved in the generalization of the measure of information to the continuous case. See also Khinchin (1957).

[23] Note that if the particles were not confined to be within the limits of a box, there would be no equilibrium state. The particles would expand their "volume" indefinitely.

no limits on the possible range of velocities a particle can attain.[24] However, in an isolated system the total energy is constant, and at equilibrium the average kinetic energy of say, mono-atomic particles is also constant. In the three cases discussed below, we shall evaluate the probability densities which maximize H, to which we shall refer to as the equilibrium density.

3.2.4.1 *The uniform distribution of locations*

Consider a particle that is confined in a one-dimensional "box" of length L. We seek the maximum of H defined in (3.2.53), but with limits $(0, L)$, subject to the condition that

$$\int_0^L f(x)dx = 1. \tag{3.2.54}$$

Applying the Lagrange method of undetermined multipliers (or the calculus of variation), we find that the equilibrium probability density $f_{eq}(x)$ that maximize H, in (3.2.53), subject to the condition (3.2.54) must satisfy the equality

$$-1 - \log f_{eq}(x) + \lambda = 0. \tag{3.2.55}$$

From (3.2.54) and (3.2.55), we obtain

$$1 = \int_0^\infty f_{eq}(x)dx = e^{\lambda-1}\int_0^L dx = e^{\lambda-1}L \tag{3.2.56}$$

or equivalently

$$f_{eq}(x) = \frac{1}{L}. \tag{3.2.57}$$

The MI associated with this density is

$$H = -\int_0^L f_{eq}(x)\log f_{eq}(x)dx = -\frac{1}{L}\log\frac{1}{L}\int_0^L dx = \log L. \tag{3.2.58}$$

Thus, the equilibrium density distribution is uniform over the entire length L. The probability of finding the particle at any

[24] Although the velocity of light is the limit, in practice, a system of particles at room temperature would have a negligible number of particles having very high velocities.

interval, say, between x and $x + dx$ is

$$f_{eq}(x)dx = \frac{dx}{L}, \tag{3.2.59}$$

which is independent of x. This result is of course in accordance with our expectations: since no point in the box is preferred, the probability of being found in an interval dx is simply proportional to the length of that interval. A more general result is when the density function $f(x)$ is defined in an interval (a, b). In this case, the maximum uncertainty is obtained for the density function

$$f(x) = \frac{1}{b-a}, \quad \text{for } a \leq x \leq b \tag{3.2.60}$$

and the corresponding value of the uncertainty is $H(X) = \log(b-a)$.

By generalization to a three-dimensional system of volume V, we have

$$f_{eq}(x, y, z)dxdydz = \frac{dxdydz}{V} \tag{3.2.61}$$

and the associated MI is[25]

$$H = \log V, \tag{3.2.62}$$

where V is the volume of the system.

Clearly, the larger L is, the larger the MI, or the uncertainty, in the location of a particle within the range $(0, L)$. Let us denote this MI by $H[L]$. In anticipating the application of this result in Chapter 5, we divide the length L into n segments each of length h (h will later be the Planck constant, but here it is an arbitrary unit of length), Figure 3.8. We can use the property of consistency to express $H[L]$ as

$$H[L] = H\left[\frac{1}{n}, \ldots, \frac{1}{n}\right] + \sum_{i=1}^{n} \frac{1}{n} H[h]. \tag{3.2.63}$$

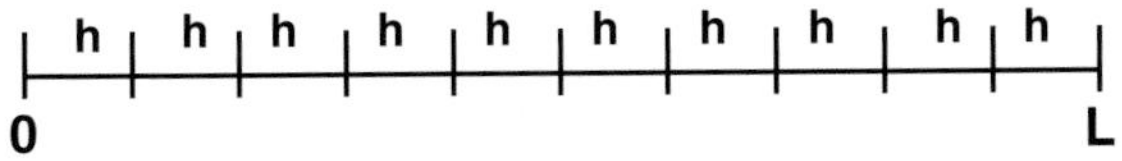

Figure 3.8. The range $(0, L)$ is divided into n segments of length h.

[25] As we shall discuss in Appendix I, we "pretend" that V is the number of states. Actually, we shall need this result only for two cases: (i) differences in H, and (ii) when in fact we cannot have infinite accuracy.

This simply corresponds to rewriting the uncertainty in two terms; first, the uncertainty with respect to which of the n boxes of size h, and the average uncertainty in the location *within* the boxes of size h.

From (3.2.50) and (3.2.63), we have

$$\begin{aligned} H[L] &= \log n + \sum_{i=1}^{n} \frac{1}{n} \log h \\ &= \log \frac{L}{h} + \log h = \log L. \end{aligned} \tag{3.2.64}$$

Now suppose that h is very small so that we do not care (or cannot care) for the location within the box of size h. All we care about is in which of the n boxes the particle is located. Clearly, the uncertainty in this case is simply the MI of a discrete and finite case, i.e.,

$$H\left[\frac{1}{n}, \ldots, \frac{1}{n}\right] = \log n = \log \frac{L}{h}. \tag{3.2.65}$$

Thus, the subtraction of $\log h$ from $H[L]$ amounts to choosing the units of length with which we locate the site of the particle, i.e., in which of the boxes the particle is located. We shall use the definition of H in the continuous case only for the *difference* in ΔH (see also Appendix I). In practice, we can never determine the location of a particle with absolute, or infinite, accuracy. There is always a short interval of length within which we cannot tell where the particle is. Hence, in all these cases, it is the discrete definition of H that applies.

3.2.4.2 *The normal distribution of velocities or momenta*

Shannon (1948) has proved the following theorem[26]:

Of all the continuous distribution densities $f(x)$ for which the standard deviation exists and is *fixed* at σ, the Gaussian (or the normal) distribution has the maximum value of H. Thus, we maximize

[26]Shannon's theorem is very general. We shall use it for a special case where σ^2 is proportional to the temperature or the average velocities of particles in an ideal gas. In this case, we obtain the Maxwell–Boltzmann distribution of the velocities.

H as defined in (3.2.53) subject to the two conditions

$$\int_{-\infty}^{\infty} f(x)dx = 1, \tag{3.2.66}$$

$$\int_{-\infty}^{\infty} x^2 f(x)dx = \sigma^2. \tag{3.2.67}$$

Using the calculus of variation, the condition for maximum H is

$$-1 - \log f(x) + \lambda_1 x^2 + \lambda_2 = 0. \tag{3.2.68}$$

The two Lagrange's constants may be obtained by substituting (3.2.68) in (3.2.66) and (3.2.67). The result is[27]:

$$f(x) = \exp[\lambda_1 x^2 + \lambda_2 - 1], \tag{3.2.69}$$

$$\begin{aligned} 1 &= \int_{-\infty}^{\infty} f(x)dx = \exp[\lambda_2 - 1] \int_{-\infty}^{\infty} \exp[\lambda_1 x^2]dx \\ &= \sqrt{-\frac{\pi}{\lambda_1}} \exp[\lambda_2 - 1], \end{aligned} \tag{3.2.70}$$

$$\begin{aligned} \sigma^2 &= \int_{-\infty}^{\infty} x^2 f(x)dx = \exp[\lambda_2 - 1] \int_{-\infty}^{\infty} x^2 \exp[\lambda_1 x^2]dx \\ &= \sqrt{\frac{\pi}{2(-\lambda_1)^3}} \exp[\lambda_2 - 1]. \end{aligned} \tag{3.2.71}$$

From the last two equations, we can solve for λ_1 and λ_2 to obtain

$$\lambda_1 = \frac{-1}{2\sigma^2}, \quad \exp[\lambda_2 - 1] = \frac{1}{\sqrt{2\pi\sigma^2}}. \tag{3.2.72}$$

Hence,

$$f(x) = \frac{\exp[-x^2/2\sigma^2]}{\sqrt{2\pi\sigma^2}} \tag{3.2.73}$$

and

$$H = -\int_{-\infty}^{\infty} f(x) \log f(x)dx = \frac{1}{2}\log(2\pi e\sigma^2). \tag{3.2.74}$$

---

[27]In this and in the next subsection, the probability density that maximizes H will be referred to as the equilibrium density.

This is an important result. We shall see that this is equivalent to the statement that the Maxwell–Boltzmann distribution of velocities is the distribution that maximizes H (or the entropy).

In the application of this result for the velocity distribution in one dimension (see Chapter 5), we have the probability distribution for v_x as

$$P(v_x) = \sqrt{\frac{m}{2\pi T}} \exp\left[-\frac{mv_x^2}{2T}\right], \tag{3.2.75}$$

where we identify the standard deviation σ^2 as

$$\sigma^2 = \frac{T}{m}, \tag{3.2.76}$$

and T is the temperature defined in units of energy. The uncertainty associated with the velocity distribution is thus[28]

$$H(v_x) = \frac{1}{2}\log(2\pi eT/m). \tag{3.2.77}$$

Assuming that the velocities along the three axes v_x, v_y, v_z are independent, we can write the corresponding uncertainty as

$$H(v_x, v_y, v_z) = H(v_x) + H(v_y) + H(v_z) = 3H(v_x). \tag{3.2.78}$$

Similarly, for the momentum distribution in one dimension, we have $p_x = mv_x$, and hence

$$P(p_x) = \frac{1}{\sqrt{2\pi mT}} \exp\left[\frac{-p_x^2}{2mT}\right],$$

and the corresponding uncertainty is

$$H(p_x) = \frac{1}{2}\log(2\pi emT). \tag{3.2.79}$$

We shall use this last expression to construct the analogue of the Sackur–Tetrode equation in Section 5.4.

The significance of the last result is that σ^2 is proportional to the temperature of the gas, and the temperature of the gas is related to the average kinetic energy of the particles in the gas. Hence, the last result means that given a temperature, or equivalently fixing the total kinetic energy of the gas molecules, the distribution of velocities v_x, v_y and v_z for which the MI is maximum, is

[28]Note that $H(v_x)$ is the MI associated with the distribution of velocities. Similarly, $H(p_x)$ is the MI associated with the distribution of momenta.

the Gaussian or the normal distribution. This is the equilibrium Maxwell–Boltzmann distribution of velocities.

3.2.4.3 *The Boltzmann distribution*

The last example we discuss here is the case where instead of fixing the variance as in (3.2.67), we fix the average of the *rv*. In this case, we look for the maximum of the quantity

$$H = -\int_0^\infty f(x) \log f(x) dx \tag{3.2.80}$$

subject to the two conditions

$$\int_0^\infty f(x) dx = 1 \tag{3.2.81}$$

and

$$\int_0^\infty x f(x) dx = a, \quad \text{with } a > 0. \tag{3.2.82}$$

Using the Lagrangian multipliers λ_1 and λ_2, we look for the maximum of the function

$$F[f(x)] = \int_0^\infty [-f(x) \log f(x) dx + \lambda_1 f(x) + \lambda_2 x f(x)] dx. \tag{3.2.83}$$

The condition for an extremum is:

$$-1 - \log f(x) + \lambda_1 + \lambda_2 = 0. \tag{3.2.84}$$

or equivalently

$$f(x) = \exp[\lambda_2 x + \lambda_1 - 1]. \tag{3.2.85}$$

Substituting this density function in the two constraints (3.2.81) and (3.2.82), we get

$$\exp[\lambda_1 - 1] \int_0^\infty \exp[\lambda_2 x] dx = 1, \tag{3.2.86}$$

$$\exp[\lambda_1 - 1] \int_0^\infty x \exp[\lambda_2 x] dx = a. \tag{3.2.87}$$

Note that λ_2 cannot be positive; otherwise the two constraints cannot be satisfied (not even for $\lambda_2 = 0$). From (3.2.86) and (3.2.87), we obtain

$$\exp[\lambda_1 - 1]\left[\frac{\exp[\lambda_2 x]}{\lambda_2}\right]_0^\infty = -\frac{\exp[\lambda_1 - 1]}{\lambda_2} = 1, \quad (3.2.88)$$

$$\exp[\lambda_1 - 1]\left[\frac{(\lambda_2 x - 1)\exp[\lambda_2 x]}{\lambda_2^2}\right]_0^\infty = \frac{\exp[\lambda_1 - 1]}{\lambda_2^2} = a. \quad (3.2.89)$$

From these two equations, we solve for λ_1 and λ_2 to obtain

$$\lambda_2 = \frac{-1}{a}, \quad (3.2.90)$$

$$\exp[\lambda_1 - 1] = \frac{1}{a}. \quad (3.2.91)$$

Hence, the density that maximizes H is

$$f(x) = \frac{1}{a}\exp\left(\frac{-x}{a}\right). \quad (3.2.92)$$

The value of H for this density is

$$H(X) = -\int_0^\infty f(x)\log f(x)dx \quad (3.2.93)$$

$$= \log a + 1 = \log(ae). \quad (3.2.94)$$

An important example of such a distribution is the barometric distribution. The number density (or the pressure) at height h, relative to the number density at height h_0, is given by the Boltzmann relation:

$$\rho(h) = \rho(h_0)\exp[-\beta mg(h - h_o)], \quad (3.2.95)$$

where $\beta = T^{-1}$ (assuming the temperature is constant throughout the column of air of length $h - h_o$), m is the mass of the particles, and g the gravitational acceleration.

Relation (3.2.95) can be converted to the pressure distribution by substituting $P = \rho T$.

3.3 The Various Interpretations of the Quantity H

After having defined the quantity H and seen some of its properties, let us discuss a few possible interpretations of this quantity. Originally, Shannon referred to the quantity he was seeking to define as "choice," "uncertainty," "information" and "entropy." Except for the last term which does not, the first three terms have an intuitive meaning. Let us discuss these for a simple case.

Suppose we are given n boxes and we are told that a coin was hidden in one and only one box. We are also told that the events, "the coin is in box k" are mutually exclusive (i.e., the coin cannot be in more than one box), and that the n events form a complete set of events (i.e., the coin is certainly in one of the boxes), and that the box in which the coin was placed was chosen at random, i.e., with probability $1/n$.

The term "choice" is easily understood in the sense that in this particular game, we have to *choose* between n boxes to place the coin.[29] Clearly, for $n = 1$, there is only one box to choose and the amount of "choice" we have is zero; we must place the coin in that box. It is also clear that as n increases, the larger n is, the larger the "choice" we have to select the box in which the coin is to be placed. The interpretation of H as the amount of "choice," for the case of unequal probabilities is less straightforward. For instance, if the probabilities of, say, ten boxes are $9/10, 1/10, 0, \ldots, 0$, it is clear that we have less choice than in the case of a uniform distribution, but in the general case of unequal probabilities, the "choice" interpretation is not satisfactory. For this reason, we shall not use the "choice" interpretation of H.[30]

The term "information" is clearly and intuitively more appealing. If we are asked to find out where the coin is hidden, it is clear, even to the lay person, that we *lack information* on "where the coin is hidden." It is also clear that if $n = 1$, we need no information, we know that the coin is in that box. As n increases, so does the amount of the information we lack, or the missing information. This interpretation can be easily extended to the case of unequal probabilities. Clearly, any non-uniformity in the distribution only increases our information, or decreases the missing information. In fact, all the properties of H listed in the previous section are consistent with the interpretation of H as the amount of missing information. For instance, for two sets of independent experiments (or games), the amount of the missing information is the sum of the

[29]Or to "choose" the box where the coin was hidden, or in general, to make a choice between n possibilities.

[30]It seems to me that although Shannon himself named Section 6 of his article "Choice, Uncertainty and Entropy," he did not use the term "choice" when interpreting the properties of the function H.

missing information of the two experiments. When the two sets are dependent, the occurrence of an event in one experiment affects the probability of an outcome of the second experiment. Hence, having information on one experiment can only reduce the amount of missing information on the second experiment.

The "information" interpretation of H is also intuitively appealing since it is also equal to the average number of questions we need to ask in order to *acquire* the missing information (see Section 3.5). For instance, increasing n will always require more questions to be asked. Furthermore, any deviation from uniformity of the distribution will decrease the average number of questions. Thus, whenever a distribution is given, the most plausible measure of the amount of MI is the quantity H. For this reason, the interpretation of H as the amount of *missing information* will prevail in this book.

There are two more interpretations that can be assigned to H which are useful. First, the meaning of H as the amount of *uncertainty.* This interpretation is derived from the meaning of probability. When $p_i = 1$, and all other $p_j = 0$ $(j \neq i)$, we are *certain* that the event i occurred. If, on the other hand, $p_1 = 9/10, p_2 = 1/10, p_3 \dots p_n = 0$, then we have more uncertainty of the outcome compared with the previous example where our uncertainty was zero. It is also intuitively clear that the more uniform the distribution is, the more uncertainty we have about the outcome, and the maximum uncertainty is reached when the distribution of outcomes is uniform. One can also say that H measures the average uncertainty that is *removed* once we know the outcome of the *rv* X. The "uncertainty' interpretation can be applied to all the properties discussed in the previous section.

Both "missing information" and "uncertainty" may have subjective quality. However, when used to interpret Shannon measure, they are objective quantities.

A slightly different but still useful interpretation of H is in terms of likelihood or expectedness. These two are also derived from the meaning of probability. When p_i is small, the event i is less likely to occur, or its occurrence is less expected. When p_i approaches one, the occurrence of i becomes more likely or more expected. Since

$\log p_i$ is a monotonous increasing function of p_i, we can say that the larger $\log p_i$, the larger the likelihood or the larger the expectedness of the event i; only the range of numbers is now changed: instead of $0 \leq p_i \leq 1$, we now have $-\infty \leq \log p_i \leq 0$. The quantity $-\log p_i$ is thus a measure of the *unlikelihood* or the *unexpectedness* of the event i.[31] Therefore, the quantity $H = -\sum p_i \log p_i$ is a measure of the *average* unlikelihood, or unexpectedness, of the entire set of the events.

There are several other interpretations of the term H. We shall specifically refrain from using the meaning of H as a measure of "entropy," nor as a measure of disorder. The first is an outright misleading term (see Preface). The second, although very commonly used, is very problematic. First, because the terms order and disorder are fuzzy concepts. Order and disorder, like beauty, are very subjective terms and lie in the eye of the beholder. Second, many examples can be given showing that the amount of information does not correlate with what we perceive as order or disorder (see also Section 1.2 and Chapter 6).[32]

Perhaps the most important objection to the usage of order and disorder for the entropy, or for H is that the concept of order or disorder, unlike information or uncertainty, does not have the properties that H has. When Shannon sought for a measure of information, or uncertainty, he posed several plausible requirements that such a measure should fulfill. None of these can be said to be a plausible requirement for the concept of order or disorder. Certainly, the additivity and the consistency properties cannot be assigned to the concept of disorder.[33] Also, the concepts of conditional information and mutual information cannot be claimed to be *plausible* properties of disorder.

Brillouin has coined the term "neg-entropy" for the quantity H. In my opinion, this suggestion amounts to a corruption of the meaningful term "information" into a vague term. It is one of the main

[31] The quantity $-\log p_i$ is sometimes called "self-information." We shall not use this term in this book.

[32] For a qualitative discussion of this aspect of entropy, see Ben–Naim (2007).

[33] It is not clear why the "disorder" of two systems should be the "sum" of the "disorder" of each system.

aims of this book to replace the vague term "entropy" by the more meaningful term "neg-information," or better yet, "missing information." As we have discussed in the Preface, the term "entropy" as coined by Classius was an unfortunate choice. It did not convey the meaning it was meant to communicate. This unfortunate choice can be understood retrospectively since it was made many years before the concept that Claussius named "entropy" was understood on a molecular level.

In this book, we shall mainly use the term information, or missing information, to replace the quantity currently called entropy. Occasionally, we shall also use the terms uncertainty or unlikelihood. It should be noted, however, that all these terms refer to the whole set of outcomes, not to a single outcome. It is a property of the entire distribution of events and not of any specific event. Furthermore, we consider this quantity as an objective quantity that "belongs" to the distribution. Sometimes, people have misinterpreted the term uncertainty or information as implying that this quantity is subjective and depends on our ignorance of the system. It is true that information, in its general sense, may or may not be subjective. However, the fact that H may be interpreted as missing information or as uncertainty does not make it a subjective quantity.

Another interpretation of the quantity $-\sum p_i \log p_i$ is as average surprisal. The quantity $-\log p_i$ may be interpreted as a measure of the extent of *surprise* associated with the specific outcome i. The smaller p_i, the larger our surprise to find that the outcome i has occurred. Hence, H is an average of the surprise associated with all the possible outcomes. We shall not use this interpretation of H.

We stress again that the amount of uncertainty or unlikelihood is a property of the *distribution* of the events, and does not depend on the *values* of the events. The following examples should clarify this statement.

Consider the following events and their probabilities:

$$A = \textit{The criminal is hiding in London},$$

$$B = \textit{The criminal is hiding in Oxford},$$

$$C = \textit{The criminal is hiding in Dover},$$

with the probabilities $P(A) = P(B) = P(C) = 1/3$. And consider another set of events:

$$D = \textit{The criminal is hiding in London},$$
$$E = \textit{The criminal is hiding in Paris},$$
$$F = \textit{The criminal is hiding in Tokyo},$$

with the probabilities $P(D) = P(E) = P(F) = 1/3$.

Clearly, the uncertainty or the unlikelihood or the MI, as measured by H, are the same for the two examples. Yet, we feel that in the first example, we have more information, i.e., we know for sure that the criminal is in England, even without asking any question. In the second example, we have no idea in which country the criminal is. The point is that the MI as defined on the distribution $(\frac{1}{3}, \frac{1}{3}, \frac{1}{3})$ does not address the question as to which country the criminal is in, but only to the three cities, London, Oxford and Dover in the first case, and London, Paris and Tokyo in the second. It does not matter how close or far the outcomes are; what matters is only the distribution of the three events.

As a second example, suppose we are given the following information on two drugs, X and Y. It is known that administering drug X to a patient with disease D, has the following outcomes:

$$\Pr[\text{the patient will recover from } D \text{ if } X \text{ is administered}] = 1/2,$$
$$\Pr[\text{the patient will die if } X \text{ is administered}] = 1/2.$$

On the other hand, for drug Y to be administered to the same disease D, we have:

$$\Pr[\text{the patient will recover fully, with absolutely no recurrence of the disease}] = 1/2,$$
$$\Pr[\text{the patient will recover, but will have some very infrequent recurrence of } D] = 1/2.$$

In both cases, our uncertainty regarding the two events is the same. However, we clearly feel that in the first case, we do not know whether the patient will survive or not. In contrast, in the second case we know for sure that the patient will survive. The point to be emphasized is that survival or non-survival are the two events in the first case, but *not* in the second case. In the second case, the uncertainty does not apply to the question of survival, but to the question of whether or not the disease will recur.

If we use the average surprisal interpretation, it is clear that in both cases, the average surprisal should be the same — although we might be extremely surprised if the patient dies upon taking the drug Y in the second example.

In this book, we shall use the definition H in (3.2.1) for any kind of information. It is always a dimensionless, non-negative quantity. When we apply the quantity H for a thermodynamic system, we shall also use the same quantity, as defined in (3.2.1), but change the notation from H to S. This is done to conform with the traditional notation for the entropy. However, unlike thermodynamic entropy, the quantity which we shall denote by S is defined as in (3.2.1) with $K = 1$ and is dimensionless.

3.4 The Assignment of Probabilities by the Maximum Uncertainty Principle

Up to this point, we have assumed that the distribution $p_1, \ldots, p_n$ is *given* and we have examined the properties of the function H *defined* on this distribution. We now turn to the "inverse" problem. We are given some information on some averages of the outcomes of a random variable, and we want to find out the "best" distribution which is consistent with the available information.[34]

Clearly, having an average value of some *rv* does not uniquely determine the distribution. There are a infinite number of distributions that are consistent with the available average quantity.

As an example, suppose we are interested in assigning probabilities to the outcomes of throwing a single die without doing any counting of frequencies, and having no information on the die.

What is the best guess of a distribution if the only requirement it must satisfy is $\sum_{i=1}^{6} p_i = 1$? There are many possible choices; here

[34]The method used in this section is due to Jaynes, based on Shannon's definition of information.

are some possible distributions:

$$
\begin{aligned}
&1, 0, 0, 0, 0, 0,\\
&\frac{1}{2}, \frac{1}{2}, 0, 0, 0, 0,\\
&\frac{1}{4}, \frac{1}{4}, \frac{1}{4}, \frac{1}{4}, 0, 0,\\
&\frac{1}{6}, \frac{1}{6}, \frac{1}{6}, \frac{1}{6}, \frac{1}{6}, \frac{1}{6}.
\end{aligned}
\tag{3.4.1}
$$

The question now is which is the "best" or the most "honest" choice that is consistent with the available information.

Clearly, the first choice is highly biased. There is nothing in the available information that singles out only one outcome to be deemed the certain event. The second choice is somewhat less biased but still there is nothing in the given information which indicates that only two of the outcomes are possible and the other four are impossible events. Similarly, the third choice is less biased than the second, but again there is nothing in the given information which states that two of the outcomes are impossible. Thus, what is left in this list is the fourth choice (of course there are more possibilities which we did not list). This is clearly the least biased choice. According to Jaynes (1957), the best distribution is obtained by maximizing the uncertainty H over all possible distributions subject to the known constraints, or the available information. We have already done this in Section 3.2.2 and found that the "best" guess for this case is the uniform distribution, i.e., the fourth choice in the list (3.4.1).

Suppose we are told that the measurements have been done and the average outcome was 4.5. In this case, the distribution must satisfy the two conditions

$$\sum_{i=1}^{6} p_i = 1, \tag{3.4.2}$$

$$\sum_{i=1}^{6} ip_i = 4.5. \tag{3.4.3}$$

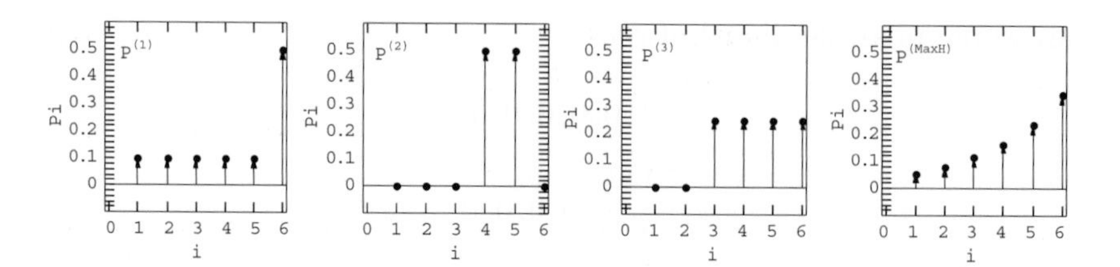

Figure 3.9. Possible distributions for a single die (3.4.4), and the distribution that maximizes H (3.4.7).

Again, it is clear that we have only two equations and six unknown quantities. Therefore, there are many solutions to this problem. Some possible solutions are (Figure 3.9).

$$\begin{aligned} p^{(1)} &= \{0.1, 0.1, 0.1, 0.1, 0.1, 0.5\}, \\ p^{(2)} &= \left\{0, 0, 0, \frac{1}{2}, \frac{1}{2}, 0\right\}, \\ p^{(3)} &= \left\{0, 0, \frac{1}{4}, \frac{1}{4}, \frac{1}{4}, \frac{1}{4}\right\} \end{aligned} \tag{3.4.4}$$

and many more. Clearly, all the distributions in (3.4.4) are consistent with the given information in (3.4.2) and (3.4.3).

The general procedure as cogently advocated by Jaynes (1957) is[35] "*in making inferences on the basis of partial information we must use the probability distribution which has the maximum entropy, subject to whatever is known.*"

The application of this principle to statistical mechanics was suggested by Jaynes (1957). It has given a new view of both the fundamental postulate, as well as of other fundamental probability distributions of statistical mechanics. It did not solve any unsolved problem. As Katz (1967) wrote: "*Information theory approach is not a miracle device to arrive at a solution of any statistical problem. It is only a general way to formulate the problem itself consistently. It*

[35] Here, Jaynes uses the term entropy as defined by Shannon, as $-K\sum_i p_i \log p_i$. Actually, Jaynes uses the meaning of "information" or "uncertainty" assigned to this quantity, not the particular quantity used in thermodynamics where K is chosen as the Boltzmann constant, and the entropy has units of energy divided by temperature.

most often happens that the 'best guess' coincided with the educated guess of those who practice guessing as an art."

This "best guess," according to Katz is the one that is consistent with the "truth, the whole truth and nothing but the truth."[36]

To find the "best" distribution, we have to maximize the uncertainty H subject to the two conditions (3.4.2) and (3.4.3). Using the method of Lagrange's multipliers, we have to find the maximum of the function

$$F = -\sum p_i \log p_i + \lambda_1 \left[1 - \sum p_i\right] + \lambda_2 \left[4.5 - \sum i p_i\right]. \quad (3.4.5)$$

The condition for maximum is

$$0 = \left(\frac{\partial F}{\partial p_i}\right)_{p_i'} = -\log p_i - 1 - \lambda_1 - \lambda_2 i = 0 \quad (3.4.6)$$

or

$$p_i = \exp[-1 - \lambda_1 - \lambda_2 i]. \quad (3.4.7)$$

The coefficients λ_1 and λ_2 can be determined from (3.4.2) and (3.4.3), i.e.,

$$1 = \sum p_i = \exp[-1 - \lambda_1] \sum i \exp[-\lambda_2 i], \quad (3.4.8)$$

$$4.5 = \sum i p_i = \frac{\sum i \exp[-\lambda_2 i]}{\sum \exp[-\lambda_2 i]} = \frac{\sum i x^i}{\sum i x^i}, \quad (3.4.9)$$

where $x = \exp[-\lambda_2]$.

These are two equations with the two unknowns λ_1 and λ_2. One can solve these equations to obtain the distribution (3.4.7).[37] This distribution, denoted $p^{(\mathrm{Max}\, H)}$, is plotted in Figure 3.9.

Next, we shall work out a simple example where at each step we add more information on the averages of the outcomes. Suppose we start with a *given* distribution, for a tetrahedral die, having four faces. The outcomes are $\{1, 2, 3, 4\}$, and we know that the

---

[36]In my view, a better way of describing what the "best guess" means is to say that it is consistent with the knowledge, the whole knowledge and nothing but the knowledge we have on the system under study. This is closer to Jaynes' original description of the merits of the maximum entropy principle.

[37]The only real solution is $x = 1.44925$ and $\exp[-1 - \lambda_1] = 0.0375$.

probabilities are: $\{\frac{1}{12}, \frac{2}{12}, \frac{3}{12}, \frac{6}{12},\}$. In this case, we can calculate all the averages we wish. In particular, we have for this case

$$\sum p_i = 1, \tag{3.4.10}$$

$$\langle i \rangle = \sum i p_i = 3.16667, \tag{3.4.11}$$

$$\langle i^2 \rangle = \sum i^2 p_i = 11, \tag{3.4.12}$$

$$\langle i^3 \rangle = \sum i^3 p_i = 40.1667. \tag{3.4.13}$$

In Figure 3.10, we plot the *known* distribution as dots. Next, we shall try to guess the distribution in steps. First, suppose we are given no information at all on this die. The only constraint on the distribution we have is (3.4.10). Using the maximum MI procedure in this case will result in

$$p_i^{(1)} = e^{-1-\lambda_1} = \frac{1}{4}. \tag{3.4.14}$$

At this level of information, our best guess is a uniform distribution (3.4.14). Next, suppose we know the average in (3.4.11). What is the best guess of the distribution? In this case, using the method of maximum MI, we get

$$p_i^{(2)} = e^{-1-\lambda_1-\lambda_2 i}. \tag{3.4.15}$$

Solving for λ_1 and λ_2 by using (3.4.10) and (3.4.11), we get the new distribution

$$p^{(2)} = \{0.0851, 0.1524, 0.273, 0.4892\}. \tag{3.4.16}$$

This distribution is also plotted in Figure 3.10. We see that we got quite a good "guess" for the distribution with only one average. Next, we assume that we have two average quantities (3.4.11) and (3.4.12). We use the same procedure and get a new distribution

$$p_i^{(3)} = e^{-1-\lambda_1-\lambda_2 i-\lambda_3 i^2}. \tag{3.4.17}$$

Solving for λ_1, λ_2 and λ_3, we get

$$p^{(3)} = \{0.0888,\ 0.1503,\ 0.2664,\ 0.4945\}, \tag{3.4.18}$$

which is almost the same as in the previous solution. The two solutions can hardly be distinguished in Figure 3.10.

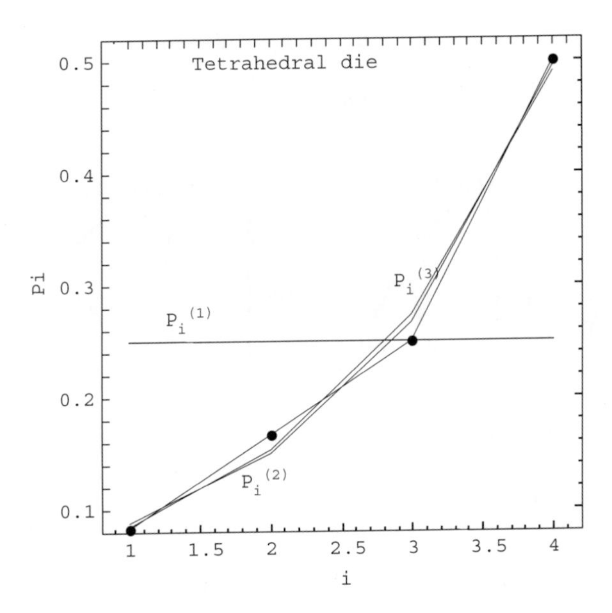

Figure 3.10. The three distributions as in (3.4.14)–(3.4.16). The dots are the exactly known probabilities.

Finally, if we have all three averages in (3.4.11)–(3.4.13), we do not need to guess anything. These equations together with (3.4.10) determine uniquely the solution for the four unknown quantities $p_1^{(4)}, p_2^{(4)}, p_3^{(4)}, p_4^{(4)}$, which are the same as the original distribution we started with, i.e., the dots in Figure 3.10.

Exercise: Start with a given distribution for an "unfair" die with distribution $\{\frac{1}{12}, \frac{1}{12}, \frac{2}{12}, \frac{2}{12}, \frac{3}{12}, \frac{3}{4}\}$. Calculate the best guess of the distribution for the cases (i) of no information and (ii) the case of knowing only the average $\sum ip_i = 4.1667$.

3.5 The Missing Information and the Average Number of Binary Questions Needed to Acquire It

We shall consider here a few examples for which we first calculate the total MI and then we ask binary questions to acquire this information. The number of questions depends on the strategy of asking questions, but the total information is the same, independent of the strategy one chooses to acquire that information. The

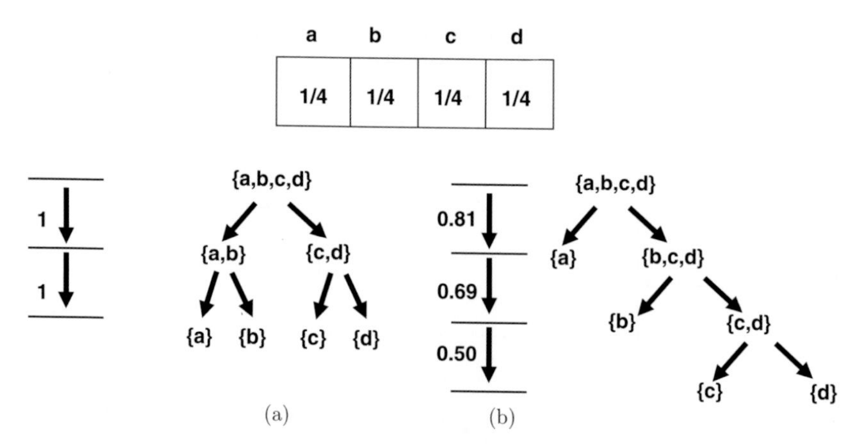

Figure 3.11. Two methods of acquiring information for the case of four boxes: (a) the smart strategy, and (b) the dumb strategy of asking questions. The reduction of MI at each stage is noted to the left-hand side of each diagram.

relation between the MI and the number of questions is the most meaningful interpretation of the quantity H. By asking questions, we acquire information. The larger the MI, the larger the average number of questions to be asked. For a mathematical proof that the minimum average number of binary questions required to determine an outcome of a random variable is between $H(X)$ and $H(X) + 1$, see Cover and Thomas (1991).

Example 1: We have four equally probable events, say a coin hidden in one of four boxes, each having probability $1/4$. Let us identify the boxes by the letters a, b, c, d, (Figure 3.11). We have to ask binary questions to locate the coin. The first method, the smartest one, is shown on the left-hand side of Figure 3.11. Split the four boxes into two halves, then again into two half-halves. The diagram in Figure 3.11a shows the method of splitting the events into groups, and the amount of information acquired at each step.

The total MI is

$$H = -\sum_{1=1}^{4} \frac{1}{4} \log_2 \frac{1}{4} = 2. \tag{3.5.1}$$

The number of steps as can be seen in the diagram is also 2:

$$
\begin{aligned}
g_1 &= H\left(\frac{1}{2}, \frac{1}{2}\right) = 1, \\
g_2 &= \frac{1}{2} H\left(\frac{1}{2}, \frac{1}{2}\right) + \frac{1}{2} H\left(\frac{1}{2}, \frac{1}{2}\right) = 1, \qquad (3.5.2) \\
g_1 + g_2 &= 2.
\end{aligned}
$$

Clearly, we have two steps. At each step, we gain 1 bit of information.

Let us calculate the probabilities for getting the MI in each step. Denote by G_i and N_i, gaining and not gaining the information on the ith step.[38] We have at the first step

$$
\begin{aligned}
P(G_1) &= 0, \\
P(N_1) &= 1.
\end{aligned} \qquad (3.5.3)
$$

This means that in the smartest strategy, there is probability zero of gaining the information in the first step. At the second step, we have

$$
P(G_2 N_1) = P(G_2/N_1)P(N_1) = 1 \times 1 = 1. \qquad (3.5.4)
$$

Thus, the average number of steps is

$$
0 \times 1 + 1 \times 2 = 2, \qquad (3.5.5)
$$

which is the same as H in (3.5.1). Note that at the second step, whatever the answer we get, we shall know where the coin is.

The second method, the dumbest one, is shown on the right-hand side of Figure 3.11. We ask questions such as: Is the coin in box a? Is it in box b? And so on. The diagram in (3.11) shows how we split the events in this case.

We see that in this case, we gain less than the maximum information at each step. Therefore, we need to ask more questions. In the first step, in contrast to the previous strategy, we *can* find the

---

[38]Note that in this and in the following examples, we shall use the phrase "gaining the information" in two senses. In one, colloquially we shall mean "gaining information" on *where* the coin is. In the second, we shall refer to gaining an amount of information in the sense of reducing the MI to zero. Also, the meaning of G_i and N_i can be slightly altered. See example in Figure 3.19.

coin in one step. The probability of finding the coin in one step is $\frac{1}{4}$. The average reduction of the MI in the first step is

$$g_1 = H\left(\frac{1}{4}, \frac{3}{4}\right) = 0.8113. \tag{3.5.6}$$

We can refer to this quantity as the *average* gain of information on the first step. On the second step, we have[39]

$$g_2 = \frac{1}{4}H(1) + \frac{3}{4}H\left(\frac{1}{3}, \frac{2}{3}\right) = 0.6887, \tag{3.5.7}$$

and on the third step, we have

$$g_3 = \frac{3}{4}\left[\frac{1}{3}H(1) + \frac{2}{3}H\left(\frac{1}{2}, \frac{1}{2}\right)\right] = \frac{1}{2}. \tag{3.5.8}$$

Clearly,

$$g_1 + g_2 + g_3 = 2. \tag{3.5.9}$$

Thus, the total MI acquired is the same. The average number of questions is different in the two methods.

The average number of questions in the second method is obtained from the following probabilities. If we ask "Is the coin in box a?" and obtain the answer "Yes", the game is ended. This happens with probability 1/4, and the "No" answer is obtained with probability 3/4. Thus,

$$P(G_1) = 1/4,$$
$$P(N_1) = 3/4.$$

To gain the information on the second step, we need to get a "No" answer at the first step, and "Yes" on the second step.[40] Hence,

$$P(G_2N_1) = P(G_2/N_1)P(N_1) = \frac{1}{3} \times \frac{3}{4} = \frac{1}{4},$$

$$P(N_2N_1) = P(N_2/N_1)P(N_1) = \frac{2}{3} \times \frac{3}{4} = \frac{2}{4}, \tag{3.5.10}$$

$$P(G_3N_1N_2) = P(G_3/N_1, N_2)P(N_1, N_2) = 1 \times \frac{2}{4} = \frac{2}{4}.$$

[39] Note that since we have a finite probability of correctly guessing the location of the coin on the first step, the quantity g_2 is the average gain of information in the second step *given* that we did not succeed in the first step.

[40] Note that in the third step, we shall know where the coin is, whether the answer is a Yes or a No.

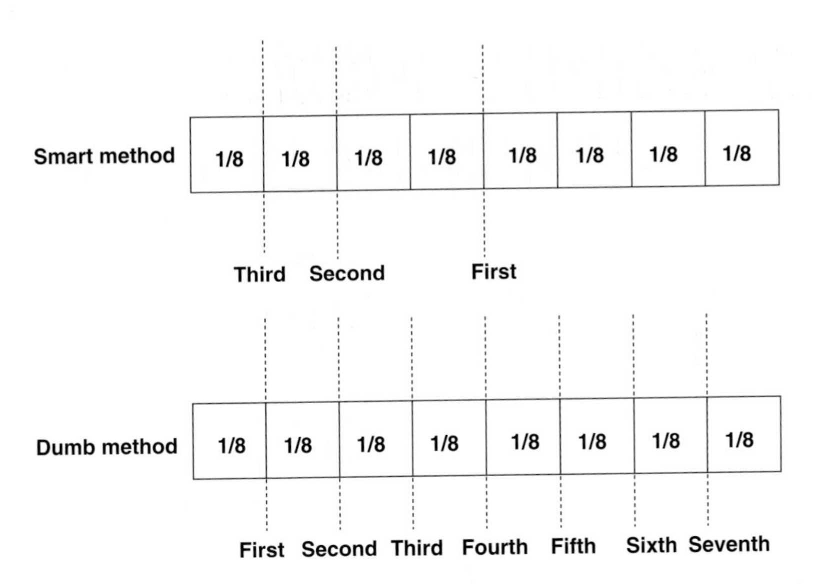

Figure 3.12. Two methods of asking binary questions for the case of eight boxes.

The average number of steps is

$$\frac{1}{4} \times 1 + \frac{1}{4} \times 2 + \frac{2}{4} \times 3 = \frac{1+2+6}{4} = \frac{9}{4} = 2.25, \tag{3.5.11}$$

which is slightly larger than the number of steps in the first method.

Example 2: We repeat the calculations as in Example 1, but with eight boxes of equal probability.

Solution: The total MI is

$$H = 8\left(\frac{1}{8}\log_2 8\right) = 3. \tag{3.5.12}$$

First method (the smartest)
Dividing each time into two equally probable parts. See Figure 3.12. The average gain of information in each step is

$$g_1 = H\left(\frac{1}{2}, \frac{1}{2}\right) = 1,$$

$$g_2 = \frac{1}{2}H\left(\frac{1}{2}, \frac{1}{2}\right) + \frac{1}{2}H\left(\frac{1}{2}, \frac{1}{2}\right) = 1,$$

$$g_3 = \frac{1}{2}\left[\frac{1}{2}H\left(\frac{1}{2},\frac{1}{2}\right) + \frac{1}{2}H\left(\frac{1}{2},\frac{1}{2}\right)\right] + \frac{1}{2}\left[\frac{1}{2}H\left(\frac{1}{2},\frac{1}{2}\right) + \frac{1}{2}H\left(\frac{1}{2},\frac{1}{2}\right)\right] = 1. \tag{3.5.13}$$

Thus, at each step, the reduction in the MI is one bit; hence, the total reduction in the MI is

$$g_1 + g_2 + g_3 = 3. \tag{3.5.14}$$

The corresponding probabilities are the following. The probability of terminating the game on the first step is zero:

$$\begin{aligned} P(G_1) &= 0, \\ P(N_1) &= 1. \end{aligned} \tag{3.5.15}$$

The probability of terminating the game on the second step is also zero:

$$\begin{aligned} P(G_2N_1) &= P(G_2/N_1)P(N_1) = 0 \times 1 = 0, \\ P(N_2N_1) &= P(N_2/N_1)P(N_1) = 1. \end{aligned} \tag{3.5.16}$$

The probability of terminating the game on the third step is one:

$$P(G_3N_1N_2) = P(G_3/N_1N_2)P(N_1N_2) = 1 \times 1 = 1. \tag{3.5.17}$$

The average number of steps is

$$0 \times 1 + 0 \times 2 + 1 \times 3 = 3, \tag{3.5.18}$$

which is exactly equal to H in (3.5.12).

The second method (the dumbest)
Choose one box at each step (see Figures 3.12 and 3.13). The total MI is the same; however, the average gain of information at each

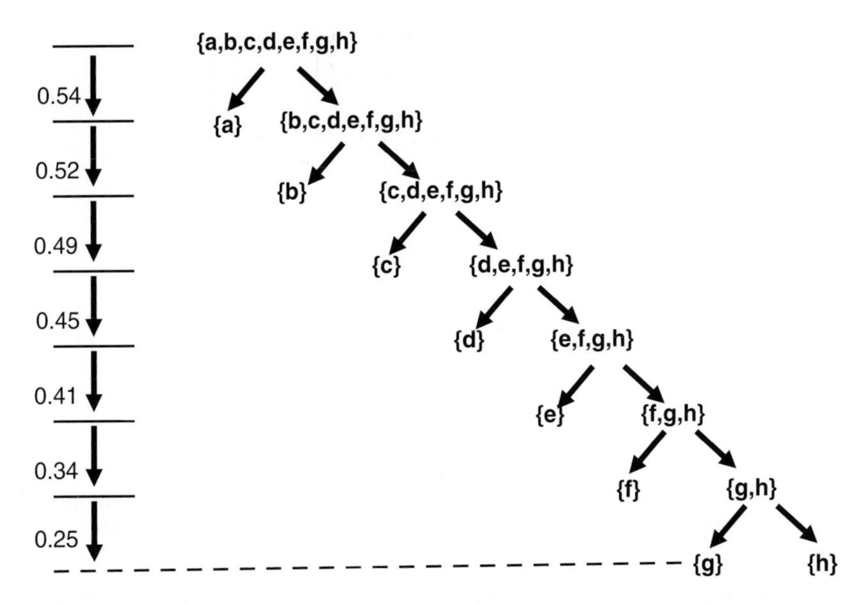

Figure 3.13. Stepwise acquisition of information for the case of eight boxes (3.5.18). The reduction of the MI is shown on the left-hand side.

step is different now:

$$
\begin{aligned}
g_1 &= H\left(\frac{1}{8}, \frac{7}{8}\right) = 0.543, \\
g_2 &= \frac{7}{8} H\left(\frac{1}{7}, \frac{6}{7}\right) = \frac{7}{8} \times 0.592 = 0.518, \\
g_3 &= \frac{7}{8} \times \frac{6}{7} H\left(\frac{1}{6}, \frac{5}{6}\right) = \frac{6}{8} \times 0.650 = 0.487, \\
g_4 &= \frac{7}{8} \times \frac{6}{7} \times \frac{5}{6} H\left(\frac{1}{5}, \frac{4}{5}\right) = \frac{5}{8} \times 0.722 = 0.451, \\
g_5 &= \frac{7}{8} \times \frac{6}{7} \times \frac{5}{6} \times \frac{4}{5} H\left(\frac{1}{4}, \frac{3}{4}\right) = \frac{4}{8} \times 0.8113 = 0.406, \\
g_6 &= \frac{7}{8} \times \frac{6}{7} \times \frac{5}{6} \times \frac{4}{5} \times \frac{3}{4} H\left(\frac{1}{3}, \frac{2}{3}\right) = \frac{3}{8} \times 0.918 = 0.344, \\
g_7 &= \frac{7}{8} \times \frac{6}{7} \times \frac{5}{6} \times \frac{4}{5} \times \frac{3}{4} \times \frac{2}{3} H\left(\frac{1}{2}, \frac{1}{2}\right) = \frac{2}{8} \times 1 = 0.25, \quad (3.5.19)
\end{aligned}
$$

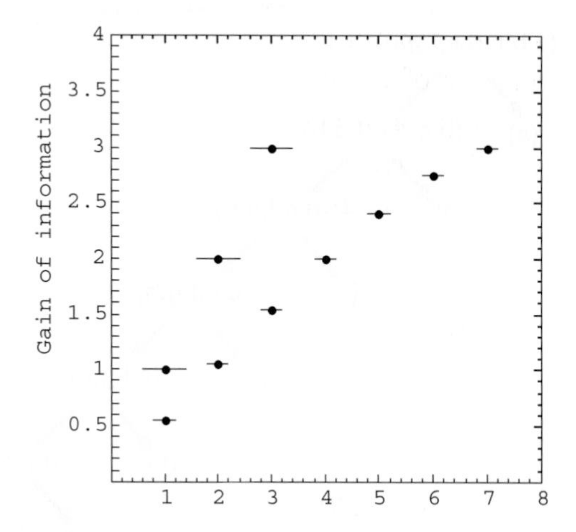

Figure 3.14. Stepwise gain of information by the two methods of asking questions. The longer bars correspond to (3.5.13), the shorter bars to (3.5.19).

Note again that in this case, in contrast to the previous method, we *can* find the coin in the first step (or the second or the third, etc.) g_i is the average gain of information on the ith step, provided we reached the ith step.

Note also that the MI left after each step increases with each step:

$$H\left(\frac{1}{8},\frac{7}{8}\right) < \left(\frac{1}{7},\frac{6}{7}\right) < H\left(\frac{1}{6},\frac{5}{6}\right) < \cdots < H\left(\frac{1}{2},\frac{1}{2}\right) = 1. \tag{3.5.20}$$

The reason is that the distribution gets more and more symmetric from the initial distribution (1/8, 7/8) to the final distribution (1/2, 1/2). On the other hand, the probabilities of reaching that step become smaller; from one to 2/8. Therefore, the average gain in information decreases at each step, i.e.,

$$g_1 > g_2 > g_3 > \cdots > g_7 = 0.25. \tag{3.5.21}$$

The sum of the average gains is

$$\sum_{i=1}^{7} g_i = 3. \tag{3.5.22}$$

In Figure 3.14, we show the gain of information for the two methods. In the smartest method (shown with longer horizontal bars), we gain 1 bit from each question, and hence we need on average a smaller number of questions. In the dumbest method, less information is acquired on average at each step (shown as shorter bars). Hence, more questions are needed on average.

The corresponding probabilities of terminating the game at each step are:

$$P(G_1) = \frac{1}{8},$$

$$P(N_1) = \frac{7}{8},$$

$$P(G_2 N_1) = P(G_2/N_1)P(N_1) = \frac{1}{7} \times \frac{7}{8} = \frac{1}{8},$$

$$P(N_2 N_1) = P(N_2/N_1)P(N_1) = \frac{6}{7} \times \frac{7}{8} = \frac{6}{8},$$

$$P(G_3 N_1 N_2) = P(G_3/N_1 N_2)P(N_1 N_2) = \frac{1}{6} \times \frac{6}{8} = \frac{1}{8},$$

$$P(N_3 N_1 N_2) = P(N_3/N_1 N_2)P(N_1 N_2) = \frac{5}{6} \times \frac{6}{8} = \frac{5}{8},$$

$$\begin{aligned} P(G_4 N_1 N_2 N_3) &= P(G_4/N_1 N_2 N_3)P(N_1 N_2 N_3) \\ &= \frac{1}{5} \times \frac{5}{8} = \frac{1}{8}, \end{aligned}$$

$$\begin{aligned} P(N_4 N_1 N_2 N_3) &= P(N_4/N_1 N_2 N_3)P(N_1 N_2 N_3) \\ &= \frac{4}{5} \times \frac{5}{8} = \frac{4}{8}, \end{aligned}$$

$$\begin{aligned} P(G_5 N_1 N_2 N_3 N_4) &= P(G_5/N_1 N_2 N_3 N_4)P(N_1 N_2 N_3 N_4) \\ &= \frac{1}{4} \times \frac{4}{8} = \frac{1}{8}, \end{aligned}$$

$$\begin{aligned} P(N_5 N_1 N_2 N_3 N_4) &= P(N_5/N_1 N_2 N_3 N_4)P(N_1 N_2 N_3 N_4) \\ &= \frac{3}{4} \times \frac{4}{8} = \frac{3}{8}, \end{aligned}$$

$$\begin{aligned} P(G_6 N_1 N_2 N_3 N_4 N_5) &= P(G_6/N_1 N_2 N_3 N_4 N_5) \\ &\quad \times P(N_1 N_2 N_3 N_4 N_5) = \frac{1}{3} \times \frac{3}{8} = \frac{1}{8}, \end{aligned}$$

$$\begin{aligned}P(N_6N_1N_2N_3N_4N_5) &= P(N_6/N_1N_2N_3N_4N_5)\\ &\quad\times P(N_1N_2N_3N_4N_5) = \frac{2}{3}\times\frac{3}{8} = \frac{2}{8},\\ P(G_7N_1N_2N_3N_4N_5N_6) &= P(G_7/N_1N_2N_3N_4N_5N_6)\\ &\quad\times P(N_1N_2N_3N_4N_5N_6) = 1\times\frac{2}{8} = \frac{2}{8}.\end{aligned} \tag{3.5.23}$$

The average number of steps in this method is

$$\frac{1}{8}\times 1+\frac{1}{8}\times 2+\frac{1}{8}\times 3+\frac{1}{8}\times 4+\frac{1}{8}\times 5+\frac{1}{8}\times 6+\frac{2}{8}\times 7=\frac{35}{8}=4\frac{3}{8}, \tag{3.5.24}$$

which is larger than the number of questions in the first method.

The general case

Suppose we have N boxes, in one of which a coin is hidden. To find out where the coin is, we ask binary questions. For simplicity, we take N to be of the form $N = 2^n$, where n is a positive integer, and also assume that the probabilities are equal. Clearly, there are many ways of asking questions (or strategies). We shall discuss the two extreme cases corresponding to the "smartest" and the "dumbest" strategies.

The total MI in this problem is

$$\text{MI} = \log_2 2^n = n. \tag{3.5.25}$$

Using the smartest strategy, the probabilities of gaining the information at the various stages and terminating the game are:

$$P(G_1) = 0, \quad P(N_1) = 1,$$
$$P(G_2) = 0, \quad P(N_2) = 1;$$

for $i < n$

$$P(G_i) = 0, \quad P(N_i) = 1;$$

for $i = n$

$$P(G_n) = 1. \tag{3.5.26}$$

Clearly, on the nth question, we are guaranteed to obtain the required information. This is a straightforward generalization of the results of the previous examples.

It is also clear that at each step, in this particular example, we obtain the maximum information of one bit. We get the total information in the minimum number of questions, on average. We can also calculate the average number of questions in this case to be

$$\sum_{i=1}^{n} P(G_i)i = n. \tag{3.5.27}$$

If N is an integer, not necessarily of the form 2^n, we cannot make the same divisions as described above. However, one can always find an integer n such that N will be in the range

$$2^n \leq N \leq 2^{n+1}. \tag{3.5.28}$$

Thus, if N is the number of boxes, we can increase the number of boxes to the closest number of the form 2^{n+1}. By doing that, the number of questions will increase by at most one. Therefore, the general dependence of the MI on the number of boxes will not change, i.e., for large N, the number of questions will go as

$$\log_2 N \approx n \log_2 2 = n. \tag{3.5.29}$$

It should be noted that the above estimates of the number of questions is for N boxes of equal probability. This gives an upper limit on the average number of questions. If the distribution of probabilities is not uniform, we can only reduce the number of questions necessary to obtain the information.

On the other hand, using the "dumbest" strategy to ask questions, we have a generalization of Example 2:

$$P(G_1) = \frac{1}{N}, \quad P(N_1) = \frac{N-1}{N},$$

$$P(G_2 N_1) = P(G_2/N_1)P(N_1) = \frac{1}{N-1}\frac{N-1}{N} = \frac{1}{N},$$

$$P(N_2 N_1) = P(N_2/N_1)P(N_1) = \frac{N-2}{N-1}\frac{N-1}{N} = \frac{N-2}{N},$$

$$
\begin{aligned}
P(G_3 N_1 N_2) &= P(G_3/N_1 N_2) P(N_1 N_2) \\
&= \frac{1}{N-2}\frac{N-2}{N} = \frac{1}{N}, \\
P(N_3 N_1 N_2) &= P(N_3/N_1 N_2) P(N_1 N_2) \\
&= \frac{N-3}{N-2}\frac{N-2}{N} = \frac{N-3}{N}, \\
P(G_{N-1} N_1 \cdots N_{N-2}) &= P(G_{N-3}/N_1 \ldots N_{N-2}) P(N_1 \cdots N_{N-2}) \\
&= 1 \times \frac{2}{N} = \frac{2}{N}.
\end{aligned}
\tag{3.5.30}
$$

At the last step, we acquire the required information. The average number of questions in this strategy is

$$
\frac{1}{N}\sum_{i=1}^{N-1} i + \frac{2(N-1)}{N} \approx \frac{N}{2}. \tag{3.5.31}
$$

Thus, the average number of questions in this case is approximately linear in N, whereas in the smartest strategy, it is linear in $\log_2 N$. This difference in behavior is shown in Figure 3.15.

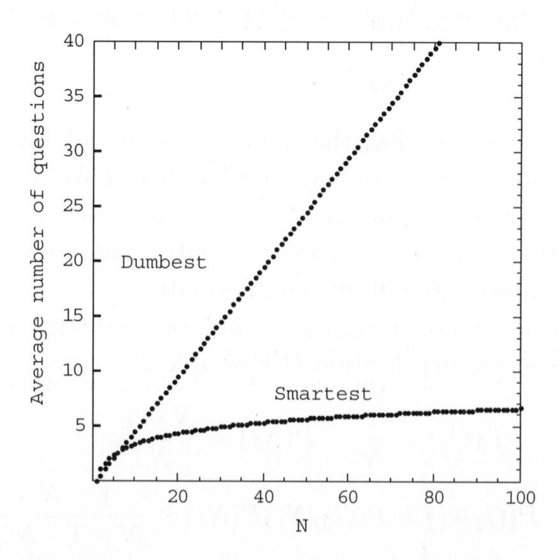

Figure 3.15. The average number of questions as a function N for the two strategies.

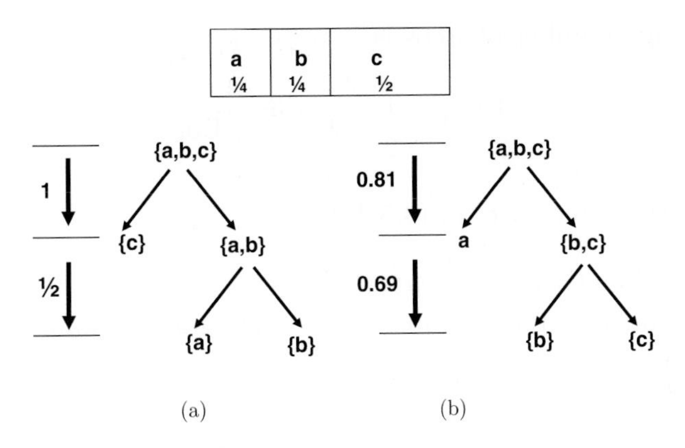

Figure 3.16. An asymmetrical case of three boxes; Example 3.

Example 3: Three boxes with unequal probabilities.
Consider the case described in Figure 3.16. The probabilities of the boxes are $\frac{1}{4}, \frac{1}{4}, \frac{1}{2}$, The total MI in this case is

$$H = -\left(\frac{2}{4}\log_2\frac{1}{4} + \frac{1}{2}\log_2\frac{1}{2}\right) = \frac{3}{2} = 1.5. \tag{3.5.32}$$

The first method (the smartest)
Divide each time into two equally probable halves. See diagram in Figure 3.16a:

$$\begin{aligned} g_1 &= H\left(\frac{1}{2}, \frac{1}{2}\right) = 1, \\ g_2 &= \frac{1}{2}H\left(\frac{1}{2}, \frac{1}{2}\right) = \frac{1}{2}, \\ g_1 + g_2 &= 1.5. \end{aligned} \tag{3.5.33}$$

This is the total amount of MI as in (3.5.32).

The probabilities of terminating the game at each step are:

$$\begin{aligned} P(G_1) &= \frac{1}{2}, \\ P(N_1) &= \frac{1}{2}, \\ P(G_2N_1) &= P(G_2/N_1)P(N_1) = 1 \times \frac{1}{2} = \frac{1}{2}. \end{aligned} \tag{3.5.34}$$

The average number of steps is

$$\frac{1}{2} \times 1 + \frac{1}{2} \times 2 = \frac{3}{2} = 1.5. \tag{3.5.35}$$

which is the same as H in (3.5.32).

The second method (the dumbest)[41]
See diagram in Figure 3.16b. The average gains of information at each step are

$$\begin{aligned} g_1 &= H\left(\frac{1}{4}, \frac{3}{4}\right) = 0.8113, \\ g_2 &= \frac{3}{4} H\left(\frac{1}{3}, \frac{2}{3}\right) = 0.6887. \end{aligned} \tag{3.5.36}$$

The total MI is

$$g_1 + g_2 = 1.5, \tag{3.5.37}$$

which is the same as in (3.5.32). However, the average number of steps is different. The probabilities, in this case are:

$$\begin{aligned} P(G_1) &= \frac{1}{4}, \\ P(N_1) &= \frac{3}{4}, \\ P(G_2 N_1) &= P(G_2/N_1)P(N_1) = 1 \times \frac{3}{4}. \end{aligned} \tag{3.5.38}$$

The average number of steps is thus

$$\frac{1}{4} \times 1 + \frac{3}{4} \times 2 = \frac{7}{4} = 1.75, \tag{3.5.39}$$

which is slightly higher than in the first method.

[41]Note that in this example the probabilities depend on which box we select first. Here, we refer to the specific selection described in Figure 3.16b.

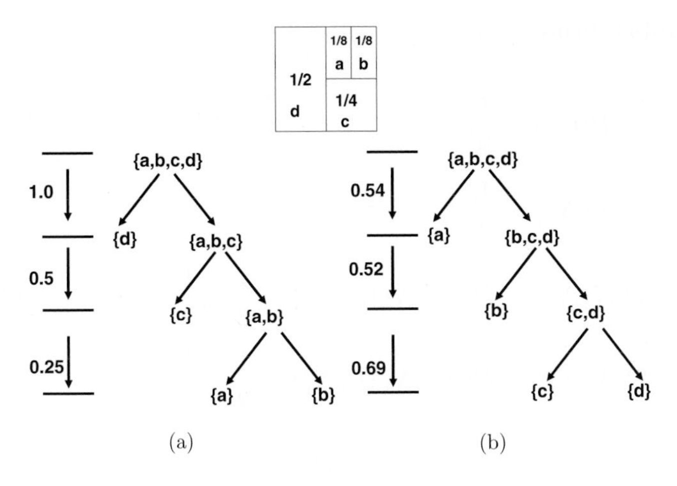

Figure 3.17. An asymmetrical case of four boxes; Example 4.

Example 4: Four boxes of unequal probabilities are shown in the upper part of Figure 3.17. The total MI is

$$H = -\left(\frac{1}{2}\log_2\frac{1}{2} + \frac{1}{4}\log_2\frac{1}{4} + \frac{2}{8}\log_2\frac{1}{8}\right) = \frac{4+4+6}{8} = \frac{7}{4} = 1.75. \tag{3.5.40}$$

The first method

The average gains of information at each step are shown in the left diagram of Figure 3.17a. The arrows show the reduction in MI at each step[42]:

$$\begin{aligned} g_1 &= H\left(\frac{1}{2},\frac{1}{2}\right) = 1, \\ g_2 &= \frac{1}{2}H(1) + \frac{1}{2}H\left(\frac{1}{2},\frac{1}{2}\right) = \frac{1}{2}, \\ g_3 &= \frac{1}{2}\left[\frac{1}{2}H\left(\frac{1}{2},\frac{1}{2}\right)\right] = \frac{1}{2}\times\frac{1}{2} = \frac{1}{4}. \end{aligned} \tag{3.5.41}$$

Hence, the total MI is

$$g_1 + g_2 + g_3 = 1.75. \tag{3.5.42}$$

[42]Note that we *can* get the information on "where the coin is" at the first step with probability $\frac{1}{2}$.

The probabilities:

$$\begin{aligned}
P(G_1) &= \frac{1}{2}, \\
P(N_1) &= \frac{1}{2}, \\
P(G_2N_1) &= P(G_2/N_1)P(N_1) = \frac{1}{2} \times \frac{1}{2} = \frac{1}{4}, \\
P(N_2N_1) &= P(N_2/N_1)P(N_1) = \frac{1}{2} \times \frac{1}{2} = \frac{1}{4}, \\
P(G_3N_1N_2) &= P(G_3/N_1N_2)P(N_1, N_2) = 1 \times \frac{1}{4}.
\end{aligned} \tag{3.5.43}$$

The average number of steps:

$$\frac{1}{2} \times 1 + \frac{1}{4} \times 2 + \frac{1}{4} \times 3 = \frac{2+2+3}{4} = 1\frac{3}{4}, \tag{3.5.44}$$

which is the same as the total MI in (3.5.40).

The second method
See Figure 3.17b.

The average gains of information at each step are:

$$\begin{aligned}
g_1 &= H\left(\frac{1}{8}, \frac{7}{8}\right) = 0.544, \\
g_2 &= \frac{7}{8} H\left(\frac{1}{7}, \frac{6}{7}\right) = 0.518, \\
g_3 &= \frac{7}{8} \times \frac{6}{7} H\left(\frac{1}{3}, \frac{2}{3}\right) = 0.689.
\end{aligned} \tag{3.5.45}$$

The sum of the total MI is

$$g_1 + g_2 + g_3 = 1.75. \tag{3.5.46}$$

The probabilities:

$$P(G_1) = \frac{1}{8},$$

$$P(N_1) = \frac{7}{8},$$

$$P(G_2N_1) = P(G_2/N_1)P(N_1) = \frac{1}{7} \times \frac{7}{8} = \frac{1}{8}, \tag{3.5.47}$$

$$P(N_2N_1) = P(N_2/N_1)P(N_1) = \frac{6}{7} \times \frac{7}{8} = \frac{6}{8},$$

$$P(G_3N_1N_2) = P(G_3/N_1N_2)P(N_1N_2) = 1 \times \frac{6}{8} = \frac{6}{8}.$$

The average number of steps is:

$$\frac{1}{8} \times 1 + \frac{1}{8} \times 2 + \frac{6}{8} \times 3 = \frac{1+2+18}{8} = 2\frac{5}{8}, \tag{3.5.48}$$

which is larger than in the first method.

Exercise: Calculate the total MI, and the gain of information at each step for the problems in Figure 3.18a.

Solution: In Figure 3.18a, we have six boxes with unequal probabilities $(\frac{1}{2}, \frac{1}{4}, \frac{1}{8}, \frac{1}{16}, \frac{1}{32}, \frac{1}{32})$. The total MI is

$$H = -\sum p_i \log_2 p_i = 1.9375. \tag{3.5.49}$$

Note that in this case, we have more boxes than in Example 1, yet the total MI is less than in that example. The reason is that, in this case, the distribution is highly non-uniform.

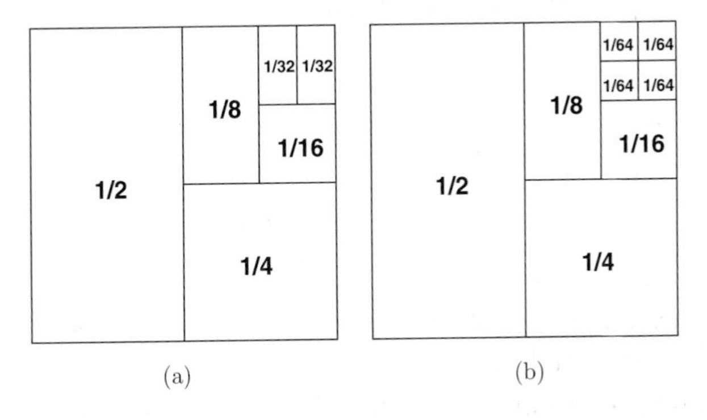

Figure 3.18. Asymmetrical cases with six and eight boxes.

Let us calculate the average number of questions we need to ask by two different routes. First, along the smart route where at each step, we divide the total number of events into equal-probability groups. Denote by $P(G_i)$ the probability of *gaining* the information on where the coin is at the ith step, and by $P(N_i)$, the probability of not gaining the information at the ith step.

The probability of gaining the information at the first step is: $P(G_1) = \frac{1}{2}$, and not gaining, $P(N_1) = \frac{1}{2}$.

The probability of gaining the information at the second step is equal to the probability of not gaining at the first step times the conditional probability of gaining at the second step, given that N_1:

$$P(G_2/N_1)P(N_1) = \frac{1}{2} \times \frac{1}{2} = \frac{1}{4}. \qquad (3.5.50)$$

The probability of gaining at the third step is equal to the probability of not gaining at the first and second steps times the conditional probability of gaining the information at the third step, given N_1, N_2:

$$P(G_3/N_1N_2)P(N_1N_2) = \frac{1}{2} \times \frac{1}{2} \times \frac{1}{2} = \frac{1}{8}. \qquad (3.5.51)$$

The probability of gaining the information at the fourth step is:

$$P(G_4/N_1N_2N_3)P(N_1N_2N_3) = \frac{1}{2} \times \frac{1}{2} \times \frac{1}{2} \times \frac{1}{2} = \frac{1}{16}. \qquad (3.5.52)$$

The probability of gaining the information on the fifth step is:

$$P(G_5/N_1N_2N_3N_4)P(N_1N_2N_3N_4) = \frac{1}{2} \times \frac{1}{2} \times \frac{1}{2} \times \frac{1}{2} \times 1 = \frac{1}{16}. \qquad (3.5.53)$$

Thus, the average number of steps is

$$\frac{1}{2} \times 1 + \frac{1}{4} \times 2 + \frac{1}{8} \times 3 + \frac{1}{16} \times 4 + \frac{1}{16} \times 5 = 1.9375, \qquad (3.5.54)$$

which is equal to the total MI.

By the second route, we ask first whether it is in the little square in the figure denoted 1/32 and continue with boxes of higher probability. The probability of gaining the information at the first step is $P(G_1) = \frac{1}{32}$, and not gaining $P(N_1) = \frac{31}{32}$. The probability of gaining the information at the second step is $P(G_2/N_2)P(N_1) = \frac{1}{31} \times \frac{31}{32} = \frac{1}{32}$.

The probability of gaining the information at the third step is:

$$P(G_3/N_1N_2)P(N_1N_2) = \frac{2}{30} \times \frac{30}{32} = \frac{2}{32}.$$

The probability of gaining the information at the fourth step is:

$$P(G_4/N_1N_2N_3)P(N_1N_2N_3) = \frac{1}{7} \times \frac{28}{30} \times \frac{30}{31} \times \frac{31}{32} = \frac{1}{8}.$$

The probability of gaining the information at the fifth step is:

$$\begin{aligned} P(G_5/N_1N_2N_3N_4)P(N_1N_2N_3N_4) &= \frac{1}{3} \times \frac{6}{7} \times \frac{28}{30} \times \frac{30}{31} \times \frac{31}{32} \\ &= \frac{1}{3} \times \frac{3}{4} = \frac{1}{4}. \end{aligned}$$

The probability of gaining the information at the sixth step is:

$$\begin{aligned} &P(G_6/N_1N_2N_3N_4N_5)P(N_1N_2N_3N_4N_5) \\ &= 1 \times \frac{2}{3} \times \frac{6}{7} \times \frac{28}{30} \times \frac{30}{32} \times \frac{31}{32} = 1 \times \frac{1}{2} = \frac{1}{2}. \end{aligned}$$

The average number of steps in this case is

$$\frac{1}{32} \times 1 + \frac{1}{32} \times 2 + \frac{1}{16} \times 3 + \frac{1}{18} \times 4 + \frac{1}{4} \times 5 + \frac{1}{2} \times 6 = 5.03. \tag{3.5.55}$$

Clearly, along this route one needs much more questions to get the required information.

Exercise: Consider eight boxes with unequal probabilities $(\frac{1}{2}, \frac{1}{4}, \frac{1}{8}, \frac{1}{16}, \frac{1}{64}, \frac{1}{64}, \frac{1}{64}, \frac{1}{64})$ (Figure 3.18b). Calculate the total MI and the average number of questions along the smartest and dumbest routes.

In all the above examples, we always had a way of dividing the total number of events into two parts with equal probabilities. This is of course, not the general case. We shall now turn to a few cases where there is no way to divide into two equally probable halves.

The simplest case is of two events only. We have seen that for the case of equal probabilities, we have Figure 3.19a:

$$H\left(\frac{1}{2}, \frac{1}{2}\right) = 1, \tag{3.5.56}$$

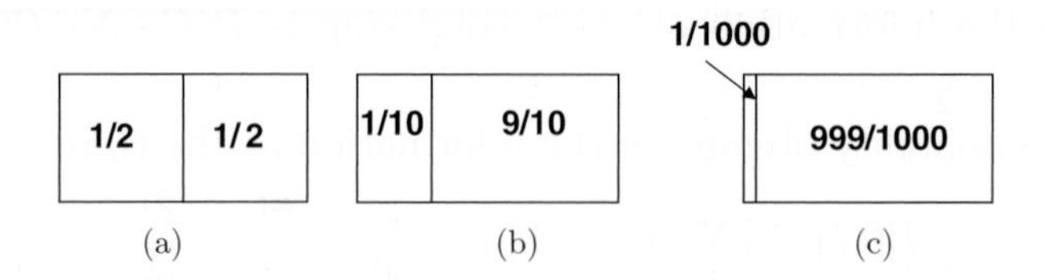

Figure 3.19. Symmetrical and asymmetrical cases with two boxes.

which means that there is one bit of information in the problem and we need to ask only one binary question. What about the unsymmetrical case where one event is much more probable than the other? Figure 3.19b shows two events with probabilities $\frac{1}{10}$ and $\frac{9}{10}$. The total MI is

$$H\left(\frac{1}{10}, \frac{9}{10}\right) = 0.49, \tag{3.5.57}$$

which means that we need to ask less than one question. But the number of questions is always an integer, and even with this example, we must ask at least *one* question. The reason is that in this example, we have no way of splitting into two equally probable halves. Yet, we certainly feel that we have more information in this case than in the previous case.

One way of expressing this additional information is the following. Instead of asking binary questions and stopping when the information is known (in which case we must ask at least one question in the two cases given above), we ask binary questions but wait until we get a "*Yes*" answer (Y). Alternatively, when we get the "Yes" answer, open the box and get the coin.

In the case of equally probable cases, we ask, is it in the right box? If the answer is "Yes," we open and take the coin. If the answer is "No," we must continue to ask, is it in the left box? Only now do we get a "Yes," answer and we shall open the box and take the coin. The probabilities in this case are (here Y_i and N_i stand for getting an answer "Yes" or "No" on the ith step, respectively):

$$\begin{aligned} P(Y_1) &= \frac{1}{2}, \\ P(N_1) &= \frac{1}{2}, \\ P(Y_2 N_1) &= P(Y_2/N_1)P(N_1) = 1 \times \frac{1}{2} = \frac{1}{2}. \end{aligned} \tag{3.5.58}$$

Thus, the average number of steps is

$$\frac{1}{2} \times 1 + \frac{1}{2} \times 2 = 1\frac{1}{2}. \tag{3.5.59}$$

Thus on average, we need 1.5 questions for this case.

On the other hand, in the second case we have the probabilities $\frac{1}{10}, \frac{9}{10}$. There is no smartest strategy in the sense discussed above. However, you should be smart enough to ask first about the box with the higher probability. Hence,

$$P(Y_1) = \frac{9}{10}, \quad P(N_1) = \frac{1}{10}, \tag{3.5.60}$$

and

$$P(Y_2 N_1) = P(Y_2/N_1)P(N_1) = 1 \times \frac{1}{10} = \frac{1}{10}. \tag{3.5.61}$$

The average number of steps is now

$$\frac{9}{10} \times 1 + \frac{1}{10} \times 2 = \frac{11}{10} = 1.1. \tag{3.5.62}$$

So we need roughly one question to gain the coin. Clearly, the larger the asymmetry between the two probabilities, the closer we shall be to a single question. For the case $(\frac{1}{1000}, \frac{999}{1000})$ (Figure 3.19c), we have the average number of questions:

$$\frac{999}{1000} \times 1 + \frac{1}{1000} \times 2 = \frac{1001}{1000} = 1.001. \tag{3.5.63}$$

In both of the last two cases, the average number of questions is smaller than in the case (3.2.59).

Exercise: Calculate the MI and the (smallest) average number of questions you need to gain the coin hidden in one of the three boxes, (Figure 3.20), knowing that the probabilities are:

$$\text{(a)} \quad \left(\frac{1}{100}, \frac{1}{100}, \frac{98}{100}\right),$$

$$\text{(b)} \quad \left(\frac{2}{200}, \frac{99}{200}, \frac{99}{200}\right).$$

Explore the number of questions using different "strategies" of choosing the order of opening the boxes.

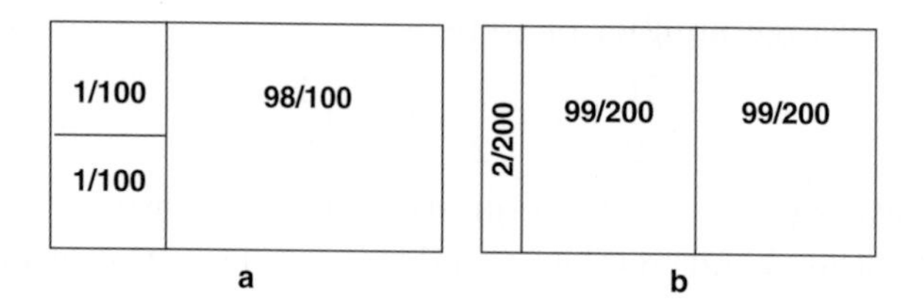

Figure 3.20. Two asymmetrical cases with three boxes.

3.6 The False Positive Problem, Revisited

Exercise: This is an extension of the problem discussed in Section 2.6. It is known that the population of carriers of a virus is $P(C) = 2/100$, i.e., two out of one hundred persons in the population are carriers. A test for the virus gives a positive (+) result if the person is a carrier in 999 out of 1,000 cases. There is also a false positive, i.e., the test is positive, for a non-carrier ($\bar{C}$), with probability 10^{-4}. Find the MI of the two experiments: X, being a carrier or not; Y, testing positive or negative. The corresponding conditional MI and the mutual information.

Solution: The probabilities in this problem are:

$$P(C) = \frac{2}{100}, \quad P(\bar{C}) = \frac{98}{100}, \tag{3.6.1}$$

$$P(+/C) = \frac{999}{1000}, \quad P(-/C) = \frac{1}{1000}, \tag{3.6.2}$$

$$P(+/\bar{C}) = 10^{-4}, \quad P(-/\bar{C}) = 1 - 10^{-4}, \tag{3.6.3}$$

$$\begin{aligned} P(+) &= P(+/C)P(C) + P(+/\bar{C})P(\bar{C}) \\ &= \frac{999}{1000} \times \frac{2}{100} + 10^{-4} \times \frac{98}{100} = 0.0201, \end{aligned} \tag{3.6.4}$$

$$\begin{aligned} P(-) &= P(-/C)P(C) + P(-/\bar{C})P(\bar{C}) \\ &= \frac{1}{1000} \times \frac{2}{100} + (1 - 10^{-4})\frac{98}{100} = 0.9799, \end{aligned} \tag{3.6.5}$$

$$\begin{aligned} P(C/+) &= \frac{P(+/C)P(C)}{P(+)} = \frac{\frac{999}{1000} \times \frac{2}{100}}{0.02} = 0.9951, \\ P(\bar{C}/+) &= 1 - P(C/+), \end{aligned} \tag{3.6.6}$$

$$P(C/-) = \frac{P(-/C)P(C)}{P(-)} = \frac{\frac{1}{1000} \times \frac{2}{100}}{0.98} = 0.0000204,$$
$$P(\bar{C}/-) = 1 - P(C/-). \tag{3.6.7}$$

The correlations between the various events are

$$g(+,C) = \frac{P(+/C)}{P(+)} = 49.756, \quad g(-,C) = \frac{P(-/C)}{P(-)} = 0.0010205, \tag{3.6.8}$$

$$g(+,\bar{C}) = \frac{P(+/\bar{C})}{P(+)} = 0.00498, \quad g(-,\bar{C}) = \frac{P(-/\bar{C})}{P(-)} = 1.02039. \tag{3.6.9}$$

The MI for X (carrier or non-carrier) is

$$H(X) = -P(C)\log P(C) - P(\bar{C})\log P(\bar{C}) = 0.1414. \tag{3.6.10}$$

The MI for Y (test positive or negative)

$$H(Y) = -P(+)\log P(+) - P(-)\log P(-) = 0.1419. \tag{3.6.11}$$

The conditional missing information are:

$$H(X/+) = -P(C/+)\log P(C/+) - P(\bar{C}/+)\log P(\bar{C}/+)$$
$$= 0.0445, \tag{3.6.12}$$
$$H(X/-) = 0.000347, \tag{3.6.13}$$
$$H(X/Y) = P(+)H(X/+) + P(-)H(X/-) = 0.001234, \tag{3.6.14}$$
$$H(Y/C) = P(+/C)\log P(+C) + P(-/C)\log P(-/C)$$
$$= 0.011408, \tag{3.6.15}$$
$$H(Y/\bar{C}) = P(+/\bar{C})\log P(+/\bar{C}) + P(-/\bar{C})\log P(-/\bar{C})$$
$$= 0.001473, \tag{3.6.16}$$
$$H(Y/X) = P(C)H(Y/C) + P(\bar{C})H(Y/\bar{C}) = 0.001671. \tag{3.6.17}$$

The joint and the mutual information are:

$$H(X,Y) = H(X) + H(Y/X) = H(Y) + H(X/Y)$$
$$= 0.14311, \tag{3.6.18}$$
$$I(X;Y) = H(X) + H(Y) - H(X,Y) = 0.14017. \tag{3.6.19}$$

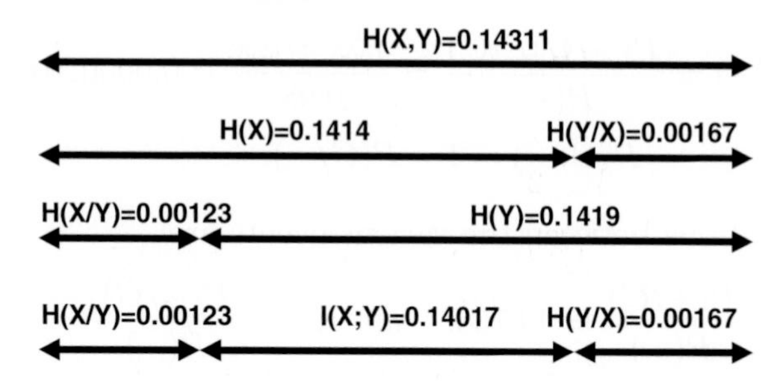

Figure 3.21. The relationships between the quantities $H(X), H(Y), H(X,Y)$ and $I(X;Y)$; Section 3.6.

Note that the correlations come in pairs, one larger than one, and one smaller than one. However, the mutual information gives larger weight to the positive correlations; hence, $I(X;Y)$ is always positive. The relative sizes of the different quantities are shown in Figure 3.21.

3.7 The Urn Problem, Revisited

Exercise: This is an extension of the problem discussed in Sections 2.5 and 2.6. In an urn, there are two white balls and two black balls. The first experiment X_I is a draw of a first ball from the urn. The drawn ball is not returned. X_{II} is the draw of the second ball from the urn after the first ball was drawn, and not returned. Calculate the MI of the two experiments, the conditional MI, and the mutual information.

Solution: We use the obvious notations; W and B for white and black:

$$P(W_I) = \frac{1}{2}, \quad P(B_I) = \frac{1}{2}, \tag{3.7.1}$$

$$P(W_{II}/W_I) = \frac{1}{3}, \quad P(B_{II}/W_I) = \frac{2}{3}, \tag{3.7.2}$$

$$P(W_{II}/B_I) = \frac{2}{3}, \quad P(B_{II}/B_I) = \frac{1}{3}, \tag{3.7.3}$$

$$\begin{aligned} P(W_{II}) &= P(W_{II}/W_I)P(W_I) + P(W_{II}/B_I)P(B_I) \\ &= \frac{1}{3} \times \frac{1}{2} + \frac{2}{3} \times \frac{1}{2} = \frac{1}{2}, \end{aligned} \tag{3.7.4}$$

$$P(B_{II}) = P(B_{II}/B_I)P(B_I) + P(B_{II}/W_I)P(W_I)$$
$$= \frac{1}{3} \times \frac{1}{2} + \frac{2}{3} \times \frac{1}{2} = \frac{1}{2}, \tag{3.7.5}$$

$$P(W_I/W_{II}) = \frac{P(W_{II}/W_I)P(W_I)}{P(W_{II})} = \frac{1/3 \times 1/2}{1/2} = \frac{1}{3}, \tag{3.7.6}$$

$$P(B_I/W_{II}) = \frac{P(W_{II}/B_I)P(B_I)}{P(W_{II})} = \frac{2/3 \times 1/2}{1/2} = \frac{2}{3}, \tag{3.7.7}$$

$$P(W_I/B_{II}) = \frac{P(B_{II}/W_I)P(W_I)}{P(B_{II})} = \frac{2/3 \times 1/2}{1/2} = \frac{2}{3}, \tag{3.7.8}$$

$$P(B_I/B_{II}) = \frac{P(B_{II}/B_I)P(B_I)}{P(B_{II})} = \frac{1/3 \times 1/2}{1/2} = \frac{1}{3}. \tag{3.7.9}$$

The correlations between the various events are

$$g(W_I, W_{II}) = \frac{P(W_I/W_{II})}{P(W_I)} = \frac{1/3}{1/2} = \frac{2}{3}, \tag{3.7.10}$$

$$g(W_I, B_{II}) = \frac{P(W_I/B_{II})}{P(W_I)} = \frac{2/3}{1/2} = \frac{4}{3}, \tag{3.7.11}$$

$$g(B_I, B_{II}) = \frac{P(B_I/B_{II})}{P(B_I)} = \frac{1/3}{1/2} = \frac{2}{3}, \tag{3.7.12}$$

$$g(B_I, W_{II}) = \frac{P(B_I/W_{II})}{P(B_I)} = \frac{2/3}{1/2} = \frac{4}{3}. \tag{3.7.13}$$

Note that two of the correlations are positive ($g > 1$) and two negative ($g < 1$). However, the mutual information which is an average of $\log g_{ij}$ is always non-negative. The reason is that the correlations come in pairs, i.e.,

$$g(W_I, W_{II}) + g(B_I, W_{II}) = 2, \tag{3.7.14}$$
$$g(W_I, B_{II}) + g(B_I, B_{II}) = 2. \tag{3.7.15}$$

Hence, when one correlation is larger than one, the second is smaller than one. However, in forming the average in $I(X_I; X_{II})$, the positive correlations get the larger weight. Therefore, the net effect is that $I(X_I; X_{II})$ is always non-negative.

The MI of the two experiments is

$$H(X_I) = -P(W_I)\log P(W_I) - P(B_I)\log P(B_I) = 1, \quad (3.7.16)$$

$$H(X_{II}) = -P(W_{II})\log P(W_{II}) - P(B_{II})\log P(B_{II}) = 1, \quad (3.7.17)$$

i.e., the MI for both X_I and X_{II} are one bit. The conditional MI are:

$$H(X_{II}/W_I) = -P(W_{II}/W_I)\log P(W_{II}/W_I) - P(B_{II}/W_I)\log P(B_{II}/W_I) = 0.918, \quad (3.7.18)$$

$$H(X_{II}/B_I) = -P(W_{II}/B_I)\log P(W_{II}/B_I) - P(B_{II}/B_I)\log P(B_{II}/B_I) = 0.918, \quad (3.7.19)$$

$$H(X_{II}/X_I) = P(B_I)H(X_{II}/B_I) + P(W_I)H(X_{II}/W_I) = 0.918, \quad (3.7.20)$$

$$H(X_I/W_{II}) = -P(W_I/W_{II})\log P(W_I/W_{II}) - P(B_I/W_{II})\log P(B_I/W_{II}) = 0.918, \quad (3.7.21)$$

$$H(X_I/B_{II}) = -P(W_I/B_{II})\log P(W_I/B_{II}) - P(B_I/B_{II})\log P(B_I/B_{II}) = 0.918, \quad (3.7.22)$$

$$H(X_I/X_{II}) = P(B_{II})H(X_I/B_{II}) + P(W_{II})H(X_I/W_{II}) = 0.918. \quad (3.7.23)$$

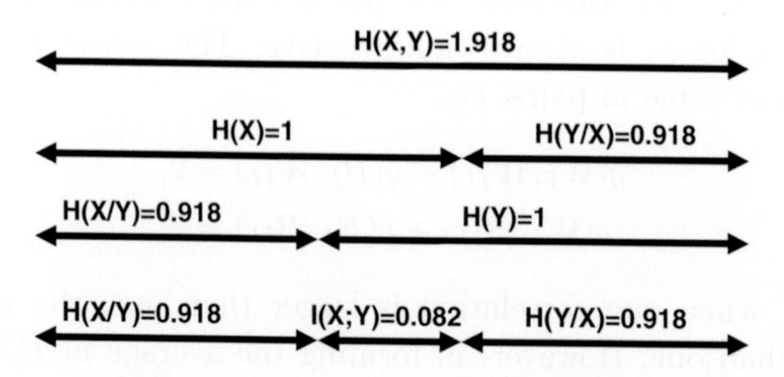

Figure 3.22. The relationships between the quantities $H(X), H(Y), H(X,Y)$ and $I(X;Y)$; Section 3.7.

Thus, all the conditional MI's are equal to 0.918. The joint and the mutual information are

$$H(X_I, X_{II}) = H(X_I) + H(X_{II}/X_I) = 1.918, \tag{3.7.24}$$

$$I(X_I; X_{II}) = H(X_I) + H(X_{II}) - H(X_I, X_{II}) = 0.082. \tag{3.7.25}$$

The relative sizes of the different quantities are shown in Figure 3.22, with $X = X_I$ and $Y = X_{II}$.

Exercise: Generalize for the case of arbitrary N_W and N_B. Examine how $H(X_I/X_{II}), H(X_{II}/X_I)$ and $I(X_I; X_{II})$ depend on the ratio N_W/N_B.

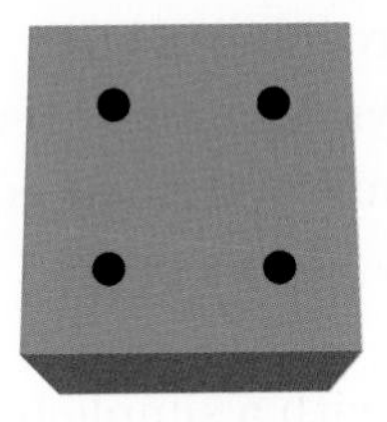

Chapter 4

Transition from the General MI to the Thermodynamic MI

In this chapter, we shall make the *transition* from the general mathematical concept of the amount of missing information applied to a general system as discussed in Chapter 3, and denoted H, to some simple thermodynamic systems.[1] Whenever we discuss thermodynamic systems at equilibrium, we shall switch notation from H to S. There are essentially four axes along which this transition is accomplished:

(i) From systems consisting of a small number of particles, not very different from the game of hiding a coin in a box, to macroscopically large systems consisting of atoms and molecules (thermodynamic systems).
(ii) From discrete to continuous space.
(iii) From independent information to dependent information.
(iv) From one type of information, such as a few particles on a lattice (locational information only), to two types of information, such as a few particles in a gaseous phase (locational and velocity information).

[1]Note that in thermodynamics, the letter H is used for the enthalpy. Here, we use H, following Shannon's notation for the MI. In this book, we shall never use the concept of enthalpy and therefore no confusion should arise. We shall refer to H as information, missing information or the amount of missing information. In all cases, we mean the Shannon measure of information.

The four transitions are intertwined: the idea is to start with the MI for very simple systems and eventually reach the MI of a macroscopic thermodynamic system at equilibrium such as binding on a surface, the MI of an ideal gas, and the analog of the Sackur–Tetrode equation.

In Section 4.1, we start with a simple binding system with a small number of sites and ligands. In Section 4.2, we take the limit of a macroscopic large system, where $M \to \infty$ and $N \to \infty$, but N/M is small and constant. The resulting systems are used as simple models for binding systems. In Section 4.3, we discuss a few simple processes: expansion, mixing and assimilation that are relevant to real binding systems. In Section 4.4, we discuss the transition from one kind (locational), to two kinds of information (locational and momentum). The culmination of all these transitions brings us to the derivation of the Sackur–Tetrode equation for the amount of missing information of an ideal gas. This is a system with a large number of particles, characterized by two types of information (location and momenta) and two correction terms (mutual information). The derivation will be done in two steps. First, we treat the locational and the momentum information as if they were independent, then introduce two corrections due to dependence; both are of quantum mechanical origin. One results from the Heisenberg uncertainty principle, and the second is due to the indistinguishability of the particles. These two corrections are cast in the form of mutual information, i.e., measuring the extent of dependence between two or more random variables. Once we achieve that, we can go on to discuss some simple processes that have been treated both by classical thermodynamics and statistical mechanics. These will be deferred to Chapter 6.

4.1 MI in Binding Systems: One Kind of Information

In this section, we start with systems of small number of binding sites and small number of ligands. The questions we shall ask are identical to the questions posed in the game of a coin hidden in one of M boxes (Chapter 3).

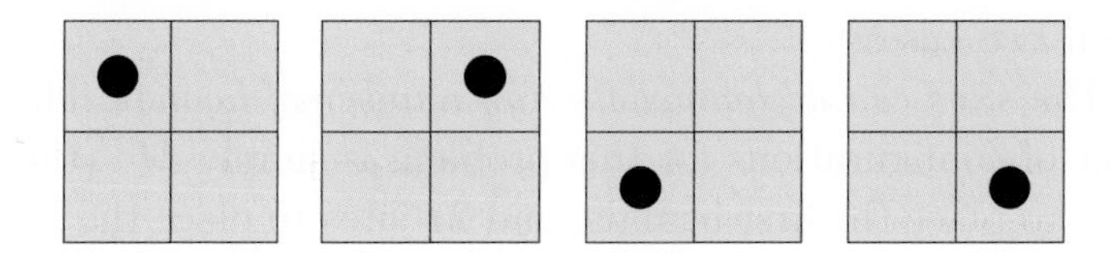

Figure 4.1. All possible configurations for $N = 1$ and $M = 4$

4.1.1 One ligand on M sites

The model is very simple. We have one ligand occupying one of M *distinguishable* binding sites (or boxes). A configuration, or an event is a complete specification of where the ligand is. Figure 4.1 shows all configurations for $N = 1$ and $M = 4$.

Since there are M equivalent, but distinguishable sites, the probability of finding the ligand in one specific site, say the jth site is simply[2]

$$P_j = \frac{1}{M}. \tag{4.1.1}$$

The corresponding MI for this system is

$$H = \log M. \tag{4.1.2}$$

Clearly, the uncertainty or the MI regarding the question "Where is the ligand?" is larger the larger M is. The more possibilities there are, the more difficult it is to find out the configuration, or the larger the number of questions required to acquire this MI (note that the number M is the analog of the number of letters in an alphabet, in connection with communication problems). In this, as well as in the next few examples, there is only one *kind* of MI, which is locational. It is presumed that the ligand does not have any other degrees of freedom, nor velocity.

4.1.2 Two *different* ligands on M sites

In this section, we assume that there are two ligands on M sites, and that there exist no ligand–ligand interactions. We distinguish

[2]We can also "derive" this distribution by starting from the general expression for the MI, $H = -\sum p_j \log p_j$, then take the maximum of MI subjected to the only constraint $\sum p_j = 1$, to get (4.1.1); see Section 3.4.

between two cases:

(a) *The sites can accommodate any number of ligands.* The total number of configurations for this problem is simply M^2 (there are M sites to place the first ligands, and M sites to place the second). Hence, the probability of finding a specific configuration (i.e., which particle is in which site) is

$$P_{ij}(1,2) = \frac{1}{M^2} = \frac{1}{M}\frac{1}{M} = P_i(1)P_j(2). \tag{4.1.3}$$

The indices i and j refers to the specific sites i and j. The number in the parentheses refer to the specific particle, 1 or 2. Clearly, in this case, the MI is[3]

$$H(1,2) = \log M^2 = \log M + \log M = H(1) + H(2), \tag{4.1.4}$$

i.e., the joint missing information on the locations of the two ligands is the sum of the MI of each ligand. There is no correlation between the events "particle 1 in site i" and "particle 2 in site j." Note that in this case, "no correlation" is a result of the assumption of no interactions between the particles, whether they are on the same or on different sites.

(b) *The sites can accommodate, at most, one ligand.* Here, the problem is slightly different from case (a) The total number of configurations in this case is $M(M-1)$ [there are M possible sites to place the first ligand, and $(M-1)$ possibilities to place the second ligand; the second cannot be placed in the occupied sites]. Thus, the probability of each specific configuration is

$$P_{ij}(1,2) = \frac{1}{M(M-1)} = P_i(1)P_j(2/1), \tag{4.1.5}$$

where $P_i(1)$ is the probability of finding particle 1 in site i, and $P_j(2/1)$ is the conditional probability of finding particle 2 in site j, given that particle 1 occupies a site,$i (i \neq j)$.

In this case, though we still presume no intermolecular interactions between the ligands, there exists correlation between the two events "particle 1 in site i," and "particle 2 in site j." We can

[3] $H(1,2)$ means the MI regarding the locations of particles 1 and 2.

express this correlation by defining the correlation function

$$g_{ij}(1,2) = \frac{P_{ij}(1,2)}{P_i(1)P_j(2)} = \frac{M^2}{M(M-1)} = \frac{1}{1-\frac{1}{M}}. \quad (4.1.6)$$

Instead of (4.1.5), we write

$$\begin{aligned} P_{ij}(1,2) &= P_i(1)P_j(2)g_{ij}(1,2) \\ &= \frac{1}{M} \times \frac{1}{M} \times \left(\frac{1}{1-\frac{1}{M}}\right). \end{aligned} \quad (4.1.7)$$

Compare this with (4.1.3). Here, the joint probability is not equal to the product of the probabilities of the two events. Note that this correlation is due to the finiteness of the number of the sites. It is easy to understand the origin of this correlation. The probability of placing the first particle in any site, say i, is as in (4.1.1). However, for the second particle, there are now only $(M-1)$ available sites, i.e., the mere occupation of one site causes a change in the probability of occupying another site.[4]

In terms of MI, we have to modify (4.1.4) to take into account the correlation. We define the mutual information for this case as

$$\begin{aligned} I(1;2) &= H(1) + H(2) - H(1,2) \\ &= \log M + \log M - \log M(M-1) \\ &= \log \frac{M^2}{M(M-1)} = \log[g(1,2)] = -\log\left(1 - \frac{1}{M}\right). \end{aligned} \quad (4.1.8)$$

Note that in (4.1.8), we omitted the subscripts i, j since this quantity is independent of the specific sites i and j $(i \neq j)$. We retain the indexes of the particles 1 and 2, to stress the fact that the particles are *different*.

Note that $I(1;2)$ is always positive; this means that knowing where particle 1 is gives us some information on where particle 2 is. Alternatively, the MI regarding the location of 2 (or 1) is *reduced* after having located particle 1 (or 2) — a result that makes sense and is in accordance with the meaning of the mutual information.

---

[4]The restriction of single occupancy can be interpreted as an infinite repulsive interaction between two ligands occupying the same site.

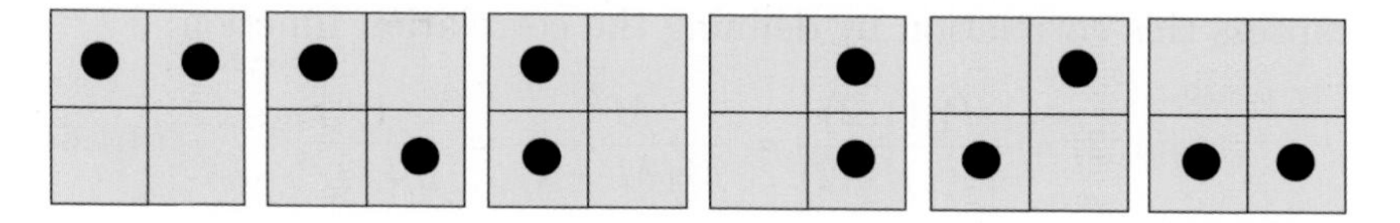

Figure 4.2. All possible configurations for $N = 2$ and $M = 4$.

We shall never be interested in this kind of correlation. We shall always assume that M is very large, where the correlation (4.1.6) will tend to unity.

4.1.3 Two *identical* ligands on M sites

In this example, we introduce another kind of correlation. To highlight this new effect, we shall eliminate the effect of the finiteness of the binding system. Before doing so, we should note that in the present problem we need not discuss cases (a) and (b) as we did in Section 4.1.2. We shall assume for simplicity that there is only a single occupancy, and that $M \gg N > 1$ (N is the number of ligands and M is the number of sites). The number of configurations for $N = 2$ and any M is:

$$W = \frac{M(M-1)}{2} \xrightarrow{M\to\infty} \frac{M^2}{2}. \tag{4.1.9}$$

Figure 4.2 show all the configurations for $N = 2$ and $M = 4$, which are given by

$$W = \frac{4(4-1)}{2} = 6. \tag{4.1.10}$$

This case is fundamentally different from the previous case. Because of the indistinguishability of the particles, we have two different singlet distributions: one, when we ask for the probability of finding a *specific* particle in site i, and the second is the probability of finding *any* particle in site i.[5] The first is simply $1/M$, the second is $2/M$. In this chapter, we shall be interested in the correlation due to the indistinguishability of the particles. Therefore,

[5]In the theory of liquids, the only significant distributions are the so-called *generic* distributions. Here, we first assume that particles are labeled, then unlabel them to find out the emergent correlations due to the indistinguishability of the particles.

we shall use the first probability to calculate the correlation. Thus, the number of configurations when a single particle is in the system is M. When there are two particles in the system, the number of configurations is $\frac{M^2}{2}$. Hence, the probabilities are:

$$P_i(1) = \frac{1}{M}, \quad P_i(2) = \frac{1}{M}, \quad P_{ij}(1,2) = \frac{2}{M^2}. \tag{4.1.11}$$

The correlation function for this case is

$$g_{ij} = \frac{P_{ij}(1,2)}{P_i(1)P_j(1)} = 2. \tag{4.1.12}$$

This means that the conditional probability of finding a particle in site j is larger than the (unconditional) probability of finding a particle at site j. This correlation may be expressed as mutual information:

$$\begin{aligned} I(1;2) &= H(1) + H(2) - H(1,2) \\ &= \log M + \log M - \log \frac{M^2}{2} = \log 2 > 0. \end{aligned} \tag{4.1.13}$$

Comparing this case with the case of Section 4.1.2, we see that if we have two distinguishable ligands, the number of states is M^2. When we un-label the ligands, i.e., make the particles indistinguishable, the number of configurations *decreases*. This is somewhat counterintuitive. We conceive the process of un-labeling as a process where we *lose* information. Here, we have erased the identity of the particles. We feel that something is lost, yet the number of configurations *decreases*, and the corresponding joint MI is smaller than the joint MI of the labeled particles. We shall return to this point in Chapter 6, and Appendix M.

4.1.4 Generalization to N ligands on M sites

We assume that the sites can accommodate at most one ligand. We have N ligands and $M > N$ sites.

For N *different* (D) ligands (or coins in M boxes), the number of configurations is

$$\begin{aligned} W^D &= M(M-1)(M-2)\cdots(M-(N-1)) \\ &= \frac{M!}{(M-N)!}. \end{aligned} \tag{4.1.14}$$

The probability of each specific configuration is $1/W^D$. This is the same as the joint probability

$$P_{i_1 i_1 \cdots i_N}(1, 2, 3, \ldots, N) = \left[\frac{M!}{(M-N)!}\right]^{-1} = \left(\frac{1}{M}\right)^N g(1, 2, \ldots, N), \tag{4.1.15}$$

where the correlation function is given by[6]

$$\begin{aligned} g(1, 2, \cdots, N) &= \frac{M^N (M-N)!}{M!} \\ &= 1 \times \frac{1}{1-\frac{1}{M}} \times \frac{1}{1-\frac{2}{M}} \times \cdots \times \frac{1}{1-\frac{N-1}{M}}. \end{aligned} \tag{4.1.16}$$

Clearly, for $M \gg N$, this correlation tends to unity:

$$g(1, 2, \ldots, N) \xrightarrow{M \to \infty} 1. \tag{4.1.17}$$

However, for an N *indistinguishable* (ID) particles, we have instead of (4.1.14)

$$W^{ID} = \frac{M!}{(M-N)!N!} = \binom{M}{N}. \tag{4.1.18}$$

This is simply the number of ways of placing N *identical* objects in M distinguishable boxes.

The probability of each specific configuration is $\binom{M}{N}^{-1}$. The corresponding correlation function is now

$$g(1, 2, \ldots, N) = \frac{P_{i_1 i_2 \cdots i_N}(1, 2, \ldots, N)}{[P_i(1)]^N} = \frac{\binom{M}{N}^{-1}}{\left(\frac{1}{M}\right)^N} \xrightarrow{M \to \infty} N! > 1. \tag{4.1.19}$$

Note that in this case, as in Section 4.1.3, there *exists* a correlation due to the indistinguishability of the particles. This correlation is different from correlations due to intermolecular interactions.

---

[6] Again, we note that this correlation is between labeled particles, and is different from correlations used in the theory of liquids. Here, we first assume that the particles are distinguishable, then introduce the correlation and the corresponding mutual information due to indistinguishability of the particles.

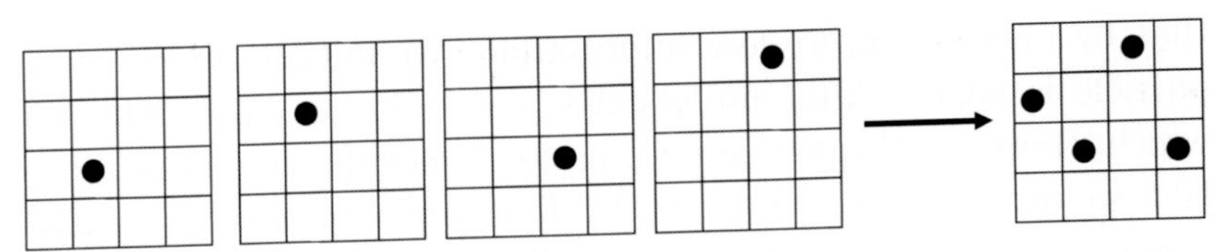

Figure 4.3. The process of assimilation four distinguishable particles. The corresponding change in MI is (4.1.20)

Consider the process depicted in Figure 4.3. The change in the MI in this process for any $N \ll M$ is:

$$\Delta H = H(1, 2, \ldots, N) - \sum_{i=1}^{n} H(1) = \log \binom{M}{N} - N \log M$$
$$= \log \frac{M! M^N}{N!(M-N)!} \xrightarrow{M \to \infty} -\log N!. \qquad (4.1.20)$$

The mutual information in this case is

$$I(1; 2; \ldots; N) = \log N!. \qquad (4.1.21)$$

This is the mutual information due to the indistinguishability of the N particles. At this stage, it is instructive to calculate the change in the MI for the following process. We start with N distinguishable (D) particles distributed on M distinguishable sites. We assume $M \gg N$, so that it does not matter if we allow only single or multiple occupation of each site. We next *un-label* the particles, i.e., we make the particles indistinguishable (see also Appendix M). The change in MI is

$$H^{ID}(1, 2, \ldots, N) - H^{D}(1, 2, \ldots, N)$$
$$= \log \binom{M}{N} - \log \frac{M!}{(M-N)!} \xrightarrow{M>>N} \log \frac{1}{N!} = -\log N!, \qquad (4.1.22)$$

i.e., in this process, the amount of MI decreases. This is again a counterintuitive result. We expect that by erasing the labels, information will be lost. Loss of information should increase the uncertainty or the MI. Here, we find that the uncertainty or the MI has actually *decreased*. The reason is that in (4.1.22), we calculated the MI associated with the *number* of *states*, or the number of configurations, and this number has decreased upon un-labeling. It is true that by un-labeling, we lose information. We initially knew which particle is which. After un-labeling, we no longer know which is which. However, this loss of information is not the kind of information we are calculating in (4.1.22). This apparent counterintuitive conclusion probably led both Maxwell and Gibbs to reach an erroneous conclusion. We shall further discuss this later in Chapter 6.

We summarize what we have found so far in this model system. The process of labeling and un-labeling is relevant for macroscopic particles, say billiard balls. In actual molecular systems, the atoms and molecules are *ID* by *nature*. We have discussed here *one kind* of information, locational: where the particles are. We have seen that there are two kinds of correlations or dependence between these events. One is due to the finiteness of the system. We saw that we can get rid of this correlation by taking the limit of $M \to \infty$. The second correlation is due to the indistinguishability of the particles. This correlation stays with us whenever we want to use information or MI about the location of molecular particles. This correlation is not removed by letting $M \to \infty$, and cannot be removed, in principle. This correlation is always positive and increases with N.

Before generalization to two kinds of information in Section 4.3, we discuss a few processes on this model. For reasons that will be clear later, we shall always assume that $M \gg N \gg 1$, i.e., there will be no correlations due to the finiteness of the system. The number of particles is very large, yet very small compared with M. (Hence, it does not matter whether the sites can be accommodated by one or more particles.)

4.2 Some Simple Processes in Binding Systems

We assume the following model. A surface consists of M sites, each site can accommodate one atom of type A, and there are no interactions between the particles.

In this section, there is *one* energy level which, by construction of the system, is simply $N\varepsilon$, where ε is the interaction energy between the atom and the site. The degeneracy of this energy level is simply the number of ways (or the number of configurations, or the number of microstates) of distributing N indistinguishable atoms on M distinguishable sites (or boxes). This number is

$$W = \binom{M}{N}. \tag{4.2.1}$$

The corresponding missing information (MI) is[7]

$$S(E, M, N) = \log W. \tag{4.2.2}$$

We have written the MI as a function of the three variables. The total energy E, the total number of sites (which in Chapter 5 will be replaced by the volume), and the total number of atoms. In this particular system, E is determined by N. We assume that all of the W states of the system are accessible.[8] We stress from the outset that the MI is a *state function*, in the sense that for any specific values of the macroscopic variables E, M, N, the value of the MI is uniquely determined when the system is at equilibrium.

We also note that the MI is an extensive quantity, i.e., it is a homogenous function of order one. This means that for any positive number $\alpha > 0$:

$$S(\alpha E, \alpha M, \alpha N,) = \alpha S(E, M, N) \tag{4.2.3}$$

as can be verified directly from (4.1.2) and (4.2.2). Next, we calculate the change in MI for some simple processes.

4.2.1 The analog of the expansion process

Consider the process depicted in Figure 4.4. We have two compartments each having the same number of M sites and N particles. Initially, the two compartments are disconnected (or are separated by a barrier that precludes the passage of ligands from one system to the other).

Clearly, the total number of configurations of the combined systems, L and R is

$$W_{\text{total}} = W_L W_R \tag{4.2.4}$$

---

[7]We use now S instead of H. We shall do so whenever we deal with a model for a real thermodynamic system at equilibrium.

[8]We could have constructed some mechanism that induces transitions between different configurations. However, in this simple model, we simply assume that all the configurations are attainable.

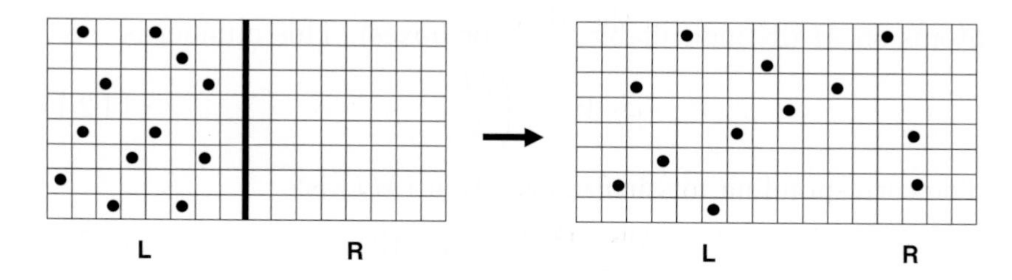

Figure 4.4. The analog of the expansion process, but on a lattice model.

and

$$\begin{aligned} S_{\text{total}} &= \log W_L W_R = \log W_L + \log W_R \\ &= S_L + S_R. \end{aligned} \tag{4.2.5}$$

Thus, the total MI of the combined (but separated) systems is simply the sum of the MI of the right (R) and the left (L) systems. This is consistent with the meaning of S as missing information as discussed in Chapter 3.

To simulate an "expansion" process, suppose that we start initially with two compartments as in the left-hand side of Figure 4.4, where the right compartment R, is empty ("vacuum").

In this case, the initial (i) number of configurations is

$$W_i = W_L W_R = W_L \tag{4.2.6}$$

and

$$S_i = S_L. \tag{4.2.7}$$

We next remove the partition that separates the two compartments. The new system is now characterized by the new variables $(E, 2M, N)$. Note that E and N do not change, only the number of sites changes from M to $2M$. The new number of configuration in the final system (f) is

$$W_f = \binom{2M}{N} \tag{4.2.8}$$

and the corresponding MI is

$$S_f = \log W_f = \log \binom{2M}{N}. \tag{4.2.9}$$

Therefore, the change in the MI in this process is

$$\Delta S = S_f - S_i = \log \frac{\binom{2M}{N}}{\binom{M}{N}} > 0. \tag{4.2.10}$$

Clearly, the number of configurations has increased by the removal of the partition. Hence, the MI is larger in the final state compared with the initial state. This is true for any N and M (of course with $N \leq M$). Let us look at one limiting case $1 \ll N \ll M$, i.e., the ligands are very "diluted" on the surface. Equivalently, the "mole fraction" of the occupied sites is very small

$$x = \frac{N}{M} \ll 1. \tag{4.2.11}$$

In this case, we have

$$\begin{aligned}
\Delta S &= \log \frac{\binom{2M}{N}}{\binom{M}{N}} = \log \frac{(2M)!(M-N)!}{(2M-N)!M!}. \\
&= \log \frac{(2M-N+1)(2M-N+2)\cdots 2M}{(M-N+1)(M-N+2)\cdot M} \\
&= \log \left[\frac{(2-\frac{N}{M}+\frac{1}{M})(2-\frac{N}{M}+\frac{2}{M})\cdots 2}{(2-\frac{N}{M}+\frac{1}{M})(2-\frac{N}{M}+\frac{2}{M})\cdots 1} \right] \\
&\xrightarrow{1\ll N\ll M} \log 2^N = N \log 2 > 0.
\end{aligned} \tag{4.2.12}$$

Note that this result is for the case in which N and M are large compared with 1 and that $N \ll M$. The significance of this result is the following. Suppose we are interested only in the locations of the particles in either R or in L.[9] With this choice, we have *all* the information on the initial system, i.e., we know that all particles are in L. Hence, $S_i = 0$. On the other hand, in the final system, each particle can be either in L or in R, hence there are altogether 2^N possibilities, the corresponding MI is $S_f = \log 2^N$. In terms of the number of questions we need to ask to acquire the MI, we take the logarithm to the base 2, hence $\Delta S = S_f - S_i = \log_2 2^N = N$. Thus, exactly N questions are needed to locate the N particles — *locate* in the sense of being either in L or in R. This example already shows

[9]Note that this is one possible *choice*, of the amount of information we are *interested* in. We shall discuss several other possibilities in Section 4.2.4.

that the MI depends on our *choice* of characterizing the system. However, the *change* in MI is independent of that choice. We shall discuss this in greater details in Section 4.2.4.

The conclusion reached here is the following: whenever we remove a barrier separating the two compartments, the amount of MI always increases. This is the analog of the expansion of an ideal gas as will be discussed in Section 6.1.

The quantity W in (4.2.1) can be given another informational interpretation. We assume that N and M are very large so that we can apply Stirling approximation (see Appendix E) to get

$$\begin{aligned}\ln W &= M \ln M - N \ln N - (M - N)\ln(M - N) \\ &= -M[x \ln x + (1 - x)\ln(1 - x)],\end{aligned} \tag{4.2.13}$$

where $x = \frac{N}{M}$ is the mole fraction of occupied sites and $1 - x$ is the mole fraction of empty sites. Thus, (4.2.13) can be viewed as M times the MI associated with each site, x and $1 - x$ are the probabilities for two possible events of each site, occupied and empty, respectively.

4.2.2 A pure deassimilation process

In the expansion process discussed in Section 4.2.1, we started with a system initially characterized by E, M, N and finally characterized by $E, 2M, N$. Only one parameter has changed, i.e., $M \to 2M$ (this is the analogue of the expansion process where the volume changes, say $V \to 2V$; see Section 6.1).

We now describe another process, where again only one parameter changes, this time the number of indistinguishable particles. The process is depicted in Figure 4.5. For reasons that will be clarified later, we refer to this process as deassimilation, i.e., the reverse of the process of assimilation. For the moment, we shall not elaborate on the meaning of this term. We shall do that in Section 6.3. Also, we shall not be interested in the way we get from the initial to the final state. All that matters is the characterization of the initial and the final states.

We are interested in the change in the MI between the initial and the final state. Initially, we have N particles on M sites, and in the final states we have split the system into two parts each

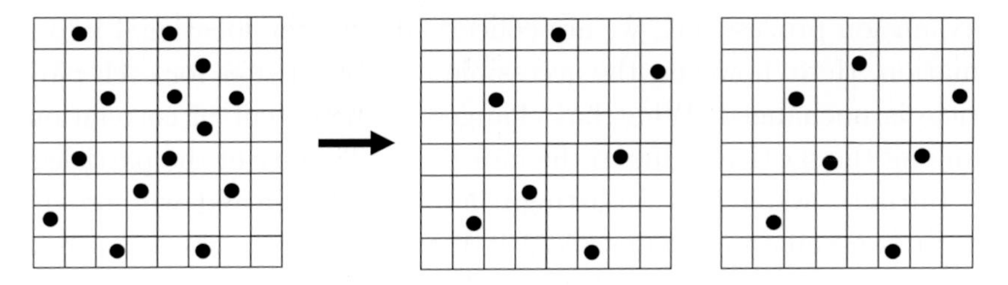

Figure 4.5. A pure process of deassimilation.

having M sites and exactly $N/2$ particles.[10] As before, we assume for simplicity, that $1 \ll N \ll M$. The change in MI in this process is

$$\begin{aligned}
\Delta H = H_f - H_i &= \log \frac{\binom{M}{N/2}\binom{M}{N/2}}{\binom{M}{N}} \\
&= \log \frac{N!}{[(N/2)!]^2} + \log \frac{(M-N)!M!}{(M-N/2)!(M-N/2)!} \\
&= \log \frac{N!}{[(N/2)!]^2} + \log \frac{\left(M - \frac{N}{2} + 1\right)\left(M - \frac{N}{2} + 2\right)\cdots M}{(M-N+1)(M-N+2)\cdots\left(M - \frac{N}{2}\right)} \\
&= \log \frac{N!}{[(N/2)!]^2} + \log \frac{\left(1 - \frac{N}{2M} + \frac{1}{M}\right)\cdots 1}{\left(1 - \frac{N}{2} + \frac{1}{M}\right)\cdots\left(1 - \frac{N}{2M}\right)} \\
&\xrightarrow{N \ll M} \log \frac{N!}{[(N/2)!]^2} \to \log \binom{N}{N/2} > 0. \qquad (4.2.14)
\end{aligned}$$

This is always positive for any N. Furthermore, for large N (strictly for $N \to \infty$), we can use the Stirling approximation to obtain

$$\Delta S = \log \binom{N}{N/2} = \log 2^N = N \log 2. \qquad (4.2.15)$$

Thus, for any finite N (and $N \ll M$), ΔH is always positive. For macroscopic N, ΔH turns into ΔS, which is exactly of the same magnitude as in (4.2.12) for the expansion process. Note that the type of information that we consider here is the same as in the

[10]The requirement of *exactly* $N/2$ particles in each side is not essential. When N is very large, we need only require that in each compartment, we have about $N/2$.

expansion process, i.e., we are concerned with the locational information. Here, however, the accessible number of sites for each particle is unchanged. What has changed is the extent of correlation among the particles, due to the indistinguishability of the particles.

Initially, we had N indistinguishable particles and we end up with two separate groups: $N/2$ indistinguishable particles in one box, and another $N/2$ indistinguishable particles in another box. Note that in the final state, the two groups are distinguishable by the fact that they are in different boxes. All the N particles are identical. However, $N/2$ in R are distinguishable from the $N/2$ in L.[11] This is a subtle point that should be considered carefully. Another point that should be considered is that though the amount of MI has increased from the initial to the final state, the process in Figure 4.5 is not a spontaneous process, i.e., a system of N identical particles do not spontaneously split into two separate boxes, each of which contain $N/2$ particles.

We now describe an equivalent process where the same amount of informational change occurs spontaneously. We shall briefly describe this process here, but we shall elaborate on this in greater detail in Chapter 6.

Suppose the ligands molecules have a chiral center, so there exists two isomers (ennantiomers) of the molecule. This will be referred to as d and l forms. The two isomers are identical in all aspects except that they are mirror images of each other. They are distinguishable, and in principle also separable. Figure 4.6 shows schematic examples of such molecules.

As in Figure 4.5, we shall start again with N molecules on M sites (Figure 4.7), but now all particles are in, say, the d form. The N molecules are indistinguishable. We place a catalyst (alternatively, we can assume that the binding surface is itself the catalyst) which initiates transitions from d to l and *vice versa*.[12]

[11] We shall further discuss this property in Chapter 6 and in Appendix J.

[12] Note that this catalyst does not use any "knowledge," as in the case of Maxwell's demon. It allows molecules to pass in *both* directions, from l to d, as well as from d to l. Here, the catalyst is the analog of removing the partition in the expansion process.

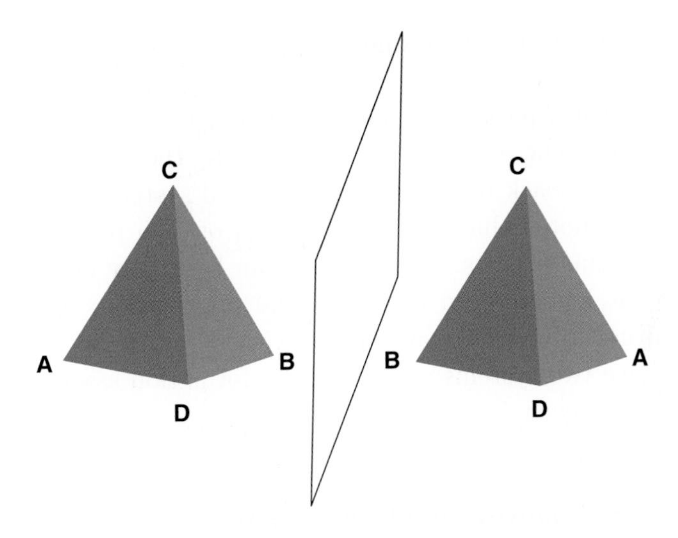

Figure 4.6. A molecule and its mirror image.

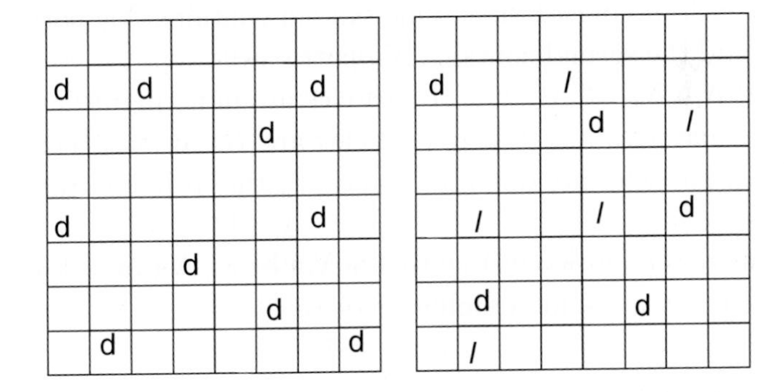

Figure 4.7. A spontaneous process of deassimilation.

After reaching equilibrium, we shall have $N/2$ of the d form and $N/2$ of the l form. The same calculation as done in (4.2.14) applies here. The resulting change in MI is

$$\Delta S = \log \binom{N}{N/2} = N \log 2. \tag{4.2.16}$$

The reason for the increase in MI here is exactly the same as in (4.2.15). In fact, the processes in Figures 4.5 and 4.7 are thermodynamically equivalent. In both, we started with N indistinguishable molecules and ended up with two groups, $N/2$ of one kind and $N/2$

of another kind. The two kinds of molecules are distinguishable. The reason for the increase in the amount of information is that the $N/2$ of the d particles acquire a new identity. This is a spontaneous process where the MI has increased by the amount (4.2.16). We shall come back to this process in a real system in Chapter 6. It should be noted that the locational MI of each particle has not changed in this process: each particle can be in any one of the M sites. What has changed is the mutual information; initially, we have mutual information associated with N particles, and at the end we have two systems with mutual information associated with $N/2$ particles each.

4.2.3 Mixing process in a binding system

We discuss here the simplest mixing process. We start with two systems each having M sites. One system contains N_A particles of type A, and the second contains N_B particles of type B. We assume that $1 \ll N_A, N_B \ll M$. The A particles are indistinguishable (ID) among themselves and the B particles are ID among themselves, but the A particles are different, hence, distinguishable from the B particles.

The process is shown in Figure 4.8. We bring N_A and N_B onto a system having the same number of sites.[13]

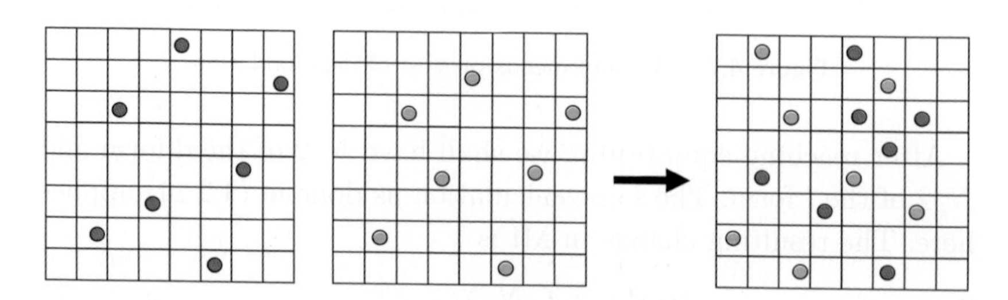

Figure 4.8. A pure process of mixing.

[13]We shall not be concerned here with the method of accomplishing this process. It can be done by employing semi-permeable partitions. Here, we are interested only in the difference between the final and the initial states. See Section 6.2.

Assuming that the site can have at most a single occupancy, we have

$$\Delta H = \log \binom{M}{N_A}\binom{M-N_A}{N_B} - \log \binom{M}{N_A} - \log \binom{M}{N_B}$$
$$\xrightarrow{M>>N_A,N_B} 0, \tag{4.2.17}$$

and for multiple occupancy

$$\Delta H = \log \frac{M^{N_A}}{N_A!}\frac{M^{N_B}}{N_B!} - \log \frac{M^{N_A}}{N_A!} - \log \frac{M^{N_B}}{N_B!} = 0. \tag{4.2.18}$$

Thus, the change in MI in this mixing process is zero. Note, however, that if the two systems contain the same kind of particles, say all of the A kind, then the change in MI is, say in the case of multiple occupancy,

$$\begin{aligned}\Delta H &= \log \frac{M^{N_A}M^{N_B}}{(N_A+N_B)!} - \log \frac{M^{N_A}}{N_A!} - \log \frac{M^{N_B}}{N_B!}\\ &= \log \frac{N_A!N_B!}{(N_A+N_B)!} < 0.\end{aligned} \tag{4.2.19}$$

ΔH (4.2.19) is always negative. This follows from the fact that the number of ways of choosing N_A out of $N_A + N_B$ is larger than one, unless $N_A = 0$, or $N_A = N_A + N_B$:

$$\binom{N_A+N_B}{N_A} = \frac{(N_A+N_B)!}{N_A!N_B!} \geq 1. \tag{4.2.20}$$

In particular, when $N_A = N_B = N$, we have

$$\Delta H = \log \frac{(N!)^2}{(2N)!} < 0, \tag{4.2.21}$$

and when $N \to \infty$, this quantity tends to[14]

$$\Delta S \to -\log 2^N = -N \log 2. \tag{4.2.22}$$

This process is clearly the reversal of the process described in Section 4.2.2. We refer to this process involving identical particles as assimilation. We see that assimilation of any two quantities of particles of the same species always involves a decrease in MI. We shall return to this kind of process in an ideal gas system in Section 6.3.

[14]Note again that ΔS is the same as ΔH, but we use ΔS whenever we deal with thermodynamic systems.

4.2.4 The dependence of MI on the characterization of the system

We have already seen in Section 4.1.1 that the MI itself depends on how we choose to view the system, but the change of MI does not depend on this choice, provided that the choice is not changed in the process. We shall further elaborate on this point since it has caused considerable confusion. We discuss here two kinds of such characterizations.

(a) *Precision with respect to the locations of the particles*
First let us extend the case of Section 4.1.1. Suppose for concreteness, we have $M = 64$ sites, and we choose to group the sites into four groups, say the sites have different forms or different colors, or simply, are at different corners of the binding system (Figure 4.9). For one atom distributed on the $M = 64$ sites, the MI is $H = \log_2 64 = 6$. However, if we are not interested in the details of at which site each particle is located, but only in which one of the four cells; upper left (UL), upper right (UR), lower left (LL) and lower right (LR), then the MI is different. In this characterization, we have altogether four configurations, among which we care to distinguish. The amount of MI in this view is simply

$$H = \log_2 4 = 2. \tag{4.2.23}$$

Clearly, the finer the description of the location of the particles, the larger the MI. More generally, suppose we divide the M sites into m groups, each containing $v = M/m$ sites. We have N particles distributed over these sites, with $N \ll M$, and we are interested only in locating the particles in the various groups. We consider again the process of expansion as in Figure 4.4. The MI in the

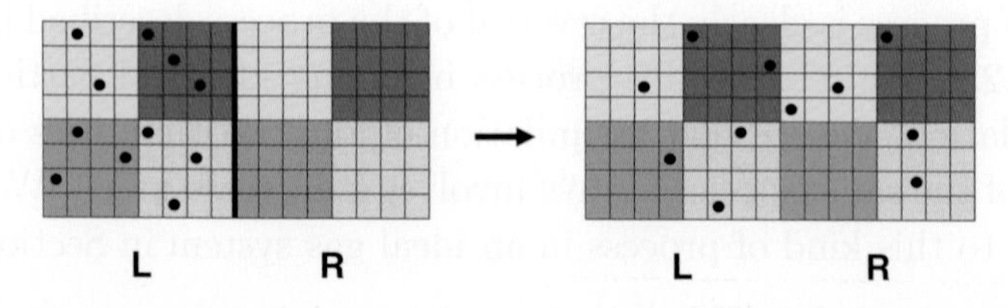

Figure 4.9. The analog of the expansion process, but considering each group of eight sites as one cell.

initial and the final states is[15]

$$H_i = \log \frac{m^N}{N!}, \quad H_f = \log \frac{(2m)^N}{N!}, \tag{4.2.24}$$

and the change in the MI in the process is

$$\Delta H = H_f - H_i = \log \frac{(2m)^N}{m^N} = N \log 2. \tag{4.2.25}$$

Note carefully that H_i and H_f *depend* on the number of groups, m or on the size of each group $v = M/m$. Clearly, if we choose smaller cells we need more information to locate the particles. In Section 4.2.1, we have chosen the extreme case where there are only two groups, or the two compartments L and R. In both cases, the *change* in MI does not depend on v. It is the same MI here, $N \log 2$ as in Section 4.2.1. We conclude that the MI depends on the precision we choose to describe the system, but changes in the MI are not dependent on this choice.

(b) *Precision with respect to the composition of the system*
We next turn to a different *type* of description of the system. Suppose there are two kinds of isotopes, say N_1 atoms of one kind and N_2 atoms of another kind, all on M sites, and we *do* care to distinguish between them, say one kind are white balls while the others are gray balls (Figure 4.10a). We are interested in the total number of configurations of this system, which is simply

$$W_a = \binom{M}{N_1}\binom{M - N_1}{N_2}. \tag{4.2.26}$$

Clearly, this is different from having $(N_1 + N_2)$ *identical* particles on M sites. The MI in the first case is

$$H_{a,1} = \log(W_{a,1}) = \log \binom{M}{N_1}\binom{M - N_1}{N_2}, \tag{4.2.27}$$

and the MI for the second case is

$$H_{a,2} = \log W_{a,2} = \log \binom{M}{N_1 + N_2}. \tag{4.2.28}$$

[15] We assume for simplicity that each site can accommodate any number of particles, and that there are no interactions between particles occupying either the same or different sites.

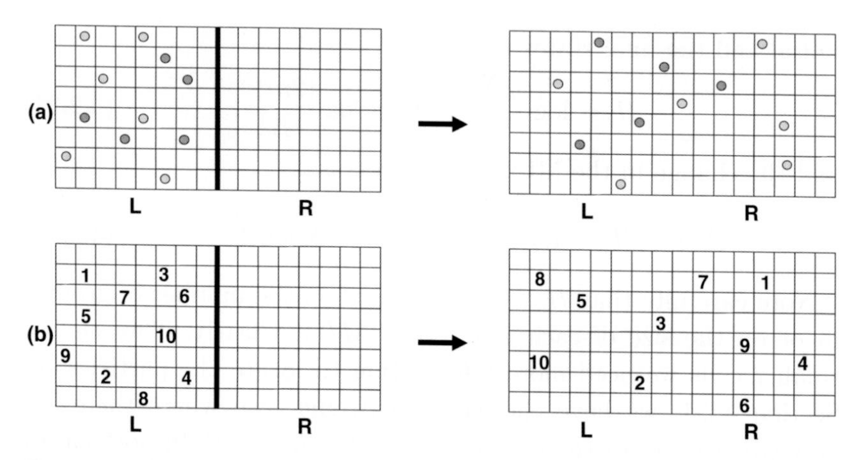

Figure 4.10. (a) The analog of the expansion process, but with mixture of two isotopes (depicted as two shades of grey). (b) The analog of the expansion process, but with labeled particles.

The difference in the MI in the two cases is

$$H_{a,1} - H_{a,2} = \log \binom{M}{N_1}\binom{M-N_1}{N_2} - \log \binom{M}{N_1+N_2}$$
$$= \log \frac{(N_1+N_2)!}{N_1!N_2!} \geq 0. \tag{4.2.29}$$

For macroscopic systems, i.e., when N_1, and N_2 are very large, this difference tends to

$$H_{a,1} - H_{a,2} = -N \log[x_1 \log x_1 + x_2 \log x_2], \tag{4.2.30}$$

where $N = N_1 + N_2$ and x_1 and x_2 are the mole fractions of the two isotopes. The expression on the right-hand side of (4.2.30), when multiplied by k_B, is very often referred to as the entropy of mixing. Clearly, this term has nothing to do with a mixing process. When proceeding from one case to the other, we simply choose to *care for*, or to distinguish between, the two isotopes, or the two colors of the balls. If we are color-blind, this term does not appear. It only depends on *our* willingness to distinguish or not to distinguish between the two species.

Thus, as in the previous example, different levels of descriptions, lead to different measures of MI. However, changes of MI, as well as the changes in the thermodynamic MI, in a physical process, are independent of that choice, provided the composition does not change in the process. We demonstrate that for the simple process of expansion.

First, we do the expansion recognizing that the two components are different (i.e., we distinguish between the white and the gray colored balls) (Figure 4.10a). The change in MI is

$$\begin{aligned}
&\Delta H(\textit{distinguishing between the colors}) \\
&\quad = \log \binom{2M}{N_1}\binom{2M-N_1}{N_2} - \log \binom{M}{N_1}\binom{M-N_1}{N_2} \\
&\quad \xrightarrow{N_1+N_2 \ll M} \log 2^{N_1+N_2} = (N_1+N_2)\log 2. \qquad (4.2.31)
\end{aligned}$$

Next, we do the same process but when we are unaware of the existence of the two isotopes (we are color-blind):

$$\begin{aligned}
&\Delta H(\textit{not distinguishing between the colors}) \\
&\quad = \log \binom{2M}{N_1+N_2} - \log \binom{M}{N_1+N_2} \\
&\quad \xrightarrow{N=N_1+N_2 \ll M} \log 2^{N_1+N_2} = (N_1+N_2)\log 2. \qquad (4.2.32)
\end{aligned}$$

Clearly, the *change* in the MI is the same, whether we do or we do not care to distinguish between the two species. This is true as long as the number of each kind of molecule does not change in the process. The result above simply means that the $(N_1 + N_2)$ molecules have expanded their accessible "volume" from M to $2M$.

We can conclude that removing the color, or deleting the labels on the two species, always lowers the MI. It is easy to show that this is true for any number of species, say if we start with N_1 molecules of one kind (say oxygen molecules), N_2 of another kind (say nitrogen molecules), N_3 of another kind (say carbon dioxide molecules), etc. The MI is different whether we distinguish or do not distinguish (voluntarily or involuntarily) between the different species. However, the *change* in MI in the process of expansion would be the same whether we care or do not care to distinguish between the species.

Let us go to the extreme case. Suppose that we could have labeled all the particles, say by numbering them $1, 2, \ldots, N$ (or color all the balls differently from each other) (Figure 4.10b).

The initial number of configurations is

$$W_i = M(M-1)(M-2)\cdots(M-N+1) = \frac{M!}{(M-N)!}. \quad (4.2.33)$$

Note that we have not divided by $N!$ since we have N distinguishable particles. The number of configurations in the final state is

$$W_f = \frac{(2M)!}{(2M-N)!}. \quad (4.2.34)$$

The change in the MI, in the process of expansion is thus [see (4.2.12)]

$$\Delta H = H_f - H_i = \log \frac{(2M)!(M-N)!}{(2M-N)!M!}$$
$$\xrightarrow{1 \ll N \ll M} N \log 2, \quad (4.2.35)$$

which is exactly the same as in the previous expansion of N indistinguishable particles. The reason is that the identities of the particles did not change in the process. What has changed is that each particle had initially M accessible sites and finally $2M$ accessible sites. Note, however, that the MI itself is very different when the particles are distinguishable or indistinguishable. For this particular example, we have

$$H(\mathit{labeled}) = \log \frac{M!}{(M-N)!}, \quad H(ID) = \log \binom{M}{N} = \log \frac{M!}{(M-N)!N!}. \quad (4.2.36)$$

Thus, the process of labeling the particles changes the amount of MI by

$$H(\mathit{labeled}) - H(ID) = \log N! \quad (4.2.37)$$

This is a very important result that should be remembered. By labeling the particles, the MI increases by the amount $\log N!$, where $N!$ is the number of ways one can label N particles. We shall return to this point in connection with the process of assimilation of ideal gases in Section 6.3 and Appendix M.

Exercise: Calculate the number of configurations for a system of $N = 2$ particles on $M = 6$ sites, where each site can accommodate at most one particle, once when the particles are labeled (say 1 and 2), and once when the particles are unlabeled. Calculate the change in the MI for the process of labeling or un-labeling the particles.

4.3 MI in an Ideal Gas System: Two Kinds of Information. The Sackur–Tetrode Equation

In this section, we discuss the MI of a system of N particles in a box of volume V. In doing so, we extend the treatment of Section 4.1 in two senses: from a discrete to a continuous system, and from one kind to two kinds of information.

The building blocks of these generalized MI were introduced by Shannon in his original work on the theory of communication. The relevant measures of information were discussed in Section 3.2. Note also that whenever we handle MI in a continuous space, we should be careful in applying the concept of MI, not to a single state but to a difference in MI between two states of the system (see also Appendix I).

4.3.1 The locational MI

For a single particle moving along a segment of length L, the equilibrium probability density is uniform (Section 3.2.4):

$$f(x) = \frac{1}{L}, \tag{4.3.1}$$

and the corresponding MI is

$$H(X) = \log L. \tag{4.3.2}$$

Note that in (4.3.2), X is the random variable, the values of which are the x coordinates. We ignore here the question of the units of length we use for L; we shall be interested only in differences in the MI, not in the MI itself.

For one particle in a box of volume V (for simplicity assume that the box is a cube, the length of each edge being L). The corresponding MI, assuming that the locations along the three axes are independent, is

$$\begin{aligned} H(\boldsymbol{R}) &= H(X) + H(Y) + H(Z) = 3\log L \\ &= \log V. \end{aligned} \tag{4.3.3}$$

For N *different*, or distinguishable (D) and non-interacting particles $1, 2, \ldots, N$, the corresponding MI is simply the sum of the MI for each particle, i.e.,

$$H^D(\boldsymbol{R}^N) = \sum_{i=1}^{N} H(\boldsymbol{R}_i) = N \log V. \tag{4.3.4}$$

Here, $\boldsymbol{R}^N$ is a shorthand notation for the configuration $\boldsymbol{R}_1, \ldots, \boldsymbol{R}_N$, of the particles.

Consider the process depicted in Figure 4.11a, which is similar to the process in Figure 4.3 except that the lattice points are replaced by the volume V. The difference in the MI in this case is

$$H^D(\boldsymbol{R}^N) - \sum_{i=1}^{N} H(\boldsymbol{R}_i) = 0. \tag{4.3.5}$$

Initially we have N boxes of equal volume V, each containing one *labeled* particle. All these are brought to one box of the same volume V. The change in MI is zero. (Note that we are discussing here totally non-interacting particles; this is a hypothetical system of point particles. Any real particles would have some finite molecular volume, and hence the same process as in Figure 4.11a would entail a finite change in MI.)

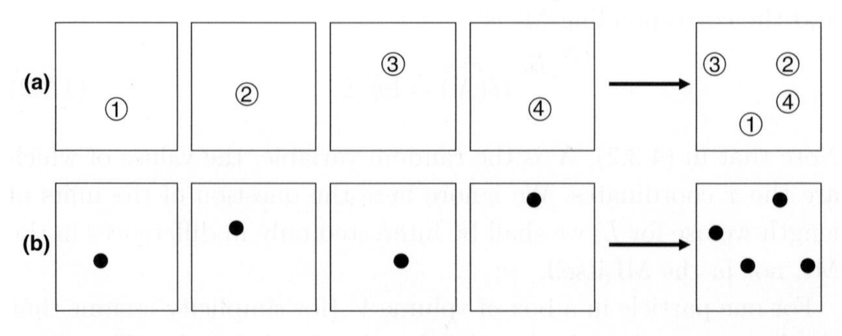

Figure 4.11. (a) The process of assimilating four labeled particles. (b) The process of assimilating four identical particles.

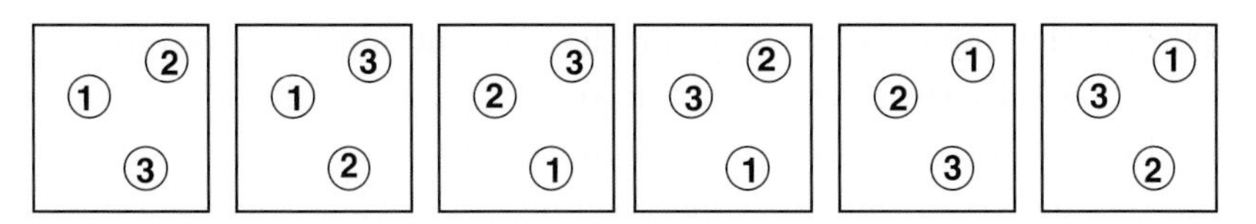

Figure 4.12. Six different configurations become indistinguisable when the labels are erased.

Next, consider the same process as in Figure 4.11a with *identical*, but *un-labeled* particles (Figure 4.11b). As long as each particle is in a different box, they are distinguishable from each other. We can say this particle is in box 1, that particle is in box 2, etc. (the boxes are distinguishable). Once they are together in a single box having volume V, they become indistinguishable. This process produces a change in the MI.

To see this, consider three labeled particles in a configuration as in Figure 4.12. Clearly, for labeled particles, these are distinguishable configurations. When counting the total number of possible configurations in (4.3.4), we have assumed that each particle can occupy any point $\boldsymbol{R}_i$ independently of the occupancy of other points in V. Thus, all of these six configurations are counted as different configurations in (4.3.4). For *ID* particles, we cannot distinguish between these six configurations. This is true for any configuration of the three particles. Thus, in (4.3.4), we have over-counted the number of configurations. To correct for this over-counting, we must divide the total number of configurations by six, i.e., instead of V^3, we must take $V^3/3!$, and for N particles, we must take $V^N/N!$. Thus, in this case we have instead of (4.3.4)

$$\begin{aligned} H^{ID}(\boldsymbol{R}^N) &= \log \frac{V^N}{N!} \\ &= H^D(\boldsymbol{R}^N) - \log N! \\ &= \sum_{i=1}^{N} H(\boldsymbol{R}_i) - \log N!. \end{aligned} \tag{4.3.6}$$

Hence, in the process of Figure 4.11b, we have

$$H^{ID}(\boldsymbol{R}^N) - \sum_{i=1}^{N} H(\boldsymbol{R}_i) = -\log N!. \tag{4.3.7}$$

This is the same as we had in the case of binding system (see Section 4.1.4). Here, we do not have to take the limit $M \to \infty$, as the finite volume V already has infinite "sites" compared to N. We can conclude that the indistinguishability of the particles introduces dependence, or correlations, between the locational information of the particles. The mutual information in this case is

$$I(\boldsymbol{R}_1; \boldsymbol{R}_2; \ldots; \boldsymbol{R}_N) = \sum_{i=1}^{N} H(\boldsymbol{R}_i) - H^{ID}(\boldsymbol{R}_1, \ldots, \boldsymbol{R}_N) = \log N!. \tag{4.3.8}$$

4.3.2 The momentum MI

Real particles are described classically by both their locations and their momenta. A single classical particle in an isolated box of volume V must have a single velocity, hence an exact kinetic energy.[16]

We therefore proceed directly to an isolated system of N non-interacting particles.[17]

Following Shannon, the maximum MI of a continuous random variable, having a fixed variance σ^2, is obtained for the normal distribution (see Section 3.2.4). We also know from the Maxwell–Boltzmann theory that at equilibrium, the distribution of the velocities is the normal distribution. The variance is simply related to the temperature by

$$\sigma_{v_x}^2 = \frac{T}{m}, \tag{4.3.9}$$

where T is the absolute temperature in units of energy (or $k_B T$ in the conventional absolute temperature), and m in the mass of each particle.

[16]We assume here that the walls of the box are not made of particles, being distributed over different momenta. The wall in this idealized isolated system is considered as being an infinite potential barrier that confines the particles to the interior of the box.

[17]A system of strictly non-interacting particles does not exist. In reality, if the gas is dilute enough, one can neglect the interaction energy among the particles compared with the total energy of the system, which in this case is the total kinetic energy of the particles. Note that collision between the particles is necessary to obtain the equilibrium distribution of velocities, or momenta.

Assuming that the motion along the three axes are independent, we write for a system of N independent particles the corresponding MI associated with the velocity distribution (see Section 3.2.4)

$$H^D(\boldsymbol{v}^N) = \frac{3N}{2}\log(2\pi eT/m), \tag{4.3.10}$$

where $\boldsymbol{v}^N = \boldsymbol{v}_I, \cdots, \boldsymbol{v}_N$. Note that $\boldsymbol{v}_i$ is the velocity vector v_x, v_y, v_z of the ith particle.

The corresponding MI for the momentum is (see Section 3.2.4)

$$H^D(\boldsymbol{p}^N) = \frac{3N}{2}\log(2\pi emT). \tag{4.3.11}$$

Here, we took the result from Section 3.2.4 multiplied by three for the three axes, and by N for the N independent particles.

4.3.3 Combining the locational and the momentum MI

In Section 4.3.1, we have seen that the joint locational MI of N particles is not the sum of the locational MI of each particle. Indistinguishability introduced correlations among the locations of the particles.

In combining the MI of the locations and of momenta, a new correlation enters. We cannot write the MI of a system of N particles as a sum of the form

$$H^D(\boldsymbol{R}^N, \boldsymbol{p}^N) = H^D(\boldsymbol{R}^N) + H^D(\boldsymbol{p}^N). \tag{4.3.12}$$

The reason is that the location and the momentum of each degree of freedom (i.e., along each axis) are not independent. The variation in the location and the momentum for each pair of coordinates x and p_x are connected by the Heisenberg uncertainty principle. As we have seen in Section 3.2, this amounts to the division of the entire phase space for each pair of x, p_x into cells of size h, where h is the Planck constant.[18]

As we have seen in Section 3.2, the passage from the continuous to the discrete space requires one to change the MI from $\log L$ to

[18]In Section 3.2, we also use the letter h to denote the size of each cell. Here h is the size of the so-called action, having the units of the product of length and momentum.

$\log \frac{L}{h}$, which transforms the "volume" of the continuous space into the number of cells in a discrete space.

We do the same for each degree of freedom of the particles. We subtract $\log h$ (here h is the Planck constant) for each axis, hence $3N \log h$ for N particles, each moving along three axes. Thus, instead of (4.3.12), we write

$$H^D(\boldsymbol{R}^N, \boldsymbol{p}^N) = H^D(\boldsymbol{R}^N) + H^D(\boldsymbol{p}^N) - 3N \log h. \tag{4.3.13}$$

This correction can be cast in the form of mutual information which we denote as

$$I(qm\ uncertainty) = 3N \log h. \tag{4.3.14}$$

We now combine all the four contributions to the MI of an ideal gas; the locational MI (4.3.4), the momentum MI (4.3.11), the correlation due to the indistinguishability of the particle, and the quantum mechanical correction (4.3.14). The result is

$$\begin{aligned} H^{ID}(1, 2, \ldots, N) &= H^D(\boldsymbol{R}^N) + H^D(\boldsymbol{p}^N) - I(qm\ uncertainty) \\ &\quad - I(indistinguishability) \\ &= N \ln V + \frac{3N}{2} \ln(2\pi emT) - 3N \ln h - \ln N! \\ &= N \ln \left[\frac{V}{N} \left(\frac{2\pi mT}{h^2} \right)^{3/2} \right] + \frac{5N}{2}. \end{aligned} \tag{4.3.15}$$

In (4.3.15), we recognize the Sackur–Tetrode equation for the MI (multiply by the Boltzmann constant k_B to obtain the entropy[19]). We shall revert back to this equation in Section 5.4 and in Appendix L. Here, we emphasize that this equation for the MI of an ideal gas was obtained from informational considerations. We have recognized two types of classical information (locational and momentum), and two types of correlations.

Note that the correlation due to the indistinguishability of the particles is different from the correlations discussed in the theory

[19] Note that we have changed to natural logarithm ln to conform with the familiar Sackur–Tetrode equation.

of liquids. In the latter, the correlation is between the densities of already indistinguishable particles. We shall discuss this type of correlation in Section 5.8.

Note carefully that both the locational and the momentum distribution employed in this section are the equilibrium distributions derived in Section 3.2. It is only for macroscopic systems at equilibrium that we identify the MI, H, with the thermodynamic MI, S. It is easy to envisage a system with initial conditions such that it will not evolve towards a thermodynamic equilibrium, and for which the Second Law of thermodynamics does not apply. For an explicit example, see Ben-Naim (2006).

4.4 Comments

(i) It is tempting to interpret $\ln N!$ as information associated with the information *about* the identity of the particles. This interpretation is potentially misleading. In fact, we have seen that there are two kinds of information; one *about* the location (i.e., where is particle i?) and one *about* the velocity or the momentum (i.e., how fast does particle i move?). We do not count information *about* the identity of the particle. Therefore, when performing any process, one should consider only the changes in the number of states. In particular, for the hypothetical or the thought experiment of labeling the particles, we find that the number of states increases. Hence, also the MI increases in contradiction with our intuitive feeling that the act of labeling should decrease the MI or the uncertainty. This sense of losing information upon un-labeling is probably the reason for an erroneous conclusion reached by Gibbs. We shall further discuss that in Chapter 6.

(ii) It is a remarkable fact that the Sackur–Tetrode expression that applies to *macroscopic* systems contains two elements from the laws of quantum mechanics — laws that govern the microscopic world. It is even more remarkable that the corrections to the (purely) classical partition function was made before the development of quantum mechanics to obtain agreements between the experimental and calculated entropy of a macroscopic system.

(iii) In this chapter, we discussed systems consisting of non-interacting particles (or nearly non-interacting). However, correlations did feature even when interactions were absent. It should be noted, however, that in most cases where correlations are studied, they originate from intermolecular forces. These are the most studied type of correlations especially in the theory of liquids. We shall briefly discuss these correlations in Chapters 5 and 6. More on these for liquids and liquid mixtures can be found in Ben-Naim (2006).

(iv) We have seen that changing the size of the cells, or the level of details we choose to describe the system, *changes* the value of the MI. It is a remarkable fact that the MI (or the entropy) of many gases, calculated by the choice of the particular cell size h (the Planck constant), is in agreement with the MI (or the entropy) measured by calorimetry. This adds additional confirmation for the validity of the Bolzmann expression for the MI (or the entropy). It also shows that Nature imposes on us a specific lower bound on the cell size, beyond which we have no freedom to choose the level of precision to describe the configuration of the system. It is also a remarkable fact that in most cases when such agreement is not found, one can explain the discrepancy by the existence of degeneracy in the state of the system near the absolute zero temperature — the so-called residual entropy.

(v) This chapter was devoted to develop the transition from a measure of MI in a simple system to a thermodynamic system. However, many of the results obtained in this chapter are relevant to experimental systems. It is instructive to derive one equation which is very useful in binding systems. In Section 4.1.4, we derived the expression for the MI of a system of N ligands on M sites:

$$S = \ln W = \ln \binom{M}{N}. \tag{4.4.1}$$

The chemical potential of the ligand can be obtained from the thermodynamic relation

$$\beta\mu = \left(\frac{\partial S}{\partial N}\right)_{E,M}. \tag{4.4.2}$$

From (4.4.1), and the stirling approximation, we get

$$\beta\mu = \frac{\partial \ln W}{\partial N} = \ln\left(\frac{M-N}{N}\right) = \ln\frac{x_0}{x_1}, \tag{4.4.3}$$

where $x_1 = N/M$ and $x_0 = (M-N)/M$ are the "mole fractions" of occupied and empty sites, respectively. In binding systems, x_1 is referred to as the fraction of the occupied sites. We also introduced the absolute activity defined by

$$\beta\mu = \ln\lambda. \tag{4.4.4}$$

From (4.4.3) and (4.4.4), we get

$$x_1 = \frac{\lambda}{1+\lambda}. \tag{4.4.5}$$

This result is well-known as the "binding isotherm" of the system, i.e., the fraction of the occupied sites as a function of the absolute

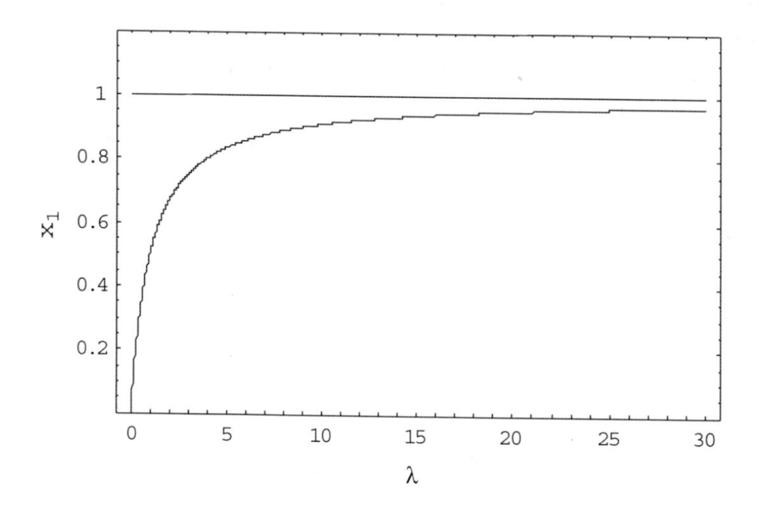

Figure 4.13. The Langmuir isotherm (4.4.5).

activity of the ligands [the latter is proportional to the partial pressure at low pressures; see Ben-Naim (2001)] (Figure 4.13).

Thus, although we have started with an extremely simple system to calculate the MI (4.4.1), the result obtained (4.4.5) is very relevant to a real system of binding of ligands on discrete sizes.

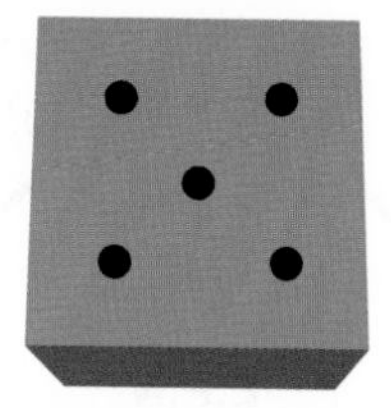

Chapter 5

The Structure of the Foundations of Statistical Thermodynamics

This chapter is devoted to presenting the foundations of statistical thermodynamics. The material of this chapter is standard and is contained in most textbooks on statistical mechanics. Therefore, the presentation will be brief, sketchy and unconcerned with details. Our aim is to reach for the applications of statistical mechanics as quick as possible. The reader who wishes to study the fundamentals in more detail should consult the bibliography at the end of the book. In this chapter, we shall also sacrifice accuracy, replacing rigor by plausibility. We shall also restrict ourselves to equilibrium systems. The field of non-equilibrium statistical mechanics is vast and requires an entire text.

We shall present both the "conventional" method, as well as Jaynes' maximum-uncertainty method, to obtain the fundamental distributions in the various ensembles. The only "new" element in this presentation is the elimination of the Boltzmann constant which renders the MI (or the entropy) dimensionless. It should be stressed from the outset that the traditional and the maximum-uncertainty methods are equivalent. We do not deem the former to be more "objective" than the latter, as some authors have claimed. In both cases, the number of microstates of a given macrostate is a physical, objective quantity; it is only the interpretation of $\ln W$ in terms of MI (or uncertainty) that has the flavor of being "subjective." In any case, either the traditional entropy defined by $k_B \ln W$,

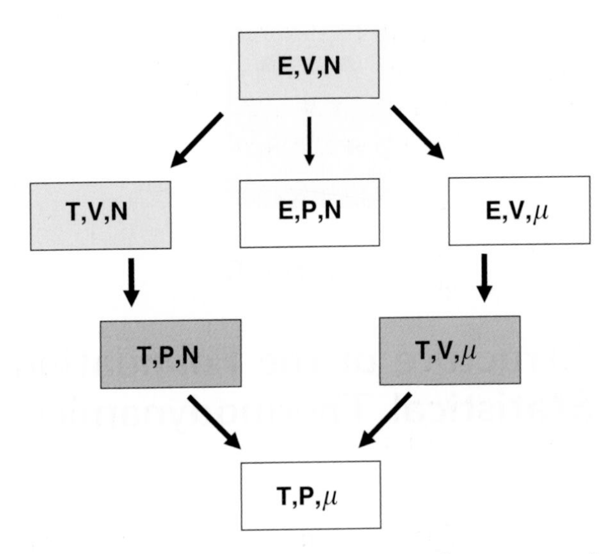

Figure 5.1. Schematic chart of the transformation from one set of variables to other sets of variables.

or the MI defined for an isolated system as $\ln W$, need ultimately be identified with the thermodynamic entropy as introduced by Clausius, with or without the Boltzmann constant.

The general structure of the theory is schematically described in Figure 5.1. The starting system under consideration is an isolated system, having constant energy E, volume V and number of particles N. It is on this system that the fundamental postulates of statistical mechanics are applied.

In the traditional approach, one uses the postulate of equal probability distribution for all the quantum mechanical states of the system (see Section 5.1). In the Jaynes' approach, the maximum MI principle is invoked as a fundamental postulate to reach the equal probability distribution of states.

Starting from the isolated system characterized by the independent variables E, V, N, we can proceed to construct other ensembles (an ensemble is a large collection of systems all characterized by the same thermodynamic variables, such as E, V, N or T, V, N, etc.). These are obtained by changing the independent variable, say from E, V, N to T, V, N to T, P, N, etc. This procedure corresponds to changing the physical boundaries of the system in the ensemble — from athermal, to thermal, to permeable boundaries, etc. In the diagram of Figure 5.1, the most useful systems are shown with shaded rendition. In Section 5.2, we transform to the variables T, V, N to obtain the canonical ensemble. A special and useful case of a canonical system is the classical limit discussed in Sections 5.3, 5.4 and 5.6. The latter leads to the Sackur–Tetrode expression, which we have already encountered in Chapter 4. In Section 5.6, we proceed to the grand-ensemble characterized by the variables T, V, μ. These systems are very useful for the study of open systems, such as osmotic systems. We shall conclude this chapter with a brief discussion of the role of intermolecular interactions in modifying the MI of a system.

5.1 The Isolated System; The Micro-Canonical Ensemble

The simplest system to start with is a system of fixed energy E, number of particles N and volume V. We shall refer to this as an E, V, N system. We shall also assume that the system is completely isolated, with no interaction with the rest of the universe. Of course, such a system does not exist; it is an idealized system that is convenient for laying the foundations of statistical thermodynamics. Even if such a strictly isolated system could exist, it would not be of interest. No experiments, nor measurements, not even observations, could be performed on such a system. Any observation necessarily will require interaction with the system. In practice, when the energy of the system is nearly constant, and the interactions with the surroundings are negligible, we treat such a system as being effectively isolated. The whole universe may be viewed as a single isolated system. However, this system is clearly far from equilibrium and therefore cannot be treated within the framework of equilibrium statistical thermodynamics.

The classical description of the microstates of an isolated system requires the specification of the locations and the velocities (or momenta) of all the N particles (assumed to be spherical). Classically, these variables could change continuously. However, quantum mechanics imposes a limit on the accuracy of the element of "action," $dxdp_x$, which is of the order of the Planck constant, h.

From quantum mechanics, we know that a system of particles characterized by the variable E, V, N can have many quantum mechanical states or microstates. These are the time-independent, or the stationary solutions of the Schrödinger equation. It is always assumed that quantum mechanics is in principle the *provider* of these solutions; statistical mechanics uses this information as *input* in the formalism to calculate average quantities. These calculated average quantities can then be compared with the corresponding experimentally measurable quantities.

It is an experimental observation that an isolated system when left undisturbed for a sufficiently long period of time reaches a state of equilibrium.[1] In this state, the thermodynamic system is fully characterized by a very few thermodynamic quantities such as E, V, N. This is a remarkable fact. The microscopic state of the system is characterized by myriads of parameters such as locations and momenta of all the particles in the system. On the other hand, the macrostate is described by a very small number of parameters such as volume, pressure, energy, etc.

There are essentially two fundamental postulates in statistical mechanics.[2]

First postulate
The time average of a dynamical quantity A in a macroscopic system is equal to the ensemble average.

The quantity we actually measure is a time average of some quantity $A(t) = A[\boldsymbol{R}^N(t), \boldsymbol{p}^N(t)]$,[3] i.e.,

$$\langle A(t) \rangle = \frac{1}{t} \int_0^t A[\boldsymbol{R}^N(t')\boldsymbol{p}^N(t')]dt', \tag{5.1.1}$$

[1]This is true for most systems studied within equilibrium statistical mechanics, such as gases and simple liquids. Some systems do not reach equilibrium for very long periods, and therefore cannot be studied by equilibrium statistical mechanics.

[2]In the old literature, the idea of the atomic constitution of matter was also taken as a fundamental postulate. See Fowler and Guggenheim (1939).

[3]We shall use lower case p_i for probability, but the vector $\boldsymbol{p}$ for the momentum with components p_x, p_y, p_z. Here, we assume that the quantity $A(t)$ is a function of t, only through the functions of $\boldsymbol{R}(t)$ and $\boldsymbol{p}(t)$ of all the particles.

where t is a finite length of time, very large compared with time-scale of molecular events. The first postulate says that

$$\lim_{t\to\infty}\frac{1}{t}\int_0^t A(t')dt' = \lim_{M\to\infty}\sum_{i=1}^{M} A_i p_i, \tag{5.1.2}$$

where A_i is the value of the function A for a system in the state i, and p_i is the probability, or the mole fraction of the system in the ensemble in state i.

The reason for adopting the not-so-obvious postulate (5.1.2) can be better understood with the following analogy. Suppose we throw a single die many times and record the frequency of occurrence of the different outcomes. With these frequencies, we can calculate average quantities pertaining to this particular die (note here that we do not require the die to be fair). Clearly, we feel that if the consecutive throws of the die are independent, then the average quantities calculated from these frequencies will be the same as the average quantities calculating from an *ensemble* of dice; all have the same characterization as the single die. In the ensemble of the die case, we throw M (M very large) simultaneously, and record the fraction of dice showing the different possible outcomes. If all the dice are independent, i.e., an outcome on one die does not affect the probability of occurrence of an outcome on a second die, then it is very plausible that the sequence (time) average will be equal to the ensemble average.

This plausible reasoning when translated into the language of our macroscopic system involves one serious difference. We cannot claim that the events "being in a specific microstate" at different times are independent. In fact, if we know the positions and velocities at one time, the equations of motion *determine* the microstate of the system at any other time, yet in spite of this dependence between the "events" at different times, we accept this postulate for a macroscopic system. We cannot *prove* this postulate. The only "proof" is in the agreement between calculated averages based on this postulate, and the corresponding measured quantities.

The second postulate

In an isolated system (micro-canonical ensemble), all possible micro-states are equally probable.

Using again the analogy with dice, we assume that if the dice are fair, then the probability of occurrence of each outcome should be the same. The reason involved is symmetry (see Section 2.3). While the symmetry argument is compelling for a fair die, it cannot be carried over straightforwardly to the macroscopic system. It is not clear whether or not the symmetry argument can be applied in this case. Nevertheless, this postulate is supported by a more powerful argument based on information theory; we shall discuss this argument below.

The two postulates cannot be proven. One can provide arguments of plausibility but not proof of these postulates. They are called postulates rather than axioms because they are less obvious than what we expect from axioms. The ultimate truth or validity of these postulates is in the agreement one obtains between computed results based on these postulates, and experimentally measurable quantities. For over a hundred years, this kind of agreement was confirmed in a myriad of cases.

In the classical description, how the microscopic state of an isolated system changes with time is easily visualized. In the quantum mechanical formulation, it is not easy to see how the system can access all states without any interaction with the surroundings. However, as in the case of dice, we can accept the idea that without any other information, the best guess is to assign equal probabilities for all the quantum states of the system characterized by E, V, N.

We start by assuming that a system characterized by the macroscopic variables E, V, N, at equilibrium has W states. All of these are assumed to be accessible to the system, within very short times, i.e., time much shorter than the typical time length of an experiment. The missing information (MI), (or the uncertainty) associated with the probability distribution for this system is

$$S = -\sum p_i \ln p_i. \tag{5.1.3}$$

In the traditional foundation of statistical mechanics, one *assumes* as a postulate, a uniform probability of all the W

micro-states,[4] i.e.,

$$p_i = \frac{1}{W}, \tag{5.1.4}$$

where p_i is the probability of finding the system in a specific microstate.

Then one defines the thermodynamic MI (or the entropy) by the relationship

$$S = -\sum_{i=1}^{W} \frac{1}{W} \ln \frac{1}{W} = \ln W. \tag{5.1.5}$$

This relationship, when multiplied by the Boltzmann constant ($k_B = 1.3807 \times 10^{23}$ J/K) is the familiar relationship between the entropy and the total number of microstates of the system characterized by the variables E, V, N. Here, we use the same quantity, denoted H by Shannon, but when applied to thermodynamics system, we use the notation S instead of H. We do that in order to preserve the traditional notation for the entropy S. However, note that S in (5.1.5) is the entropy divided by the Boltzmann constant k_B. Therefore, S is dimensionless and can be expressed in units of bits. In the traditional definition of the Boltzmann entropy in (5.1.5), one has to show that this entropy is identical to the thermodynamic entropy of the system. In fact, one needs to show that *differences* in the Boltzmann entropies are equal to differences in the thermodynamic entropy. The same applies to the MI defined in (5.1.3).

In the Jaynes formalism, one obtains the uniform distribution (5.1.4) from the principle of maximum MI (originally the maximum entropy principle), i.e., one maximizes MI in (5.1.3) subject to the condition $\sum p_i = 1$, to obtain the distribution (5.1.4). As we have seen, having only this information on the microstates of the system, the distribution (5.1.4) is the best guess, or the least biased distribution. Thus, Jaynes' procedure provides plausibility — not

[4]In the classical version of (5.1.4), one assumes equal *a priori* probabilities for equal regions on a constant energy surface, in an abstract phase space. It is interesting to note that in actual applications, the classical limit of statistical mechanics was found to be more useful. However, the statement of the postulate is easier to make in quantum mechanical language.

a proof — for one of the fundamental postulates of statistical mechanics.

Having the distribution (5.1.4) still leaves open the question of how to calculate W for a specific macroscopic system. We have seen one simple example of W in Chapter 4. In general, this quantity is very difficult to calculate. It should be made clear however, that the calculation of W is not part of the formalism of statistical mechanics. W is presumed to be *given* as input into the formalism. Thus, for each set of variables E, V, N, the quantity W is uniquely determined, and hence the MI is also defined for such a system, i.e.,

$$S(E, V, N) = \ln W. \tag{5.1.6}$$

Although we do not know how to calculate W for the general macroscopic system, it is believed that for macroscopic systems, W increases very sharply with the energy, such as

$$W(E) \approx E^f, \tag{5.1.7}$$

where f is the number of degrees of freedom of the system and where E the energy measured relative to the energy of the ground state, E_0, i.e., the lowest possible value of the energy of the system. Since f is of the order of the Avogadro number $N_{\mathrm{AV}} \approx 10^{23}$, W grows very steeply with E as well as with N.

Once we identify $S(E, V, N)$ in (5.1.6) as the MI of the system, and that the changes of the MI are equal to the changes in the thermodynamic entropy of the system (in dimensionless units), then one can use thermodynamic identities to obtain all the thermodynamic quantities of the system. The most important derivatives of S are obtained from the relation

$$TdS = dE + PdV - \mu dN. \tag{5.1.8}$$

Hence,

$$\beta = \frac{\partial \ln W}{\partial E}, \tag{5.1.9}$$

$$\beta P = \frac{\partial \ln W}{\partial V}, \tag{5.1.10}$$

$$-\beta\mu = \frac{\partial \ln W}{\partial N}, \tag{5.1.11}$$

where P is the pressure, μ is the chemical potential of the system and β is T^{-1}.

Relation (5.1.6) is the fundamental cornerstone of statistical mechanics. It is a remarkable relation between a quantity Clausius named entropy, which was defined in terms of heat transfer and temperature on the one hand, and the total number of states of an isolated system characterized by the variables E, V, N on the other. These two quantities seem to be totally unrelated. The first is defined in terms of physically measurable quantities (heat transfer and temperature); the second is based on counting of states, which seems to be almost devoid of any physical reality. There is no formal proof that the two quantities are identical. The validity of the relationship between the two quantities ultimately rests on the agreement between the calculated values of S, based on (5.1.6), and experimental data based on Clausius' definition.

Another remarkable property of this relation, due to the development of statistical mechanics by Gibbs, is the realization that the *function* $S(E, V, N)$ encapsulates all the thermodynamics of the system, e.g., by taking the first derivatives of (5.1.6), one can get the temperature, pressure and chemical potential,[5] and from the second derivative the compressibility, heat capacity, etc.

For these reasons, (5.1.6) is referred to as the fundamental relationship in statistical thermodynamics. The function $S(E, V, N)$ has the same status as the canonical partition function $Q(T, V, N)$ or the grand canonical partition function $\Xi(T, V, \mu)$, discussed in the next sections.

The relation between entropy and the number of states, although well understood and aggressively promoted by Boltzmann, was never explicitly published by Boltzmann himself. It seems that Planck was the first to publish this relationship in 1923, and proclaimed it to be a definition of the absolute entropy. Boltzmann was also the one who tried hard to convince his contemporaries that entropy is related to probabilities. Although probability does

[5]The chemical potential, the most important quantity in chemical thermodynamics, was introduced by Gibbs. We have already seen one application of (5.1.11) in Section 4.4.

not explicitly feature in (5.1.5), it does implicitly. We shall discuss that aspect in relation to the Second Law in Section 6.12.

It should also be emphasized that early in the 20th century, probability arguments were foreign to physics. The first who introduced probability to physics was James Clerk Maxwell. The idea that entropy is related to probabilities also rendered the Second Law less absolute, as compared with other laws of physics. In the original, non-atomistic formulation of the Second Law by Rudolf Clausius, it was proclaimed to be absolute. In Boltzmann's atomistic formulation, exceptions were permitted. This was a revolutionary idea in the pre-quantum physics era, when probabilistic arguments were only starting to diffuse into physics.

5.2 System in a Constant Temperature; The Canonical Ensemble

As we have noted, the actual calculation of W is usually extremely complicated: only a very few cases are computationally feasible, such as the example discussed in Chapter 4.

To overcome this difficulty, one proceeds to remove the constraint of either one or more of the constant variables E, V or N (see Figure 5.1). We shall start with the case of a system in contact with a thermal, or heat, reservoir. Traditionally, one envisages a very large *ensemble* of systems all conforming to the same macroscopic variables E, V, N but are completely isolated from each other. We now bring the systems into thermal contact (Figure 5.2), i.e., the athermal boundaries between all the isolated systems are replaced by diathermal (or heat conducting) boundaries. In the new ensemble, heat (or thermal energy) can flow from one system to another. We know from thermodynamics (the Zeroth Law) that any two macroscopic systems in thermal contact at equilibrium will have the same temperature. The energy of the systems which were initially fixed at E will now fluctuate and be replaced by an average energy of $\bar{E}$. Note that the whole ensemble is isolated.

There are several ways to proceed from the system characterized by the variables E, V, N to a system characterized by the variables T, V, N. One simple way is to take one system in the ensemble

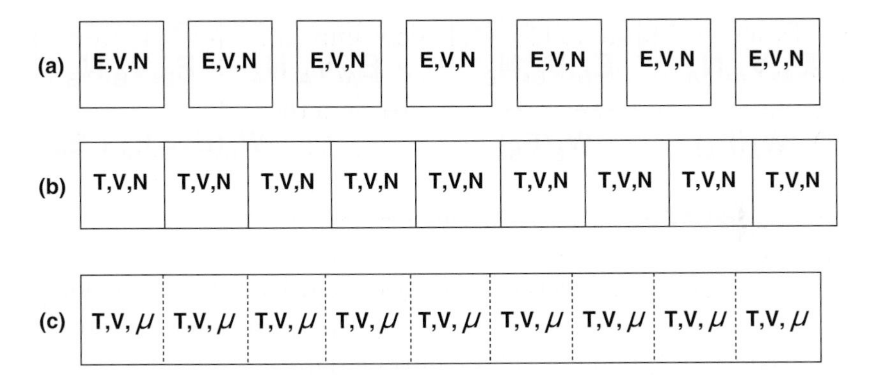

Figure 5.2. (a) An ensemble of isolated systems, all with the same E, V, N. (b) An ensemble of systems in thermal equilibrium, all with the same T, V, N. (c) An ensemble of systems in thermal and material equilibrium, all with the same T, V, μ.

(Figure 5.2) and view all the other systems as a heat reservoir (or a thermal bath). The main property of the heat reservoir is that it is so large that even when it exchanges thermal energy with the system, its own thermal energy is almost unaffected, and hence its temperature remains very nearly constant.

We next turn to the question of finding the probability distribution for the energy levels. The traditional method of calculating this distribution is discussed in Appendix K. As will be shown there, it is essentially equivalent to the procedure we shall use here due to Jaynes. The latter is somewhat more general than the former. Let E_i be the energy level, measured relative to the ground state which may conveniently be chosen to be zero. Let p_i be the probability of finding the system in any one of the states, having the energy level E_i. According to Jaynes' maximum uncertainty principle, we write the MI for this system as

$$S = -\sum_i p_i \log p_i. \tag{5.2.1}$$

Find the maximum of MI subject to the two conditions

$$\sum p_i = 1, \tag{5.2.2}$$

$$\sum p_i E_i = \bar{E}. \tag{5.2.3}$$

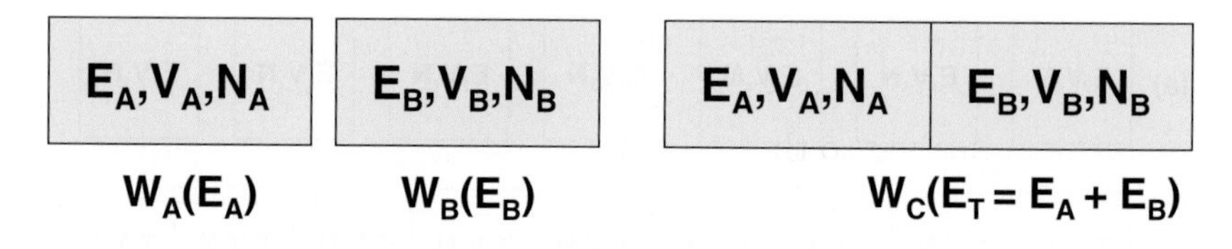

Figure 5.3. Two systems brought to thermal equilibrium.

Using the Lagrange method of undetermined multipliers, we maximize S, subject to the two conditions (5.2.2) and (5.2.3), and get the result (see Section 3.4 and Appendix G)

$$p_i = \frac{\exp[-\lambda_2 E_i]}{Q}, \tag{5.2.4}$$

where

$$Q = \sum_i \exp[-\lambda_2 E_i]. \tag{5.2.5}$$

The sum in (5.2.5) is carried out over all possible states of the system. There are many states corresponding to the energy level E_i. These are degeneracies of the energy levels. We denote by $W(E_k, V, N)$ the number of states (or the degeneracy) having the same energy level E_k and rewrite Q as

$$Q(\lambda_2, V, N) = \sum_k W(E_k, V, N) \exp[-\lambda_2 E_k]. \tag{5.2.6}$$

Now the summation is over all the energy levels, rather than over all quantum mechanical states of the system as in (5.2.5). What we have achieved so far is the general *form* of the distribution (5.2.4) which is referred to as the Boltzmann distribution (see also Section 3.2).[6]

The quantity $Q(\lambda_2, V, N)$ is known as the canonical partition function.[7] This partition function turns out to be the most useful one in applications of statistical mechanics.

[6]In the traditional formalism, we apply the postulates of statistical mechanics to the ensemble of systems at thermal contact. This is essentially equivalent to the Jaynes prescription. See Appendix K.

[7]The term "canonical" was introduced by Gibbs meaning *standard*. The quantum states are the eigenstates of the energy operator with eigenvalue E.

Next, we turn to identify the physical meaning of the parameter λ_2. There are various ways of doing that.[8] Here, we shall identify λ_2 by appealing to thermodynamic relations. We shall return to this identification of λ_2 in connection with ideal gases in Section 5.5.

Consider two macroscopic systems A and B that are characterized by the variables (E_A, V_A, N_A) and (E_B, V_B, N_B), respectively. When the two systems are isolated, the corresponding number of states are $W(E_A)$ and $W(E_B)$, respectively.

We suppress the dependence of W on the volume and on the number of particles since these will be held constant in the following discussion.

We now bring the two systems into thermal contact. Thermal energy can flow from one system to the other. The volume and the number of particles of each of the system are unchanged. The combined system C is presumed to be isolated. Let W_C be the total number of states of the combined systems A and B at thermal equilibrium. Since the total energy of the combined system is fixed at say E_T, the number of states of the combined system is

$$W_C(E_T) = \sum_{E_A} W_A(E_A) W_B(E_T - E_A), \tag{5.2.7}$$

where we assume for simplicity that the energies are discrete. The summation in (5.2.7) is over all possible energy levels E_A of A. Since E_T is presumed constant, the energy of B is determined by E_A.

The probability of finding the system A with energy E_A is simply

$$P(E_A) = \frac{W_A(E_A) W_B(E_T - E_A)}{W_C(E_T)}, \tag{5.2.8}$$

where we have used the classical definition of the probability (Section 2.3). Recall that the combined system A and B is isolated. Therefore, the probability of each state of the combined system is $1/W_C(E_T)$. $W_C(E_T)$ is the *total* number of states, or the events of C, whereas in the numerator, we have the total number of events consistent with the requirement that system A will have energy E_A (and system B must have energy $E_T - E_A$). We now

[8] For details, see for example, Rushbrooke (1949) and Hill (1960).

use the fact that $W_A(E_A)$ increases extremely rapidly with E_A. Hence, $W_B(E_T - E_A)$ decreases extremely rapidly with E_A. Therefore, $P(E_A)$ has a sharp peak at some value E_A^*. The condition for maximum value of $P(E_A)$ is the same condition for the maximum of $\ln P(E_A)$. Hence,

$$\frac{1}{P(E_A)}\frac{\partial P(E_A)}{\partial E_A} = \frac{\partial \ln P(E_A)}{\partial E_A} = \frac{\partial \ln W_A(E_A)}{\partial E_A} + \frac{\partial \ln W_B(E_T - E_A)}{\partial E_A}$$

$$= \frac{\partial \ln W_A(E_A)}{\partial E_A} - \frac{\partial \ln W_B(E_B)}{\partial E_B} = 0, \qquad (5.2.9)$$

where we have denoted $E_B = E_T - E_A$ and used the equality $dE_A = -dE_B$.

Thus, the condition for maximum value of $P(E_A)$ is the same as the condition for the equilibrium value of E_A,[9] i.e., at equilibrium we have

$$\left[\frac{\partial \ln W_A(E_A)}{\partial E_A}\right]_{eq} = \left[\frac{\partial \ln W_B(E_B)}{\partial E_B}\right]_{eq}. \qquad (5.2.10)$$

This equality means that at thermal equilibrium, the derivative of the MI with respect to the energy is the same for A and B. We know from thermodynamics that any two systems in thermal equilibrium must have the same temperature. Also, we know from thermodynamics that the derivative of the MI with respect to the energy is

$$\left(\frac{\partial S}{\partial E}\right)_{V,N} = \frac{1}{T}, \qquad (5.2.11)$$

where T is the absolute temperature. Since we identify the thermodynamic MI, with $\ln W$ for each system, following from (5.2.10):

$$\left(\frac{\partial \ln W_A(E_A)}{\partial E_A}\right)_{eq} = \frac{1}{T_A} = \frac{1}{T_B} = \frac{1}{T}, \qquad (5.2.12)$$

where T is the absolute temperature measured in units of energy.

[9] The average value, the most probable and the equilibrium value of E_A are all different concepts. However, because of the extreme sharpness of $P(E)$, they are practically the same.

Next, we go back to (5.2.6), where we have defined the quantity Q as a normalization constant in (5.2.4). We now apply the same argument as above regarding the behavior of the two factors $W(E_k)$ and $\exp[-\lambda_2 E_k]$ as a function of E_k. The first is a sharply increasing function of E_k; the second a steeply decreasing function of E_k (for $\lambda_2 \geq 0$; see below). Hence, one can replace the sum in (5.2.6) by the maximal term,[10] the condition for which is

$$\frac{\partial \ln W(E_k)}{\partial E_k} - \lambda_2 = 0$$

or equivalently

$$\lambda_2 = \frac{\partial \ln W(E_k)}{\partial E_k}. \tag{5.2.13}$$

Comparing (5.2.13) and (5.2.12), we can conclude that the parameter λ_2 is simply the inverse of the temperature. We shall henceforth use the notation

$$\beta = \lambda_2 = \frac{1}{T}. \tag{5.2.14}$$

In Section 5.5, we shall see that the parameter β is related, for a classical ideal monoatomic gas, to the average kinetic energy of the atoms.

With this identification of λ_2, the quantity Q defined in (5.2.6) is now a function of the variables T, V, N. It is referred to as the canonical partition function. Having the partition function Q, one can obtain all the thermodynamic quantities from Q and its derivatives with respect to T, V and N.

Let E^* be the energy value for which the quantity $W(E_k, V, N)$ $\exp[-\beta E_k]$ is maximum. Define the new quantity $A(\beta, V, N)$ by

$$\begin{aligned} \exp[-\beta A(\beta, V, N)] &= W(E^*, V, N) \exp[-\beta E^*] \\ &= \exp[S(E^*, V, N) - \beta E^*] \\ &= \exp\{-\beta[E^* - TS(E^*, V, N)]\}. \end{aligned} \tag{5.2.15}$$

[10]This replacement is quite often done in statistical mechanics. We shall discuss one such case in more detail in Appendix N.

Hence, from (5.2.15) and (5.2.6), and replacing the sum in (5.2.6) by its maximal term, we get the fundamental relationships

$$\begin{aligned} A(T,V,N) &= E^* - TS(E^*,V,N) \\ &= -T\ln Q(T,V,N), \end{aligned} \tag{5.2.16}$$

where

$$T^{-1} = \left.\frac{\partial S}{\partial E}\right|_{E=E^*}. \tag{5.2.17}$$

We can immediately identify $A(T,V,N)$ as the Helmholtz energy of the system. Note that the probability distribution $P(E)$ is extremely sharp. Therefore, one can also identify E^* with the average energy of the system characterized by the variables T, V, N.

Having the general relationship between the canonical partition function and the Helmholtz energy $A(T,V,N)$, we can obtain all the thermodynamic quantities using thermodynamic identities. For instance, for the differential of A, we have[11]

$$dA = -SdT - PdV + \sum \mu_i dN_i. \tag{5.2.18}$$

We get

$$\begin{aligned} S &= \frac{-\partial A}{\partial T} = \ln Q + T\frac{\partial \ln Q}{\partial T}, \\ P &= \frac{-\partial A}{\partial V} = T\frac{\partial \ln Q}{\partial V}, \\ \mu_i &= \frac{\partial A}{\partial N_i} = -T\frac{\partial \ln Q}{\partial N_i}. \end{aligned} \tag{5.2.19}$$

Having identified λ_2, we can now rewrite the probability p_i for the state i in (5.2.4) as

$$p_i = \frac{\exp[-\beta E_i]}{Q}, \tag{5.2.20}$$

and the probability of finding the system in an energy level E as

$$P(E) = W(E)\exp[-\beta E]/Q. \tag{5.2.21}$$

[11]For a multi-component system, N is replaced by the vector $\boldsymbol{N} = (N_1, N_2, \ldots, N_c)$, where N_i is the number of molecules of species i.

The transformation from the function $S(E, V, N)$ to $A(T, V, N)$ within thermodynamics is known as the Legendre transformation of the function $S(E, V, N)$, where we replace the fixed energy E by a fixed temperature T. In statistical mechanics, the corresponding transformation is the discrete analog of the Laplace transform for the discrete variable E [for a continuous variable E, we need to replace the sum in (5.2.6) by an integral].

It is instructive to reproduce the Clausius original definition of the entropy change, here in terms of MI, as an amount of heat transferred at a fixed temperature.

Consider a macroscopic system in contact with a heat reservoir at a constant temperature T having average energy $\bar{E}$. We assume that a very small amount of heat ΔQ_{rev} is transferred to the system, such that $\Delta Q_{rev} \ll \bar{E}$. As a result of this energy change, the MI of the system will change. For very small ΔQ_{rev}, we take the first order expansion of MI about $\bar{E}$, to obtain

$$\begin{aligned} S(\bar{E} + \Delta Q_{rev}) &= \ln W(\bar{E} + \Delta Q_{rev}) \\ &= \ln W(\bar{E}) + \left.\frac{\partial \ln W}{\partial E}\right|_{E=\bar{E}} \Delta Q_{rev} + \cdots . \end{aligned} \tag{5.2.22}$$

Thus, for very small ΔQ_{rev}, we have

$$\Delta S = S(\bar{E} + \Delta Q_{rev}) - S(\bar{E}) = \beta \Delta Q_{rev}. \tag{5.2.23}$$

The original definition of the entropy change for a reversible transfer of heat ΔQ_{rev} to a system at constant temperature T, as defined by Clausius, was

$$\Delta S = \frac{\Delta Q_{rev}}{T}. \tag{5.2.24}$$

Originally, ΔS in (5.2.24) was defined in units of energy divided by temperature (in units of Kelvin, K). As we discussed in Chapter 1, had the temperature been defined in units of energy, then Clausius' entropy would have been rendered dimensionless. This would make the Clausius entropy change in (5.2.24) identical to the change in the MI in (5.2.23).

5.3 The Classical Analog of the Canonical Partition Function

There is a standard procedure to obtain the classical analogue of the quantum mechanical partition function.[12] Here, we shall use an heuristic argument based on analogy.

The classical analogue of the quantum mechanical canonical partition function can be written as

$$Q_{quant} = \sum_i \exp[-\beta E_i] \to Q_{class} = \int \cdots \int d\boldsymbol{R}^N d\boldsymbol{p}^N \exp[-\beta H_N]. \tag{5.3.1}$$

Here, the sum over i extends over all possible states of the system. By purely formal analogy, we write the (purely) classical equivalent of the quantum mechanical partition, replacing the sum over all states, by the integration over all possible classical states of the system of N, simple spherical particles in a volume V and at temperature T. $\boldsymbol{R}_i$ is the locational vector for the center of the ith particle and $\boldsymbol{p}_i$ is the corresponding momentum vector. $\boldsymbol{R}^N$ and $\boldsymbol{p}^N$ are shorthand notations for the vectors $\boldsymbol{R}^N = \boldsymbol{R}_1, \ldots, \boldsymbol{R}_N$ and $\boldsymbol{p}^N = \boldsymbol{p}_1, \ldots, \boldsymbol{p}_N$, and similarly for the product $d\boldsymbol{R}^N = d\boldsymbol{R}_1 \cdots d\boldsymbol{R}_N$ and $d\boldsymbol{p}^N = d\boldsymbol{p}_1 \cdots d\boldsymbol{p}_N$. H_N is the classical Hamiltonian of the system, which in this case constitutes the kinetic energies of all particles and the total potential energy of interactions among the particles. Thus, we write

$$H_N = \sum_i \frac{\boldsymbol{p}_i^2}{2m} + U_N(\boldsymbol{R}_1, \ldots, \boldsymbol{R}_N), \tag{5.3.2}$$

where $\boldsymbol{p}_i$ is the momentum of particle i, i.e., $\boldsymbol{p}_i$ is the vector[13]

$$\boldsymbol{p}_i = (p_{x_i}, p_{y_i}, p_{z_i}), \tag{5.3.3}$$

and the mass m is assumed to be the same for all particles.

[12]Kirkwood (1933) following Wigner showed how to obtain the limiting expression for the quantum mechanical partition function as a power series in Planck constant h. The leading term in this expansion is now referred to as the classical PF. More appropriately this should be referred to as the classical limit of the quantum mechanical PF [see also Hill (1956)].

[13]Note that $\boldsymbol{R}_i$ and $\boldsymbol{p}_i$ are vectors, but $d\boldsymbol{R}_i$ and $d\boldsymbol{p}_i$ are elements of "volume," i.e., $d\boldsymbol{R}_i = dx_i dy_i dz_i$ and $d\boldsymbol{p}_i = dp_{x_i} dp_{y_i} dp_{z_i}$. In this section, we use the letter H for the Hamiltonian function. In the rest of the book H is used for the MI.

U_N is the total interaction energy among all the particles of the system. The integration over all of the momenta is well known. The result is

$$\int d\boldsymbol{p}_1 \cdots d\boldsymbol{p}_N \exp\left[-\sum_{i=1}^{N} \frac{\beta \bar{p}_i^2}{2m}\right] = \left\{\int dp_1 \exp\left[-\frac{\beta \bar{p}_1^2}{2m}\right]\right\}^N$$
$$= \left\{\int_{-\infty}^{\infty} dp_x \exp\left[-\frac{\beta p_x^2}{2m}\right]\right\}^{3N}$$
$$= (2\pi m T)^{3N/2}. \quad (5.3.4)$$

It is easy to show that had we defined the Helmholtz energy as

$$A_{class} = -T \ln Q_{class} = -T \ln V^N (2\pi m T)^{3N/2}, \quad (5.3.5)$$

we would have obtained the correct equation of state of an ideal gas [i.e., for $U_N \equiv 0$ in (5.3.2)]

$$P = -\left(\frac{\partial A}{\partial V}\right)_{N,T} = \frac{NT}{V}, \quad (5.3.6)$$

and also the correct average energy of the system

$$\bar{E} = \frac{3}{2} NT. \quad (5.3.7)$$

However, it was known for over a hundred years that the entropy, or the MI derived from the classical partition function, would have come out wrong, namely

$$S_{class} = N \ln V + \frac{3}{2} N \ln(2\pi m T) + \frac{3}{2} N. \quad (5.3.8)$$

Clearly, if we replace N by $2N$ and V by $2V$, the MI will not be doubled. This result is sometimes referred to as the Gibbs paradox (see Appendix O).

Similarly, the chemical potential of an ideal gas would have the form

$$\mu_{class} = \left(\frac{\partial A_{class}}{\partial N}\right)_{T,V} = T \ln \frac{(2\pi m T)^{3/2}}{V}, \quad (5.3.9)$$

which has the incorrect dependence on the density $\rho = N/V$ [see (5.3.17) below].

Indeed, it was found that the "classical" partition function, as defined in (5.3.1), must be corrected by two factors to get the *classical limit* of the quantum mechanical partition function. Instead of (5.3.1), we need to make the correspondence

$$Q_{\text{quant}} \to Q_{class} = \frac{1}{h^{3N}N!}\int\cdots\int d\boldsymbol{R}^N, d\boldsymbol{p}^N \exp[-\beta H_N]. \tag{5.3.10}$$

Both of the new factors h^{3N} and $N!$ have their origins in quantum mechanics. Both can be interpreted as mutual information (see Section 4.3). Both can be viewed as corrections due to over-counting.

The first factor h arises from counting too many configurations in the "phase space." The uncertainty principle dictates that we cannot distinguish between configurations for which the element of action $dxdp_x$ is smaller than the Planck constant h. Therefore, we must divide by h for each degree of freedom — hence the factor h^{3N} in the denominator of (5.3.10). This is the same type of mutual information we encountered in Section 3.2.4.

The second factor $N!$ is also due to correction for over-counting configurations in the purely classical partition function. When integrating over all locations, we have tentatively *labeled* the particles $1, 2, \ldots, N$, then integrated for each particle over the entire volume V. In quantum mechanics, this "labeling" is not possible. What we do in practice is to first label the particles, say $1, 2, \ldots, N$. This results in an over-counting of configurations; therefore, we must correct this over-counting by dividing by $N!$. This labeling and unlabeling of particles should be done with some care. We shall further discuss this point in Appendix M.

From the "corrected" classical partition function in (5.3.10), we obtain the MI of the system of non-interacting particles

$$S = \frac{5}{2}N - N\ln\rho\Lambda^3, \tag{5.3.11}$$

where

$$\Lambda = \frac{h}{\sqrt{2\pi mT}}. \tag{5.3.12}$$

Expression (5.3.11), when multiplied by the Boltzmann constant k_{B}, is the well-known Sackur–Tetrode equation for the entropy.

We can transform this equation to express the MI in terms of the energy, volume and number of particles. For an ideal monoatomic gas, the average energy is[14]

$$E = N\varepsilon_K = \frac{3}{2}TN = N\frac{m\langle v^2\rangle}{2}, \tag{5.3.13}$$

where ε_K is the average kinetic energy per particle.

The equation of state for this system is

$$P = \rho T. \tag{5.3.14}$$

We can identify the relation between the temperature and the average energy per particle as

$$T = \frac{2}{3}\frac{E}{N}. \tag{5.3.15}$$

Hence, we can express S as a function of the variable E, V, N

$$S(E,V,N) = \ln\left[\frac{e^{5N/2}(4\pi mE)^{3N/2}}{\rho^N h^{3N}(3N)^{3N/2}}\right]. \tag{5.3.16}$$

The chemical potential for this system is

$$\mu = \left(\frac{\partial A}{\partial N}\right)_{T,V} = -T\left(\frac{\partial S}{\partial N}\right)_{E,V} = T\ln\rho\Lambda^3. \tag{5.3.17}$$

This should be compared with relation (5.3.9) for the chemical potential, obtained from the classical partition function (5.3.5).

Having obtained the MI for an ideal monoatomic gas, we can write the number of states of the system as

$$W = \exp[S] = \left[\frac{e^{5/2}}{\rho\Lambda^3}\right]^N, \tag{5.3.18}$$

where W is given as a function of T, V, N. We can also use (5.3.15) to transform into a function of E, V, N to obtain

$$W = \left[\left(\frac{4\pi m}{3}\right)^{3/2}\frac{e^{5/2}}{h^3}\frac{E^{3/2}V}{N^{5/2}}\right]^N. \tag{5.3.19}$$

[14]For simplicity of notation, we omit the bar over E.

5.4 The Re-interpretation of the Sackur–Tetrode Expression from Informational Considerations

In the previous section, we have derived the Sackur–Tetrode expression for an ideal monoatomic ideal gas of N particles in volume V and at temperature T at equilibrium. In Section 4.3.3, we have derived the same expression, but from informational considerations. In anticipation of its applications in Chapter 6, we shall summarize here the four contributions to the Sackur–Tetrode expressions:

(i) *The locational MI, S(Locations)*. In a one-dimensional box, we have seen (Section 4.3.3) that the uncertainty is[15]

$$S(Locations) = N \ln L, \tag{5.4.1}$$

where L is the length of the "box." Assuming that the locations along the three axes x, y, z are independent, we can easily generalize (5.4.1) to obtain

$$S(Locations) = N \ln V. \tag{5.4.2}$$

The meaning of this MI is clear. Each particle can wander in the entire volume V. The larger the volume V, the greater the uncertainty is, or the greater the missing information regarding the location of the particles. Assuming that the particles are independent, we have to take N times the uncertainty for each particle.

(ii) *The momentum MI, S(Momenta)*. For the momenta in one dimension, we have seen the expression for $S(Momenta)$ in Section 4.3.3:

$$S(Momenta) = \frac{1}{2} N \ln(2\pi emT). \tag{5.4.3}$$

Again, assuming that the momenta along the three axes are independent, we obtain

$$S(Momenta) = \frac{3}{2} N \ln(2\pi emT). \tag{5.4.4}$$

So far, we have accounted for the MI in the location and momenta of each particle, as if these were independent. Also, the N particles

[15]Note again that L and V must be dimensionless quantities. However, we shall use only differences in S, where units of L or V will cancel out.

are considered to be independent. Thus, we have the "classical" Sackur–Tetrode expression

$$\begin{aligned} S &= S(Locations) + S(Momenta) \\ &= N \ln V + \frac{3}{2} N \ln(2\pi emT) \end{aligned} \qquad (5.4.5)$$

[see (5.3.8)].

As we have pointed out in Section 5.3, this expression for the MI (or the original entropy) is not satisfactory. It does not lead to results consistent with experiments. There are two corrections that must be introduced:

(i) *Correction due to the quantum mechanical uncertainty.* In Section 3.2.4, we have seen how to correct the missing information when we pass from the continuous range of locations $(0, L)$ to the discrete division of the length L into cells of length h. In this process, we effectively transformed the expression for S from the continuous space $(0, L)$ to the discrete space of n cells, each of which of length h.

In a similar fashion, we recognize that in "counting" all possible points in the "phase space" of each particle, we have counted too many configurations. We must divide the phase space of each particle into cells of size h^3, thereby transforming the MI as calculated in (5.4.5) from a continuous space into a discrete space. The correction due to the quantum mechanical uncertainty principle is therefore

$$\Delta S(uncertainty\ principle) = -\ln h^{3N} = -3N \ln h. \qquad (5.4.6)$$

(ii) *Correction due to indistinguishability of the particles.* The second correction to the "classical" Sackur–Tetrode expression is more subtle. We recall that in Section 5.3, we have written the classical partition function (5.3.1), where we have performed integrations over all locations and all momenta of the particles. In the very writing of (5.3.1), we have *labeled* the particles by giving each particle a number $1, 2, \ldots, N$. This labeling of particles is not permitted by quantum mechanics. Therefore, we must correct for the labeling of the particles by un-labeling them. Here is the subtle point. Unlabeling is conceived as loss of information. Loss of information is equivalent to an *increase* in MI or an *increase* in the uncertainty. In other words, we expect that un-labeling should involve an increase in MI. However, this conclusion is incorrect, and the conception of an *increase* in MI is only an illusion. More on this in Chapter 6 and Appendix M.

The fact is that by un-labeling the particles we actually *decrease* the number of *states* of the system (see a simple illustration in Section 4.1). Decreasing the number of states is associated with a *decrease* in the MI or a decrease in the uncertainty. Hence, the necessary correction is

$$\Delta S(\textit{indistinguishability of the particles}) = -\ln N!. \qquad (5.4.7)$$

We can now collect the two contributions (5.4.2) and (5.4.4), and the two corrections to construct the Sacker–Tetrode equation for this system

$$\begin{aligned} S &= [N \ln V] + \left[\frac{3}{2} N \ln(2\pi emT)\right] \\ &\quad + [-3N \ln h] + [-\ln N!] \\ &= N \ln\left[\frac{V}{N}\left(\frac{2\pi mT}{h^2}\right)^{3/2}\right] + \frac{5}{2}N, \end{aligned} \qquad (5.4.8)$$

where we used the Stirling approximation $\ln N! \approx N \ln N - N$.

From (5.4.8), we can calculate the changes in MI associated with either the change in location or in momenta. Perhaps, the simplest way of calculating the effect of changing the locational and the momenta information is to use the well-known thermodynamic equation for the change in MI between two states (T_1, V_1) and (T_2, V_2). For one mole of ideal gas

$$S_2 - S_1 = C_V \ln \frac{T_2}{T_1} + N_{AV} \ln \frac{V_2}{V_1}, \qquad (5.4.9)$$

where C_V is the heat capacity at constant volume, and N_{AV} is the Avogadro number (in conventional thermodynamics C_V is measured in entropy units and the gas constant R appears in the second term; here the two terms are dimensionless).

Clearly, for any process where the volume changes at constant temperature, the change in the MI is only *locational* information, i.e., from V_1 to V_2. On the other hand, for a process at constant volume, the change from T_1 to T_2 is equivalent to the change in the distribution of the momenta — from one momentum distribution to another.

5.5 Identifying the Parameter β for an Ideal Gas

In Section 5.2, we have interpreted the parameter λ_2 that was used as an undetermined multiplier. In the previous Section 5.4, we have derived the classical limit of the quantum mechanical partition function, from which we can derive the distribution of momenta along say the x-axis

$$P(p_x) = \left(\frac{\beta}{2\pi m}\right)^{1/2} \exp\left[-\frac{\beta p_x^2}{2m}\right]. \tag{5.5.1}$$

From the momentum distribution function, it is easy to get the velocity distribution along the x-axis. Putting $p_x = mv_x$, we get

$$P(v_x) = \left(\frac{m\beta}{2\pi}\right)^{1/2} \exp\left[-\beta\frac{mv_x^2}{2}\right]. \tag{5.5.2}$$

The average square velocity along the x-axis is

$$\langle v_x^2 \rangle = \int_{-\infty}^{\infty} v_x^2 p(v_x) dv_x. \tag{5.5.3}$$

Assuming that the motions along the three axes are independent, we can get from (5.5.2) the distribution of velocities:

$$P(v) = \left(\frac{\beta m}{2\pi}\right)^{3/2} 4\pi v^2 \exp\left[-\frac{\beta m v^2}{2}\right]. \tag{5.5.4}$$

This is the Maxwell–Boltzmann distribution of velocities. This is the probability density for finding a particle with velocity between v and $v + dv$ from which one can obtain the average of v^2

$$\begin{aligned}\langle v^2 \rangle &= \langle v_x^2 \rangle + \langle v_y^2 \rangle + \langle v_z^2 \rangle \\ &= 3\langle v_x^2 \rangle.\end{aligned} \tag{5.5.5}$$

By integrating (5.5.4) over all possible velocities, we obtain

$$\frac{1}{2}m\langle v^2 \rangle = \frac{3}{2\beta}. \tag{5.5.6}$$

Since we already know the relation between the average kinetic energy and the temperature

$$\frac{1}{2}m\langle v^2\rangle = \frac{3}{2}T, \tag{5.5.7}$$

we can identify β as

$$\beta = \frac{1}{T}. \tag{5.5.8}$$

This is the same result we obtained in Section 5.2. Here, the result is valid for a classical ideal gas only.

The root mean square velocity is defined by

$$v_{\rm rms} = \sqrt{\langle v^2\rangle} = \sqrt{\frac{3T}{m}}. \tag{5.5.9}$$

5.6 Systems at Constant Temperature and Chemical Potential; The Grand Canonical Ensemble

We now proceed one step forward and transform from the independent variables T, V, N into the new independent variables T, V, μ where μ is the chemical potential. (Here we discuss only one component system. The generalization for a multi-component system is quite straightforward.) As we have discussed in the previous section, there are different ways of achieving this transformation of variables. On a purely thermodynamic level if we already have the Helmholtz energy $A(T, V, N)$ given by

$$A(T, V, N) = E(T, V, N) - TS(T, V, N), \tag{5.6.1}$$

we can transform from the independent variable N into the corresponding intensive variable μ, defined by

$$\mu = \left(\frac{\partial A}{\partial N}\right)_{V,N} \tag{5.6.2}$$

to define a new potential function, denoted tentatively as ψ by

$$\psi(T, V, \mu) = A(T, V, \mu) - \mu N(T, V, \mu). \tag{5.6.3}$$

Note that in this transformation, the thermodynamic variable N becomes a function of the independent variables T, V, μ.

Physically, this transformation can be viewed as a result of opening the system to a material reservoir similar to the "opening" of the isolated system to the heat reservoir (Figure 5.2c).

We can think of a large number of systems all at fixed values of T, V, N brought into contact in such a way that particles can flow from one system to another. We know from thermodynamics that any two thermodynamic systems at thermal and material contact at equilibrium must have the same temperature and the same chemical potential.

Instead of looking at a large ensemble of system, we can also focus on a single system and view all the surrounding system as a very large heat and particle reservoir at constant temperature and chemical potential. The number of particles, as well as the energy of the system, will fluctuate about the average values $\bar{N}$ and $\bar{E}$, respectively.

We wish to find the probability p_i of finding the system at a specific state i, for which the energy of the system is E_i and the number of particles N_i. In this case, we need to maximize the quantity

$$H = -\sum p_i \log p_i, \tag{5.6.4}$$

subject to the conditions

$$\sum_i p_i = 1, \tag{5.6.5}$$

$$\sum_i p_i E_i = \bar{E}, \tag{5.6.6}$$

$$\sum_i p_i N_i = \bar{N}, \tag{5.6.7}$$

where the summation is over all possible states of the system. We proceed by defining the auxiliary function

$$L = -\sum p_i \log p_i - \lambda_1 \sum p_i - \lambda_2 \sum p_i E_i - \lambda_3 \sum p_i N_i. \tag{5.6.8}$$

Take the derivatives of L with respect to p_i and requiring that all derivatives be zero, we obtain the condition for maximum L

$$-\log p_i - 1 - \lambda_1 - \lambda_2 E_i - \lambda_3 N_i = 0 \tag{5.6.9}$$

for all i, or equivalently

$$p_i = \exp[-1 - \lambda_1 - \lambda_2 E_i - \lambda_3 N_i]. \tag{5.6.10}$$

To identify the multiplier λ_i, we substitute (5.6.10) into (5.6.5) to (5.6.7) to obtain

$$\Xi \equiv \exp[-1 - \lambda_1] = \sum_i \exp[-\lambda_2 E_i - \lambda_3 N_i], \tag{5.6.11}$$

where Ξ is referred to as the Grand partition function. Before we identify the undetermined multipliers λ_2 and λ_3, we note that by taking the derivatives of $\ln \Xi$, we obtain

$$\begin{aligned} \frac{\partial \ln \Xi}{\partial \lambda_2} &= \frac{1}{\Xi} \sum (-E_i) \exp[-\lambda_2 E_i - \lambda_3 N_i] \\ &= -\sum E_i p_i = -\bar{E}, \end{aligned} \tag{5.6.12}$$

$$\begin{aligned} \frac{\partial \ln \Xi}{\partial \lambda_3} &= \frac{1}{\Xi} \sum (-N_i) \exp[-\lambda_2 E_i - \lambda_3 N_i] \\ &= -\sum N_i p_i = -\bar{N} \end{aligned} \tag{5.6.13}$$

where $\bar{E}$ and $\bar{N}$ are the average energy and average number of particles, respectively.

As in the case of Section 5.2, we can also look at two systems A and B at thermal and material equilibrium. The combined system C is isolated and has a total energy E_T and a total number of particles N_T. The total number of states of the combined system is $W_C(E_T, N_T)$. Since the combined system is isolated, all of the $W_C(E_T, N_T)$ states are equally probable. For any specific energy E_A and number of particles N_A of the system A, there are $W_A(E_A, N_A)$ number of states. For each of these $W_A(E_A, N_A)$ states of system A, there are $W_B(E_T - E_A, N_T - N_A)$ states of the system B. Therefore, the total number of states of the combined system is

$$W_C(E_T, N_T) = \sum_{N_A} \sum_{E_A} W(E_A, N_A) W(E_T - E_A, N_T - N_A). \tag{5.6.14}$$

Since all these states are considered to be equally probable events, the probability of finding the system A with energy E_A and number of particles N_A according to the classical definition of probability [see Section 2.3 and (5.2.8)] is

$$P(E_A, N_A) = \frac{W_A(E_A, N_A)W_B(E_T - E_A, N_T - N_A)}{W_C(E_T, N_T)}. \tag{5.6.15}$$

We now proceed as in Section 5.2. Since for a fixed (but macroscopic) $N_A, W_A(E_A, N_A)$ is a steeply increasing function of E_A, and $W_B(E_T - E_A, N_T - N_A)$ will be a steeply decreasing function of E_A, the product in the numerator will have a sharp maximum at the average value $\bar{E}_A$. The condition for maximum with respect to E_A has already been carried out in Section 5.2, which led to the identification of the temperature of the two systems at thermal equilibrium:

$$\left(\frac{\partial \ln W_A(E_A, N_A)}{\partial E_A}\right)_{eq} = \frac{1}{T} = \lambda_2. \tag{5.6.16}$$

Similarly, we use the condition of maximum of $P(E_A, N_A)$ but now with respect to N_A at a fixed (but macroscopic) E_A to obtain the analog of (5.2.10)

$$\frac{\partial \ln W_A(E_A, N_A)}{\partial N_A} = \frac{\partial \ln W_B(E_B, N_B)}{\partial N_B}. \tag{5.6.17}$$

We know from thermodynamics that when two systems are at thermal and material equilibrium, the chemical potentials are also equal, so we also have the thermodynamic relation

$$\left(\frac{\partial S}{\partial N}\right)_{E,V} = \frac{-\mu}{T}, \tag{5.6.18}$$

where T is the absolute temperature (in units of energy) and μ is the chemical potential. Hence, from (5.3.17) and (5.3.18), we have

$$\frac{\partial \ln W_A(E_A, N_A)}{\partial N_A} = \frac{-\mu}{T}. \tag{5.6.19}$$

Once we have identified λ_2 in (5.6.16), we can proceed to identify λ_3 using a similar argument to that used in Section 5.2. First, we rewrite (5.6.11) as

$$\begin{aligned}\Xi &= \sum_i \exp[-\beta E_i - \lambda_3 N_i] \\ &= \sum_N \exp[\lambda_3 N] \sum_E W(E,V,N)\exp[-\beta E] \\ &= \sum_N Q(\beta,V,N)\exp[-\lambda_3 N], \end{aligned} \quad (5.6.20)$$

where in the second equality, we have replaced the sum over all states i, by first fixing N and summing over all energy levels E [as in (5.2.6)], and then summing over all possible N. We next identify the sum over E with the canonical partition function Q as in (5.2.6).

Using again a similar argument as in Section 5.2, we take the maximal term in the sum (5.6.20). The condition for maximum is

$$\frac{\partial \ln Q}{\partial N} - \lambda_3 = 0. \quad (5.6.21)$$

We now use the relation between Q and the Helmholtz energy in (5.2.16) to rewrite (5.6.21) as

$$\lambda_3 = \frac{\partial \ln Q}{\partial N} = \frac{-\partial(A/T)}{\partial N} = \frac{-\mu}{T}. \quad (5.6.22)$$

Having identities λ_2 and λ_3, we can rewrite the probabilities in (5.6.10) as

$$P(E,N) = \frac{\exp[-\beta E + \beta\mu N]}{\Xi}, \quad (5.6.23)$$

and the probability of finding the system with N particles as

$$P(N) = \frac{Q(T,V,N)\exp[\beta\mu N]}{\Xi}. \quad (5.6.24)$$

Replacing the sum in (5.6.20) by the maximal term, we obtain

$$\begin{aligned}\Xi(T,V,\mu) &= Q(T,V,N^*)\exp[\mu N^*/T] \\ &= \exp[-(A(T,V,N^*) - \mu N^*)/T] \\ &= \exp[P(T,V,\mu)V/T]. \end{aligned} \quad (5.6.25)$$

Note that in (5.6.21), we view N^* that satisfies the equality (5.6.21) as a function of λ_3, T and V, or equivalently of T, V, μ. In (5.6.25), we also used the thermodynamic identity

$$P(T,V,\mu)V = A[T,V,N^*(T,V,\mu)] - \mu N^*(T,V,\mu). \quad (5.6.26)$$

Hence, the fundamental connection between thermodynamics and the Grand partition function is

$$P(T,V,\mu)V = T\ln\Xi(T,V,\mu). \tag{5.6.27}$$

Here, the pressure P is viewed as a function of the independent variables T, V, μ. From (5.6.27), one can obtain all the thermodynamic quantities of the system by taking derivatives of Ξ with respect to T, V, μ. Thus, from

$$d(PV) = SdT + PdV + \bar{N}d\mu, \tag{5.6.28}$$

and (5.6.27), we get

$$\begin{aligned} S &= \ln\Xi + T\frac{\partial \ln\Xi}{\partial T}, \\ \bar{N} &= T\frac{\partial \ln\Xi}{\partial \mu}, \\ P &= T\frac{\partial \ln\Xi}{\partial V} = T\frac{\ln\Xi}{V}. \end{aligned} \tag{5.6.29}$$

It is instructive to obtain the equation of state of an ideal gas from the Grand partition function

$$\begin{aligned} \Xi(T,V,\mu) &= \sum_{N\geq 0} Q(T,V,N)\exp[\beta\mu] \\ &= \sum_{N\geq 0}\frac{q^N}{N!}\exp[\beta\mu] = \exp(\lambda q), \end{aligned} \tag{5.6.30}$$

where λ is the absolute activity defined by

$$\lambda = \exp[\beta\mu], \tag{5.6.31}$$

and $q = q_{int}V/\Lambda^3$, with q_{int} as the internal partition function of a single molecule.

From the general results (5.6.29) and (5.6.30), we obtain for an ideal gas:

$$\bar{N} = T\frac{\partial \ln\Xi}{\partial \mu} = \lambda\frac{\partial \ln\Xi}{\partial \lambda} = \lambda q = PV/T, \tag{5.6.32}$$

which is the equation of state of an ideal gas.

Note that in (5.6.32), the *average* number of particles replaces the exact N in the canonical ensemble (5.3.6). As we have seen in Section 5.3, the equation of state of an ideal gas is universal in the sense that it does not depend on the particular gas under consideration. This is not true for other quantities such as the chemical potential, which for an ideal gas is

$$\beta\mu = \ln\lambda = \ln\bar{N}q^{-1} = \ln(\bar{N}q_{int}^{-1}\Lambda^3/V). \tag{5.6.33}$$

Clearly, different gases will have different internal degrees of freedom and different masses. For an ideal monoatomic gas with $q_{int} = 1$, this reduces to

$$\beta\mu = \ln\rho\Lambda^3 \tag{5.6.34}$$

with $\rho = \bar{N}/V$.

5.7 Systems at Constant Temperature and Pressure; The Isothermal Isobaric Ensemble

We shall briefly outline the procedure of transforming from the independent variables T, V, N to T, P, N (see Figure 5.1). We leave the detail as exercises for the reader.

Starting with an ensemble of systems, all characterized by the same values of T, V, N, we remove the constraint on fixed volume by allowing an "exchange of volume" between the systems. This means that we replace the rigid boundaries between the systems, by movable or flexible boundaries. We know from thermodynamics that for each pair of systems, connected by a movable boundary, in mechanical equilibrium, the volume would fluctuate about some average value but the pressure of the two systems will be the same.

For simplicity, we shall assume that the volume of the system can attain only discrete values, say $V_1, V_2, \ldots$. We write the MI as in the previous section

$$H = -\sum p_i \log p_i, \tag{5.7.1}$$

and find the maximum of MI subject to the conditions

$$\sum_i p_i = 1, \tag{5.7.2}$$

$$\sum_i p_i E_i = \bar{E}, \tag{5.7.3}$$

$$\sum_i p_i V_i = \bar{V}, \tag{5.7.4}$$

where $\bar{V}$ is the average volume of the system in the T, P, N ensemble. Using the procedure as in Sections 5.2 and 5.6, we can obtain the probability of each state as

$$p_i = \frac{\exp[-\beta E_i - \beta P V_i]}{\Delta(T, P, N)}. \tag{5.7.5}$$

The probability of finding the system at a specific volume V_i is

$$\Pr(V_i) = \frac{Q(T, V_i, N)\exp[-\beta P V_i]}{\Delta(T, P, N)}, \tag{5.7.6}$$

where $\Delta(T, P, N)$ is referred to as the isothermal isobaric partition function:

$$\Delta(T, P, N) = \sum_V Q(T, V, N)\exp[-\beta P V]. \tag{5.7.7}$$

The relation between $\Delta(T, P, N)$ and thermodynamics is

$$G(T, P, N) = -T \ln \Delta(T, P, N), \tag{5.7.8}$$

where $G(T, P, N)$ is the Gibbs energy.

One can also rewrite (5.7.7) as an integral over the continuous variable V, in which case $\Pr(V)$ in (5.7.6) will turn into a probability density, i.e., $\Pr(V)dV$ is the probability of finding the system having volume between V and $V + dV$.

The partition function $\Delta(T, P, N)$ is less convenient than $Q(T, V, N)$ in calculating thermodynamic quantities of theoretical models. However, when discussing processes under constant pressure and temperature, the partition function $\Delta(T, P, N)$ is the more relevant one to use.

From $\Delta(T, P, N)$, one can obtain all the thermodynamic quantities by taking the derivative of G, e.g.,

$$\begin{aligned} S &= -\frac{\partial G}{\partial T}, \\ V &= \frac{\partial G}{\partial P}, \\ \mu &= \frac{\partial G}{\partial N}. \end{aligned} \tag{5.7.9}$$

Finally, we note that if we do one further step in the diagram of Figure 5.1, and change all of the extensive variables E, V, N

into intensive variables T, P, μ, one gets the so-called "generalized partition function"

$$\Gamma(T, P, \mu) = \sum_E \sum_V \sum_N W(E, V, N) \exp[-\beta E - \beta PV + \beta\mu N]$$
$$\approx \exp[-\beta(E + PV - ST - G)] = e^0 = 1. \qquad (5.7.10)$$

Thus, the corresponding "thermodynamic potential" for these variables is zero. This is in accordance with the Gibbs–Duhem identity

$$SdT - VdP + \sum N_i d\mu_i = 0, \qquad (5.7.11)$$

i.e., the variable T, P, μ_i cannot be changed independently. We note that in spite of the fact that the three variables T, P, μ are dependent, the partition function in (5.7.10) is sometimes useful [see, for example, Ben-Naim (1992)].

5.8 The Mutual Information due to Intermolecular Interactions

In Section 4.3, we discussed two kinds of correlation among the particles. One is due to the uncertainty principle; the second is due to the indistinguishability of the particles. In this section, we discuss a third kind of correlation, which is expressed as mutual information among the particles due to intermolecular interactions.

The classical canonical partition function is written as[16]

$$Q(T, V, N) = \frac{Z_N}{N!\Lambda^{3N}}, \qquad (5.8.1)$$

where Z_N is the configurational PF of the system

$$Z_N = \int \cdots \int d\boldsymbol{R}^N \exp[-\beta U_N(\boldsymbol{R}^N)]. \qquad (5.8.2)$$

The probability density for finding the particles at a specific configuration $\boldsymbol{R}^N = \boldsymbol{R}_1, \ldots, \boldsymbol{R}_N$ is

$$P(\boldsymbol{R}^N) = \frac{\exp[-\beta U_N(\boldsymbol{R}^N)]}{Z_N}. \qquad (5.8.3)$$

[16] For simplicity, we assume that the particles are spherical and have no internal degrees of freedom.

When there are no intermolecular interactions (ideal gas), we have $Z_N = V^N$ and the corresponding partition function is reduced to

$$Q^{ig}(T,V,N) = \frac{V^N}{N!\Lambda^{3N}}. \tag{5.8.4}$$

We define the change in the Helmholtz energy due to the interactions as

$$\Delta A = A - A^{ig} = -T \ln \frac{Q(T,V,N)}{Q^{ig}(T,V,N)} = -T \ln \frac{Z_N}{V^N}. \tag{5.8.5}$$

The corresponding change in the MI is

$$\begin{aligned} \Delta S &= -\frac{\partial \Delta A}{\partial T} = \ln \frac{Z_N}{V^N} + T \frac{1}{Z_N} \frac{\partial Z_N}{\partial T} \\ &= \ln Z_N - N \ln V + \frac{1}{T} \int d\boldsymbol{R}^N P(\boldsymbol{R}^N) U_N(\boldsymbol{R}^N). \end{aligned} \tag{5.8.6}$$

We now substitute $U_N(\boldsymbol{R}^N)$ from (5.8.3) into (5.8.6) to obtain

$$\Delta S = -N \ln V - \int P(\boldsymbol{R}^N) \ln P(\boldsymbol{R}^N) d\boldsymbol{R}^N. \tag{5.8.7}$$

Note that the second term on the right-hand side of (5.8.7) has the form of MI. We can also write the first term on the right-hand side of (5.8.7) as the MI of an ideal gas. For an ideal gas $U_N(\boldsymbol{R}^N) = 0$ and $P^{ig}(\boldsymbol{R}^N) = (1/V)^N = P(\boldsymbol{R}_1)P(\boldsymbol{R}_2)\cdots P(\boldsymbol{R}_N)$.

Hence,

$$\begin{aligned} \Delta S &= \ln P^{ig}(\boldsymbol{R}^N) - \int P(\boldsymbol{R}^N) \ln P(\boldsymbol{R}^N) d\boldsymbol{R}^N \\ &= H(1,2,\ldots,N) - H^{ig}(1,2,\ldots,N) \\ &= -\int P(\boldsymbol{R}^N) \ln \left[\frac{P(\boldsymbol{R}^N)}{P^{ig}(\boldsymbol{R}^N)}\right] d\boldsymbol{R}^N \\ &= -\int P(\boldsymbol{R}^N) \ln \left[\frac{P(\boldsymbol{R}^N)}{\prod_{i=1}^N P(\boldsymbol{R}_i)}\right] d\boldsymbol{R}^N \\ &= I(1;2;\ldots;N). \end{aligned} \tag{5.8.8}$$

The last expression on the right-hand side of (5.8.8) has the form of a mutual information. The quantity

$$g(1,2,\ldots,N) = \frac{P(\boldsymbol{R}^N)}{\prod_{i=1}^N P(\boldsymbol{R}_i)} \tag{5.8.9}$$

has the form of a correlation due to the interactions between particles. Note that $P(\boldsymbol{R}_i) = V^{-1}$ is the probability density of finding a specific (labeled) particle at $\boldsymbol{R}_i$.

A special case of interest is, in the limit of very low density, when pair interactions are operative but interactions among more than two particles are rare and can be neglected. To obtain this limit, we write Z_N as

$$Z_N = \int d\boldsymbol{R}^N \prod_{i<j} \exp[-\beta U_{ij}]. \tag{5.8.10}$$

Define the so-called Mayer f *function* by

$$f_{ij} = \exp(-\beta U_{ij}) - 1 \tag{5.8.11}$$

and rewrite Z_N as

$$\begin{aligned} Z_N &= \int d\boldsymbol{R}^N \prod_{i<j}(f_{ij}+1) \\ &= \int d\boldsymbol{R}^N \left[1 + \sum_{i<j} f_{ij} + \sum_{i<i<k} f_{ij} f_{jk} + \cdots \right]. \end{aligned} \tag{5.8.12}$$

Neglecting all terms beyond the first sum, we obtain

$$\begin{aligned} Z_N &= V^N + \frac{N(N-1)}{2} \int f_{12} d\boldsymbol{R}^N \\ &= V^N + \frac{N(N-1)}{2} V^{N-2} \int f_{12} d\boldsymbol{R}_1 d\boldsymbol{R}_2, \end{aligned} \tag{5.8.13}$$

where we have integrated over $N-2$ variables $\boldsymbol{R}_3, \ldots, \boldsymbol{R}_N$.

We now identify the second virial coefficient as

$$B_2(T) = \frac{-1}{2V} \int_V \int_V f_{12} d\boldsymbol{R}_1 d\boldsymbol{R}_2 \tag{5.8.14}$$

and rewrite Z_N as

$$\begin{aligned} Z_N &= V^N - N(N-1)\, V^{N-1} B_2(T) \\ &= V^N \left[1 - \frac{N(N-1)}{V} B_2(T)\right]. \end{aligned} \tag{5.8.15}$$

The corresponding Helmholtz energy change is

$$\Delta A = A - A^{ig} = -T \ln \frac{Z_N}{V^N}$$
$$= -T \ln \left[1 - \frac{N(N-1)}{2V^2} \iint f_{12}(\boldsymbol{R}_1, \boldsymbol{R}_2) d\boldsymbol{R}_1 d\boldsymbol{R}_2 \right]. \quad (5.8.16)$$

At low densities, $\rho \to 0$, the limiting behavior of the right-hand side of (5.8.16) is

$$\Delta A = \frac{TN(N-1)}{2V^2} \iint f_{12}(\boldsymbol{R}_1, \boldsymbol{R}_2) d\boldsymbol{R}_1 d\boldsymbol{R}_2. \quad (5.8.17)$$

Let Z_2 be the configurational partition function for a system with exactly two particles. For this special case, $N = 2$, the corresponding change in the MI is obtained from (5.8.5)

$$\Delta S = -\left(\frac{\partial \Delta A}{\partial T} \right)_{V,N}$$
$$= \ln \left(\frac{Z_2}{V_2} \right) + \frac{\iint \exp[-\beta U(\boldsymbol{R}_1, \boldsymbol{R}_2)] \beta U(\boldsymbol{R}_1, \boldsymbol{R}_2) d\boldsymbol{R}_1 d\boldsymbol{R}_2}{Z_2}. \quad (5.8.18)$$

For this system, the probability of finding the two particles at the exact configuration $(\boldsymbol{R}_1, \boldsymbol{R}_2)$ is

$$P(\boldsymbol{R}_1, \boldsymbol{R}_2) = \frac{\exp[-\beta U(\boldsymbol{R}_1, \boldsymbol{R}_2)]}{Z_2}. \quad (5.8.19)$$

Hence, from (5.8.18) and (5.8.19), we get

$$\Delta S = -\ln(V^2) - \iint [P(\boldsymbol{R}_1, \boldsymbol{R}_2) \ln P(\boldsymbol{R}_1, \boldsymbol{R}_2)] d\boldsymbol{R}_1 d\boldsymbol{R}_2, \quad (5.8.20)$$

which is the analog of (5.8.18) for this special case, $N = 2$.

When the particles are independent, we have $P^{ig}(\boldsymbol{R}_1, \boldsymbol{R}_2) = V^{-2}$ and the normalization condition for the probability is[17]

$$\int \int P^{ig}(\boldsymbol{R}_1, \boldsymbol{R}_2) d\boldsymbol{R}_1 d\boldsymbol{R}_2 = \int \int P(\boldsymbol{R}_1, \boldsymbol{R}_2) d\boldsymbol{R}_1 d\boldsymbol{R}_2 = 1, \quad (5.8.21)$$

[17]Note that $P(\boldsymbol{R}_1, \boldsymbol{R}_2)$ is the probability density for two specific particles 1 and 2. For details, see Hill (1960) or Ben-Naim (2006).

we get from (5.8.20)

$$\Delta S = \iint P(\boldsymbol{R}_1, \boldsymbol{R}_2) \ln \frac{P(\boldsymbol{R}_1, \boldsymbol{R}_2)}{P^{ig}(\boldsymbol{R}_1, \boldsymbol{R}_2)} d\boldsymbol{R}_1 d\boldsymbol{R}_2 = -I(1;\ 2), \tag{5.8.22}$$

which is the analog of (5.8.8) for this special case.

For a system with N particles, but in the limit of very low densities, we have instead of (5.8.22)

$$\Delta S = -\frac{N(N-1)}{2} \iint P(\boldsymbol{R}_1, \boldsymbol{R}_2) \ln \frac{P(\boldsymbol{R}_1, \boldsymbol{R}_2)}{P^{ig}(\boldsymbol{R}_1, \boldsymbol{R}_2)} d\boldsymbol{R}_1 d\boldsymbol{R}_2. \tag{5.8.23}$$

The integral on the right-hand side of (5.8.22) has the form of mutual information per pair of interacting particles. The whole right-hand side of (5.8.23) is the mutual information for the total of $N(N-1)/2$ pairs of particles.

We now discuss two simple cases of this mutual information.

(i) *Hard spheres of diameter* σ

Hard spheres are defined by the interaction potential function:

$$U(R) = \begin{cases} \infty & \text{for } R \le \sigma, \\ 0 & \text{for } R > \sigma. \end{cases} \tag{5.8.24}$$

From (5.8.17), we get

$$\begin{aligned} \Delta A &= \frac{-TN(N-1)^{\sigma}}{2V} \int_0^{\infty} \{exp[-\beta U(R)] - 1\} 4\pi R^2 dR \\ &= \frac{TN(N-1)^{\sigma}}{2V} \left(\frac{4\pi\sigma^3}{3}\right). \end{aligned} \tag{5.8.25}$$

The corresponding change in MI is

$$\begin{aligned} \Delta S &= \frac{-N(N-1)}{2v} \left(\frac{4\pi\sigma^3}{3}\right) \\ &\approx N \ln \left[\frac{V - (N-1)\frac{4\pi\sigma^3}{6}}{V}\right] = -I(1;\ 2; \ldots; N). \end{aligned} \tag{5.8.26}$$

The interpretation of this negative change in MI is simple. When passing from an ideal to a real gas of hard spheres, the accessible volume for each particle reduces from V to $V - \frac{(N-1)}{2}\frac{4\pi\sigma^3}{3}$. Figure 5.4 illustrates the reduction in the accessible volume for the

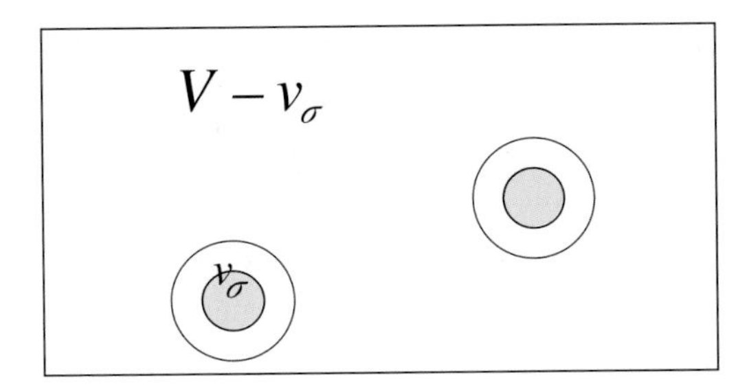

Figure 5.4. The reduction of the accessible volume for each particle from V to $V - v_\sigma$.

case of exactly two particles, $N = 2$. This mutual information is similar to what we encountered in Section 4.1. We noted there that the restriction of one ligand per site is equivalent to an infinite repulsion between two ligands occupying a single site. Here again, we have infinite repulsion due to the finite size of the hard sphere, which effectively reduces the accessible volume—hence a reduction in the MI.

(i) *Square well potential*
The square well potential is defined by

$$U(R) = \begin{cases} \infty & \text{for } R \leq \sigma, \\ -\varepsilon & \text{for } \sigma \leq R < \sigma + \delta, \\ 0 & \text{for } R \geq \sigma + \delta. \end{cases} \tag{5.8.27}$$

Next, we examine the the change in the MI when we turn on the square well potential (5.8.27). In this case, we examine only the contribution to ΔS due to the additional part of the potential function at $\sigma \leq R \leq \sigma + \delta$. For the case of exactly two particles, we have

$$\Delta A = \frac{T}{V} \int_\sigma^{R_M} (exp[-U(R)/T] - 1) 4\pi R^2 dR \tag{5.8.28}$$

$$\Delta S = \frac{1}{V} \left(\int_\sigma^{R_M} (exp[-U(R)/T] - 1) 4\pi R^2 dR + \int_\sigma^{R_M} (exp[-U(R)/T])(U(R)/T) 4\pi R^2 dR \right), \tag{5.8.29}$$

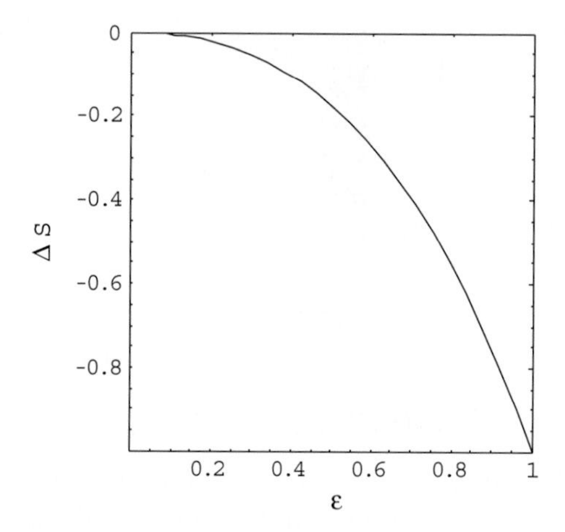

Figure 5.5. The additional decrease of the MI due to turning on the square well potential.

where for simplicity we assume that the system is spherical, with radius R_M.

Again, ΔS is negative and it is larger, the larger value of ε. The interpretation of this result in terms MI is again quite straightforward. Suppose we have only two particles and neglect the volume of each particle (this was taken into account in the previous example). In this case, the accessible volume for one particle is the entire volume V. However, the second particle is now more likely to be in the spherical shell of radius σ and width δ around the first particle. This additional information reduces the MI (see Figure 5.5) The larger $\varepsilon > 0$, the larger the reduction in the MI. The general conclusion is that whenever we introduce interactions between particles, the MI is reduced, i.e., the mutual information is positive.

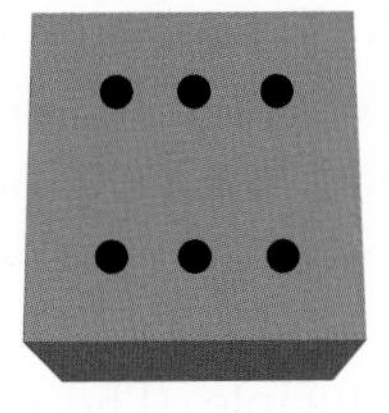

Chapter 6

Some Simple Applications

The purpose of this chapter is three-fold; first, to present a few very simple applications of statistical thermodynamics. Some of these are discussed in most text books on statistical mechanics, while some are not. All of these examples are uniformly analyzed in terms of the changes in the MI. Second, the examples are simple enough so that they can be viewed as exercises. The reader is urged to do these exercises with or without the solutions that are given as part of the text. Finally, most of these examples are designed to demonstrate the workings of the Second Law of Thermodynamics. Instead of the common formulation of the Second Law in terms of the increasing entropy, we shall show that all of the processes occurring spontaneously in an isolated system involve the increasing of the MI.

Using information as a fundamental concept makes the understanding of the Second Law much easier. It also removes the mystery that has befogged entropy and the Second Law for a long time. We shall briefly discuss the Second Law of Thermodynamics in the last section of this chapter. We shall see that the MI is the best way of describing *what* is the quantity that changes in a spontaneous process. However, the only way of understanding *why* the MI changes in one way is in terms of probability. This aspect will be discussed in Section 6.12.

As we have seen in Chapter 4, there are essentially two kinds of information: one associated with the *locations* of the particles and the second associated with the momenta of the particles. There are also corrections for over-counting. These can be cast in the form of mutual-information, which basically measures an average correlation, or dependence between the particles. For an ideal gas, we found two kinds of such corrections. One is due to the quantum mechanical uncertainty principle. The second is due to the indistinguishability of the particles. When intermolecular interactions exist, there is an additional mutual information due to these interactions. In this chapter, we shall discuss various processes where one or two of these factors change.

In all of the examples discussed in this chapter, we shall be talking about changes in MI in various *processes*. However, we shall never follow the process itself. The only quantities of interest will be differences in MI between two equilibrium states; the initial and the final states. It should be noted that the identification of the entropy S with the Shannon measure of information H is valid only at equilibrium states.

6.1 Expansion of an Ideal Gas

We start with the simplest spontaneous process in an isolated system, i.e., a system of N particles having fixed energy E and contained in volume V (Figure 6.1).

The system is an ideal gas, i.e., no interactions exist between the particles (or, if they exist, the density is very low so that the interactions between particles can be neglected). Also, we assume for simplicity that the particles are structureless, i.e., point masses with no internal degrees of freedom. The total energy of the system

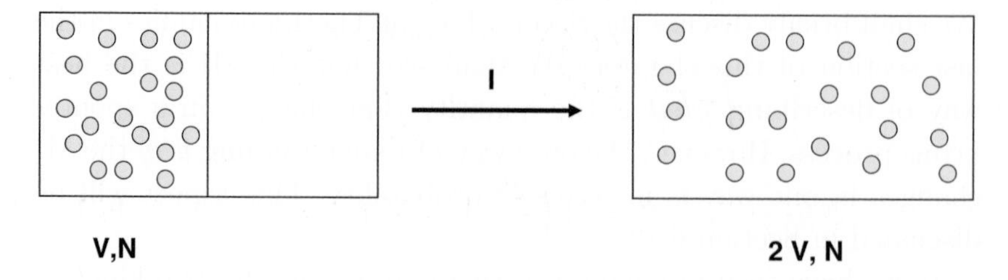

Figure 6.1. Expansion of an ideal gas, from volume V to $2V$.

is simply the sum of the translational kinetic energy of the particles. All the conclusions of these and the next few sections are valid for complex particles. As long as none of the internal degrees of freedom change in the process, the conclusion reached regarding the changes in the MI will not be affected.

The change in the MI in the process depicted in Figure 6.1 can be obtained most straightforwardly from the Sackur–Tetrode equation for the MI that we derived in Sections 4.3 and 5.4. Since the energy E and the number of particles N do not change in this process, the change in the MI in this process is due only to the change of the accessible volume for each particle, from the initial value V to the final value $2V$. The change in the MI for N particles is thus

$$\Delta S_I = N \ln \frac{2V}{V} = N \ln 2 > 0. \tag{6.1.1}$$

This is the same result one obtains from thermodynamics.[1]

Note that we could have derived this result directly from considering the change of the locational information as we have done in Section 4.3.

For an ideal gas system, constant energy is equivalent to constant temperature. Therefore, the entropy change obtained from either the Sackur–Tetrode equation or from the canonical partition function must be the same. We therefore rederive the result (6.1.1) from the canonical partition function.

We write the PF for the system in the initial (i) and in the final (f) states as

$$Q_i = \frac{V^N}{N!\Lambda^{3N}}, \quad Q_f = \frac{(2V)^N}{N!\Lambda^{3N}}. \tag{6.1.2}$$

Using the thermodynamic relationship for the MI, as the temperature derivative of the Helmholtz energy

$$S = -\left(\frac{\partial A}{\partial T}\right)_{V,N} = \frac{\partial (T \ln Q)}{\partial T}, \tag{6.1.3}$$

we get

$$\Delta S_I = S_f - S_i = \ln\left(\frac{2V}{V}\right) = N \ln 2 > 0. \tag{6.1.4}$$

---

[1] See, for example, Denbigh (1967).

It is instructive, however, to re-derive the result (6.1.1) directly by counting configurations in a discrete sample space.[2] In this manner, we illustrate again that the MI *does* depend on details we choose to describe the system. However, the *change* of the MI is independent of this choice. This is the same result we have obtained in Section 4.2.4. Suppose that we divide the volume V into m cells each of volume v, such that $V = mv$. Suppose that we are interested in the distribution of the particles only in the different cells. We assume for simplicity that each cell can contain any number of particles. We could also ignore $N!$ in this example. Since N does not change in the process, the change in MI will not depend on N. The number of configurations in the initial state is $m^N/N!$, and in the final state is $(2m)^N/N!$. Hence, the corresponding MI in the initial and final states are

$$H_i = \ln \frac{m^N}{N!}, \tag{6.1.5}$$

$$H_f = \ln \frac{(2m)^N}{N!}. \tag{6.1.6}$$

The change of the MI in the process of Figure 6.2 is

$$\Delta S_I = S_f - S_i = N \ln \frac{2m}{m} = N \ln 2. \tag{6.1.7}$$

Thus, the *amount* of MI, H_i and H_f in (6.1.5) and (6.1.6) depends on the accuracy we choose in locating the particles; the finer the size of the cell, the larger m, and the larger the MI in (6.1.7). The *difference* in the MI is however independent of the size of the cells.

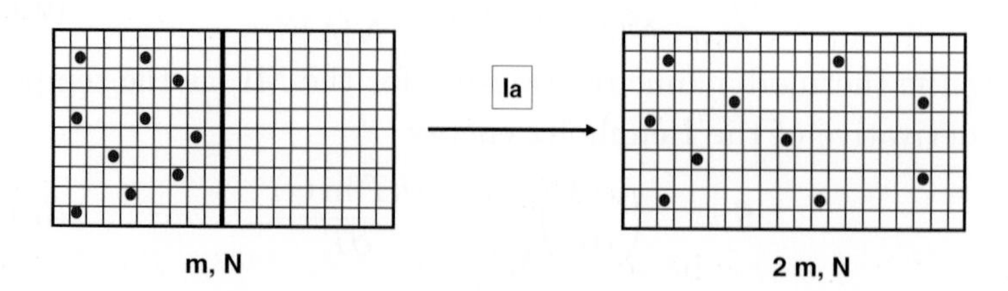

Figure 6.2. Expansion of an ideal gas, as in Figure 6.1, but from m cells to $2m$ cells.

[2]This also illustrates the validity of the result (6.1.4) even though we used the definition of S for the continuous case.

When $m \to \infty$, the MI in both the initial and the final states tends to infinity. However, if we first take the difference in MI, then let $m \to \infty$, we get a finite change of $N \ln 2$, independent of m.

Note carefully that the change in the MI in the expansion process depends only on the initial and the final volume accessible to each particle. It does not depend on the *kind* of the particles. This sounds like a trivial comment in the context of the expansion process. It is far from trivial in the context of the same process discussed in Section 6.4.

6.2 Pure, Reversible Mixing; The First Illusion

Figure 6.3 shows a pure process of mixing. It is *pure* in the sense that none of the parameters E, V or N changes in this process. Only *mixing* is observed. In Figure 6.3a, we depicted only the initial and the final states. In Figure 6.3b, we show the *process* itself. In most of this chapter, we shall not be concerned with the process itself, but only on the difference in the MI between the initial and the final states. Initially, we have N_A particles of type A in a volume V, and N_B particles of type B in a volume V. We bring the two systems into the same final volume V, and we now have a mixture of A and B.

We can calculate the change in MI in this process either from the Sackur–Tetrode equation, or from the canonical partition functions. For the initial and the final states, we have

$$Q_i = \frac{V^{N_A}}{N_A!\Lambda_A^{3N_A}} \frac{V^{N_B}}{N_B!\Lambda_B^{3N_B}}, \quad Q_f = \frac{V^{N_A+N_B}}{N_A!N_B!\Lambda_A^{3N_A}\Lambda_B^{3N_B}}. \tag{6.2.1}$$

Clearly, since $Q_i = Q_f$, we get

$$\Delta S_{II} = 0. \tag{6.2.2}$$

Note that if we are interested only in the change in the MI from the initial to the final states, all we need are the partition functions in these two states. In most textbooks on thermodynamics, one can also find a reversible process that transforms the initial to the final states. This process requires the existence of semi-permeable partitions that are permeable selectively, one to A, and the second

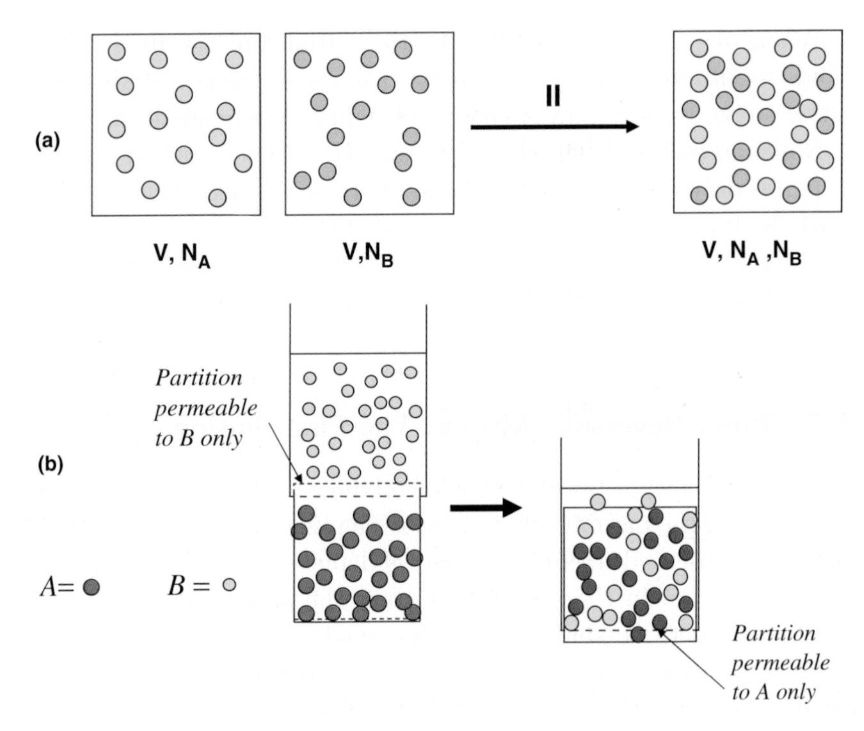

Figure 6.3. (a) Pure mixing of two components A and B. (b) Reversible process of mixing.

to B (see Figure 6.3b). However, the result (6.2.2) does not *depend* on the existence of such partitions.

The interpretation of the result (6.2.2) is simple and straightforward. The volume V accessible for each particle before and after the process did not change. Also, the temperature, and hence the velocity distribution, did not change in the process. Therefore, the net effect is strictly a zero change in the MI.

From this simple result, we conclude that mixing in ideal gas systems, in itself, has no effect on any thermodynamics process. Hence, thermodynamically speaking, mixing is a *non-process*. This is in sharp contrast to the frequent statements made in the literature, as well as in daily life, regarding mixing as an inherently irreversible process.

From the example discussed in this section, we conclude that the association of *mixing* with increase of entropy is only an illusion.

There are two reasons that contribute to this illusion. First is the common association of mixing, or "mixed-upness," with disorder, and the association of disorder with increase in entropy. As we have discussed in Chapter 1, the first part of the preceding sentence is qualitatively correct, while the second part is, in general, incorrect.

The second reason is that in daily life, we do see many mixing processes that are indeed spontaneous and involve an increase in entropy. None of these is a pure mixing process. As we shall see in Sections 6.4 and 6.5, both mixing and demixing can be made spontaneous when coupled with another process, such as expansion. In all of these processes, it is the expansion, not the mixing, which is responsible for the increase in entropy, or the MI. It just happens that mixing processes coupled with expansion are more frequently encountered (see Section 6.4). This fact creates the *illusion* that the mixing part of the process such as IV in Figure 6.6 is *responsible* for the increase in entropy.

6.3 Pure Assimilation Process; The Second Illusion

Figure 6.4 shows two processes similar to the one described in Figure 6.3, but now the particles in the two compartments are of the same type, say A.[3] We refer to this process as pure assimilation for the following reasons.[4] In process IIIa, we have N_A particles in one box of volume V, and another N_A particles in a second box. The particles in one box are distinguishable from the particles in the second box by the very fact that they are confined to two macroscopically distinguishable boxes.[5] In the final state, we have $2N_A$ indistinguishable particles in the same volume V. The volume accessible to each particle, as well as the velocity distribution of the particles, is unchanged in this process. What has changed is only

---

[3]Note again that in Figure 6.4 we do not discuss the *process* itself. All we are interested in is the difference in the MI between the initial and the final states.

[4]In this book, we shall use the terms "mixing" and "assimilation" for the two processes which Gibbs (1906) referred to as "mixing of particles of different kinds," and "mixing of particles of the same kind" (see also Section 6.6).

[5]Indistinguishability of identical particles is further discussed in Appendix J.

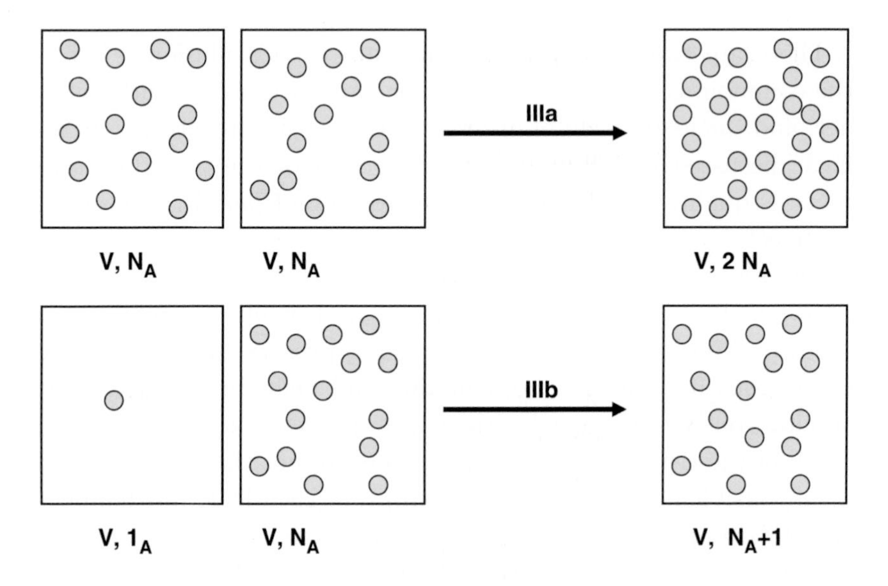

Figure 6.4. Pure process of assimilation.

the *number* of indistinguishable particles, from N_A ID particle and another N_A ID particles into $2N_A$ ID particles. Process IIIb is the same as IIIa but with one A particle in the left box.

The particles of one box when "mixed" with particles in the second box are *assimilated.* There is a strong sense of loss of information that can never be recovered. This sense of loss has already been alluded to by Maxwell in connection with the Second Law, and also by Gibbs. We shall discuss these ideas in Section 6.6 where assimilation is coupled with a process of expansion, a process that is more closely related to the ones discussed by both Maxwell and Gibbs. We shall see that this sense of loss of information has led to erroneous conclusions. It is probably a result of relating consciously or unconsciously the loss of this kind of information to increase in the MI. These conclusions are erroneous however. Before analyzing this conclusion, let us calculate the change in the MI in this process. Since these processes involve changes in the number of particles, we must discuss separately particles that obey Fermi–Dirac and Bose–Einstein statistics. As in the previous section, we shall not discuss the actual execution of the process. What matters is only the initial and the final states.

6.3.1 Fermi–Dirac (FD) statistics; Fermions

Let M be the number of molecular states available to each of the N Femions. The total number of states for the whole system, with the restriction that no more than one particle can occupy a specific molecular state, is well-known (see Appendix B):

$$W_{FD}^{(N)} = \frac{M!}{N!(M-N)!} = \binom{M}{N}. \tag{6.3.1}$$

This is the number of ways of placing N indistinguishable particles in M boxes (or molecular states) with the restriction that the occupancy of each molecular state can be either zero or one. Clearly, to satisfy this condition, M must be larger than N. The MI associated with this number of states is

$$S_{FD}^{(N)} = \ln W_{FD}^{(N)}. \tag{6.3.2}$$

In order to interpret this quantity as informational measure, we adopt the assumption that each of the $W_{FD}^{(N)}$ states has the same probability, equal to

$$P_{FD}^{(N)} = \binom{M}{N}^{-1}. \tag{6.3.3}$$

The informational measure defined on this probability distribution is

$$S_{FD}^{(N)} = \ln W_{FD}^{(N)} = -\sum_{i=1}^{\binom{M}{N}} P_{FD}^{(N)} \ln P_{FD}^{(N)}. \tag{6.3.4}$$

Note that the informational measure for this system is rendered possible only because of the fundamental postulate of statistical mechanics expressed in (6.3.3). We first examine the change in the MI process IIIb in Figure 6.4:

$$\begin{aligned}\Delta S_{FD}^{(1,N)} &= \ln \frac{W_{FD}^{(N)}}{W_{FD}^{(1)} W_{FD}^{(N-1)}} \\ &= \ln \left(\frac{1}{N} - \frac{1}{M} + \frac{1}{NM} \right) < 0.\end{aligned} \tag{6.3.5}$$

Since $M > N$, this change in MI is *negative*. Thus, upon assimilating one Fermion into $N-1$ Fermions of the same kind, we have a *decrease* in MI. In the more general case of assimilation of N_1 and N_2 Fermions, the change in the MI process IIIa (Figure 6.4):

$$\Delta S_{FD}^{(N_1,N_2)} = \ln \frac{W_{FD}^{(N_1+N_2)}}{W_{FD}^{(N_1)} W_{FD}^{(N_2)}}. \tag{6.3.6}$$

Again, this is negative (to prove that, one can simply repeat process IIIb, N_1 times, until all of the N_1 particles in process IIIa are transferred from the left box into the right box). We conclude that the process of pure assimilation involving indistinguishable Fermions always *decreases* the MI. This result is due to a *decrease* in the total number of states of the whole system.

6.3.2 Bose–Einstein (BE) statistics; Bosons

Here, we consider a system of N non-interacting Bosons. The number of states for the whole system of N indistinguishable Bosons, with no restriction on the occupation number in each molecular state, is the well-known (Appendix B):

$$W_{BE}^{(N)} = \frac{(M+N-1)!}{N!(N-1)1} = \binom{M+N-1}{N}. \tag{6.3.7}$$

Again, viewing $W_{BE}^{(N)}$ as the *total* number of states of the whole system (at fixed energy, volume and number of Bosons), we write the MI of the system as

$$S_{BE}^{(N)} = \ln W_{BE}^{(N)}. \tag{6.3.8}$$

In addition, we assume *per postulate* that all states have equal probability, i.e.,

$$P_{BE}^{(N)} = \binom{M+N-1}{N}^{-1}. \tag{6.3.9}$$

We can then interpret (6.3.8) as a Shannon-measure of information, i.e.,

$$S_{BE}^{(N)} = -\sum_{i=1}^{\binom{M+N-1}{N}} P_{BE}^{(N)} \ln P_{BE}^{(N)}, \tag{6.3.10}$$

where the sum, on the right-hand side of (6.3.10), is over all the states of the system.

The changes in the MI processes IIIb and IIIa are thus

$$\Delta S_{BE}^{(1,N)} = \ln \frac{W_{BE}^{(N)}}{W_{BE}^{(1)} W_{BE}^{(N-1)}} = \ln\left(\frac{1}{N} + \frac{1}{M} - \frac{1}{NM}\right) < 0 \tag{6.3.11}$$

and

$$\Delta S_{BE}^{(N_1,N_2)} = \ln \frac{W_{BE}^{(N_!+N_2)}}{W_{BE}^{(N_1)} W_{BE}^{(N_2)}} < 0. \tag{6.3.12}$$

Thus, we find again that in both processes IIIa and IIIb of the pure assimilation of Bosons, the MI *decreases*.

6.3.3 Maxwell–Boltzmann (MB) statistics

It is well-known[6] that in the limit of $M \gg N$, both the FD and the BE statistics tend to the Maxwell–Boltzmann statistics and the following inequality holds for any N and $M (M > N)$ (see Figure 6.5):

$$W_{BE}^{(N)} > W_{MB}^{(N)} > W_{FD}^{(N)} \tag{6.3.13}$$

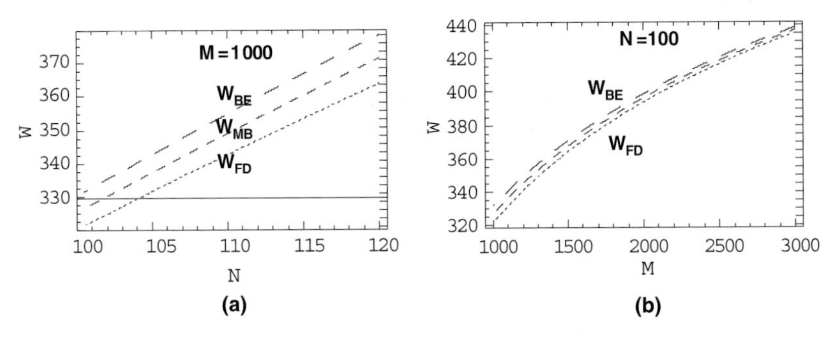

Figure 6.5. (a) W_{BE}, W_{MB}, W_{FD} as a function of N for a fixed $M = 1000$. (b) W_{BE}, W_{MB}, W_{FD} as a function of M for a fixed $N = 100$ (The curve for W_{MB} is the intermediate curve).

[6] See, for example, Hill (1960).

and

$$W_{FD}^{(N)} \to W_{MB}^{(N)} = \frac{M^N}{N!}, \quad W_{BE}^{(N)} \to W_{MB}^{(N)} = \frac{M^N}{N!}. \tag{6.3.14}$$

Thus, in the limit of $M \gg N$, the MB statistics is obtained from either the FD or the BE statistics.

The MI associated with $W_{MB}^{(N)}$ is

$$S_{MB}^{(N)} = \ln W_{MB}^{(N)}. \tag{6.3.15}$$

Assuming that each of the $W_{MB}^{(N)}$ states has equal probability, i.e.,

$$P_{MB}^{(N)} = \left(\frac{M^N}{N!}\right)^{-1}. \tag{6.3.16}$$

The corresponding MI defined on this distribution is

$$S_{MB}^{(N)} = -\sum_{i=1}^{\left(\frac{M^N}{N!}\right)} P_{MB}^{(N)} \ln P_{MB}^{(N)}. \tag{6.3.17}$$

The relevant changes in MI in the processes IIIa and IIIb are

$$\begin{aligned} \Delta S_{MB}^{(1,N)} &= \ln \frac{W_{MB}^{(N)}}{W_{MB}^{(1)} W_{MB}^{(N-1)}} \\ &= -\ln N < 0 \end{aligned} \tag{6.3.18}$$

and

$$\begin{aligned} \Delta S_{MB}^{(N_1,N_2)} &= \ln \frac{W_{MB}^{(N_1+N_2)}}{W_{MB}^{(N_1)} W_{MB}^{(N_2)}} \\ &= -\ln \frac{(N_1+N_2)!}{N_1!N_2!} < 0. \end{aligned} \tag{6.3.19}$$

The inequality (6.3.19) follows from the inequality

$$\binom{N_1+N_2}{N_1} = \frac{(N_1+N_2)!}{N_1!N_2!} > 1. \tag{6.3.20}$$

Thus, in all the processes of pure assimilation, the MI *decreases*. This is clearly a counter-intuitive result. We shall revert to this aspect of the assimilation process in Section 6.6.

As we have done in Section 6.2, we can calculate the change of MI for the MB case from the canonical PF. The canonical partition functions in the initial and final states are

$$Q_i = \frac{V^{N_1}}{N_1!}\frac{V^{N_2}}{N_2!\Lambda_A^{3(N_1+N_2)}}, \quad Q_f = \frac{V^{(N_1+N_2)}}{(N_1+N_2)!\Lambda_A^{3(N_1+N_2)}}. \tag{6.3.21}$$

The corresponding change in MI of the process is

$$\Delta S = S_f - S_i = \ln\frac{N_1!N_2!}{(N_1+N_2)!} < 0, \tag{6.3.22}$$

In agreement with (6.3.19). We stress again that this process is referred to as a pure assimilation process. It is *pure* in the sense that only the *number* of indistinguishable particles have changed; the volume accessible to each particle as well as the temperature has not changed in this process. A special case of (6.3.22) is when $N_1 = 1$ and $N_2 = N - 1$, for which we have

$$\Delta S = \ln\frac{1!(N-1)!}{N!} = -\ln N < 0. \tag{6.3.23}$$

This process is important in the study of solvation where one particle, initially distinguishable from the rest of the particles of the same type, is assimilated into the other members of the same species.[7]

It is tempting to interpret (6.3.23) directly in terms of changes in the *identity* of the particles. We write

$$\Delta S = -\ln N = \sum_{i=1}^{N}\frac{1}{N}\ln\frac{1}{N} = \sum_{i=1}^{N} p\ln p < 0. \tag{6.3.24}$$

This has the correct form of an informational measure, defined on the probability distribution $p = 1/N$. But what are these probabilities? Recall that we have started with a system of *one distinguishable* particle and $(N-1)$ indistinguishable particles, and ended with N indistinguishable particles. One interpretation of p could be the probability that a particle picked at random would be

[7]See Ben-Naim (2006).

that one that was originally separated and distinguishable from the other. Thus, we are tempted to accept the intuitively very appealing interpretation that upon assimilation, we *lose* information. Initially, we *knew* which particle was in the separate box; after assimilation, we lost that information, we lost it forever and we can never retrieve this information; we can never find out which particle we have just added.[8] This argument is so compelling that even Gibbs himself concluded that retrieving this particular particle is "entirely impossible" (see Section 6.6). Looking again at the sign of ΔS in (6.3.23), we see that ΔS is *negative*, corresponding to a negative change in MI. Negative MI is equivalent to the *gain* in information.

The sense of loss of information in either process IIIa or IIIb is genuine.[9] This sense of loss of information seems to conflict with the finding that the MI has decreased in the process of assimilation. However, this conflict is an illusion, perhaps a deeper illusion than that involved in the process of mixing as discussed in Section 6.2. The source of this illusion is that we tend to associate *any* loss of information with an increase in MI. However, as we have seen in Section 4.3, we have two kinds of information in the classical description that contribute to the MI; the locational and the momentum information. Neither of these types of information change in the assimilation process. We have also seen in Section 4.3 that the change in the number of indistinguishable particles is associated with the mutual information, i.e., with the correlation among the particles. Likewise, we have also seen that the process of unlabeling of particles does feel like a loss of information, actually it causes an increase in the number of states — hence an *increase* in the MI. We shall revert to this question of assimilation in Section 6.7 and in Appendix M.

[8]This loss of information not only applies to the addition of a particle of the same kind as described in process IIIa, but also for the change of the identity of a molecule from one kind to another. We discuss a specific example in Section 6.7.

[9]This is similar to taking a letter, putting it in an envelope, and dropping the envelope into a bag of N identical envelopes. If the envelopes cannot be opened, we can never retrieve the information on "where the letter is."

6.4 Irreversible Process of Mixing Coupled with Expansion

The process IV in Figure 6.6 is perhaps the most common process of mixing discussed in textbooks on thermodynamics. It is in this process that the erroneous conclusions are most frequently reached that mixing is essentially an irreversible process. As the title of this section says, this is a process of mixing coupled with expansion. In the literature, this is referred to as mixing at constant pressure, or simply as a mixing process.

Following Gibbs original treatment two processes IV and V are presented in connection with the process of mixing as depicted in Figures 6.6 and 6.7. For simplicity, we take $N_A = N_B = N$ and $V_A = V_B = V$, but all the arguments and conclusions are valid for any N_A, N_B, V_A and V_B. It is easily shown that in process IV *there* is a positive change of MI $\Delta S_{IV} = -N \sum x_i \ln x_i = N \ln 2 > 0$ whereas in process V, no change is observed. Hence, $\Delta S_V = 0$. Comparing these two results, the natural conclusion is that the *mixing* in IV is the *cause* for the positive change in MI. Indeed, the mixing in process IV is the only *conspicuous* difference between the two processes IV and V. As will be shown below, the *conspicuous* mixing

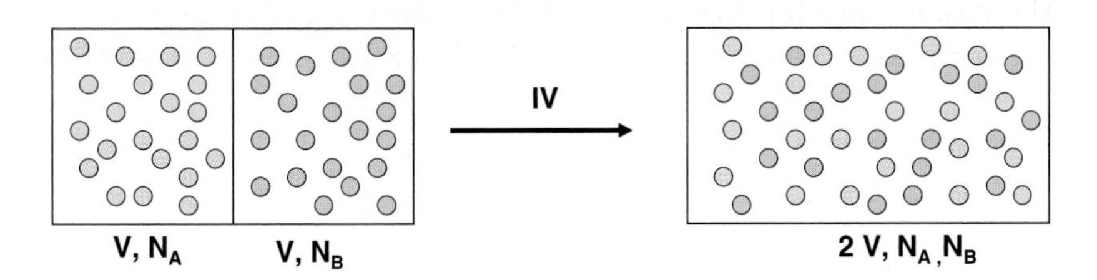

Figure 6.6. Irreversible mixing coupled with expansion.

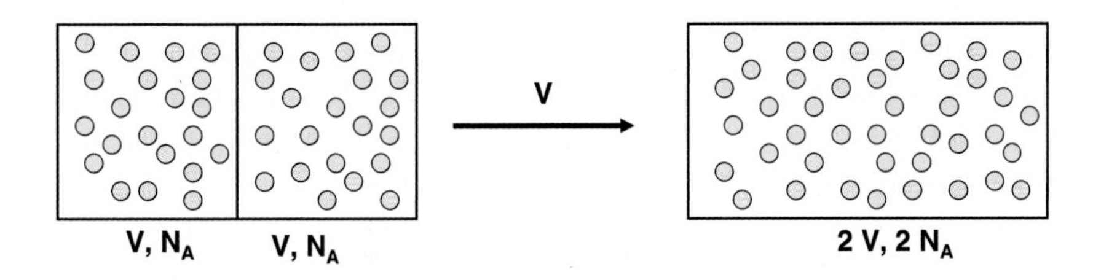

Figure 6.7. Reversible assimilation coupled with expansion.

in IV does not necessarily imply that the mixing is the *cause* for the irreversibility of the process IV. The perception that mixing is an inherently irreversible process is not only part of our basic scientific education, intimately woven into our comprehension of the Second Law of Thermodynamics, it also conforms with our everyday experience. Many processes of mixing that we encounter everyday are *indeed* spontaneous processes, involving an increase in the MI.

No doubt, this is the reason why most textbooks on thermodynamics reach the (erroneous)conclusion that *mixing* is an inherent irreversible process — hence the quantity $-\sum x_i \ln x_i$ is referred to as the *entropy of mixing*.

In this section, we analyze the origin of $\Delta S > 0$ in process IV, and in Section 6.6, we shall analyze the origin of $\Delta S = 0$ in process V. We have already seen that the *pure* mixing process involves no change in MI (see process II in Figure 6.3). In other words, a *pure mixing* process does not cause any change in MI, whereas a pure assimilation processes involves a decrease in MI.

As in the previous sections, we can calculate the change of MI in the process IV (Figure 6.6) either from the Sackur–Tetrode equation or from the canonical PF. We write for the initial and the final states the PF's as

$$Q_i = \frac{V^{N_A} V^{N_B}}{N_A! N_B! \Lambda_A^{3N_A} \Lambda_B^{3N_B}}, \quad Q_f = \frac{(2V)^{N_A} (2V)^{N_B}}{N_A! N_B! \Lambda_A^{3N_A} \Lambda_B^{3N_B}}. \tag{6.4.1}$$

The corresponding change in MI is

$$\begin{aligned} \Delta S_{IV} = S_f - S_i &= N_A \ln \frac{2V}{V} + N_B \ln \frac{2V}{V} \\ &= (N_A + N_B) \ln 2. \end{aligned} \tag{6.4.2}$$

We thus see that process IV is equivalent to two processes of the same kind as process of expansion I where the systems of N_A and N_B expand from initial volume V into $2V$. This equivalence is depicted in Figure 6.8 (i.e., process IV is equivalent to the two processes IV_1 and IV_2).

Once we realize that the *expansion* from V to $2V$ is the cause of the change in the MI, the puzzle regarding the origin in the difference between processes IV and V is passed to process V. We shall further discuss process V in Section 6.6. Here, it is important

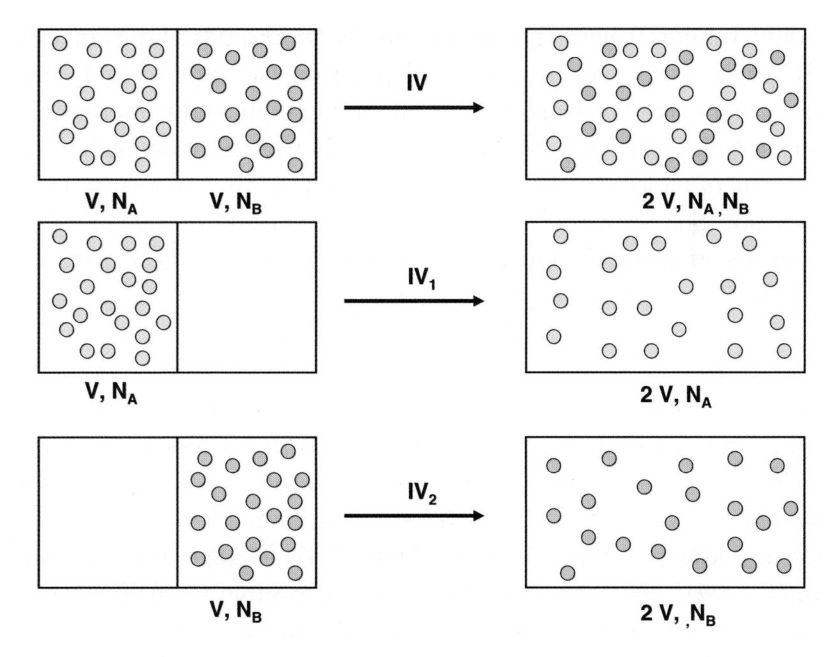

Figure 6.8. The mixing in process IV is equivalent to two processes of expansion; IV_1 and IV_2.

to realize that since the change in MI due to the expansion process is independent of the *kind* of molecules, the change in ΔS in the mixing process is also independent of the type of molecules. This conclusion was not always so obvious. Gibbs considered the fact that the value of the "entropy of mixing" is independent of the *kind of gas*, no less striking than the fact that ΔS becomes zero when the two gases become identical. On this matter, Gibbs writes[10]

> *"But if such considerations explain why the mixture of gas-masses of the same kind stands on different footing from mixtures of gas-masses of different kinds, the fact is not less significant that the increase of entropy due to mixture of gases of different kinds in such a case as we have supposed, is independent of the nature of the gases."*

Indeed, if one conceives of the *mixing* itself as the cause of the "entropy of mixing" then it is quite puzzling to find that the entropy of mixing is independent of the *kind* of the mixing molecule. It

[10]Gibbs (1906), page 167.

should be noted here that in mixing of two liquids the change in MI (or entropy) will in general, depend on the type of molecules that are mixed. The reason is that in mixing liquids, the interactions between AA, AB and BB are different, and strongly dependent on the type of molecules. However, for ideal gases, the mixing, in itself, does not play any role in determining the value of the so-called "entropy of mixing." Once we recognize that it is the *expansion*, not the *mixing*, which causes a change in the MI, then the puzzling fact that the change in MI is independent of the kind of molecules evaporates.

Clearly, we do observe *mixing* in process IV. This is a fact! But it is also clear that the mixing that occurs in process IV *does not* contribute to the change in MI. In both processes I and IV, each particle was initially confined to a smaller volume V and finally can access a larger volume $2V$. In both cases, the change in MI has the *same* source: the change in the accessible volume to each particle. The irrefutable conclusion is that the *mixing observed in process IV has no effect on the entropy of this process.* This conclusion leaves us with an uneasy feeling: if mixing in process IV does not affect the MI in the process, then what *is* the cause of the *different* results we obtain for the MI change in processes IV and V in Figures 6.6 and 6.7?

Since the only conspicuous difference in processes IV and V is the *mixing* in IV and non-mixing in V, it is almost inevitable to ascribe the change in MI to the *mixing* in IV, and the *no change* in MI in process V to the apparent non-process in V. This conclusion is erroneous, however. To clarify, and to answer the aforementioned question, we have to examine the change in MI in process V, which looks deceptively simple but is actually more complex and more difficult to understand than the change of MI in the process of mixing in IV. We shall return to this process in Sections 6.6 and 6.7.

6.5 Irreversible Process of Demixing Coupled with Expansion

While processes IV and V are well-known and well-discussed in textbooks, the following example is not as well-known. It is, however,

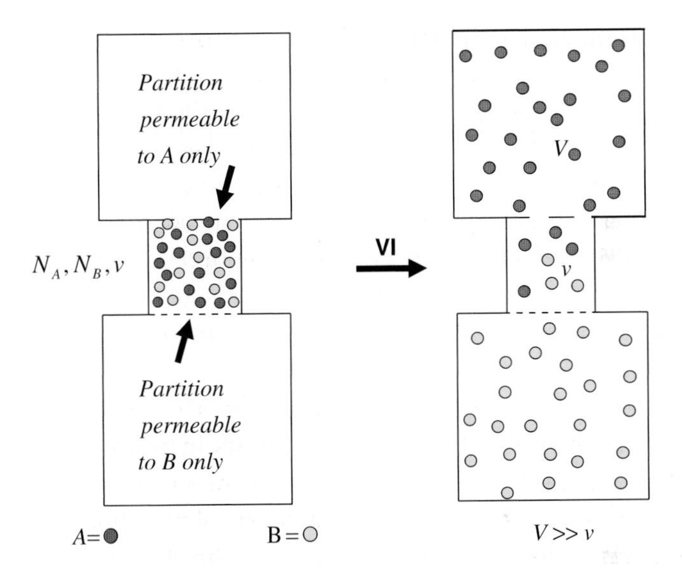

Figure 6.9. Irreversible de-mixing coupled with expansion.

an important process since it demonstrates that both mixing and demixing could be irreversible when coupled with a driving force which is an irreversible one involving increase of MI. In Section 6.11, we shall discuss another possible driving force, the solvation Gibbs energy, that can be coupled with mixing or demixing.

Process VI depicted in Figure 6.9 is a spontaneous irreversible demixing. We start with a mixture of N_A and N_B in a small volume v, and let the A molecules expand into a larger volume V through a partition permeable to A only. Similarly, we let B expand from v to V.

The change in MI for this process is[11]

$$\Delta S_{VI} = N_A \ln\left(\frac{V+v}{v}\right) + N_B \ln\left(\frac{V+v}{v}\right) > 0. \qquad (6.5.1)$$

If we choose $V \gg v$, we can achieve an almost complete demixing of A and B. If we insist, we can also demix completely the remaining mixture in v with no change in MI.[12] For $N_A = N_B = N$, we find

[11] For more details, see Ben-Naim (1987).

[12] Simply reversing the process II described in Section 6.2, but for the remaining mixture in v.

that the total change in MI is

$$\Delta S_{VI} = 2N \ln\left(\frac{V}{v}\right) - N\frac{v}{V} \approx 2N \ln\left(\frac{V}{v}\right) > 0. \tag{6.5.2}$$

The important lesson we learned from processes IV and VI is that the *observed* mixing in IV, and the observed demixing in VI, are *not* the driving force for these processes. In both processes the mixing is *coupled* to expansion — which provides the thermodynamic driving force. Clearly, by the same logic that many authors conclude from process IV, that *mixing* is inherently irreversible, we could have concluded from process VI that *demixing* is inherently an irreversible process. The two conclusions are erroneous.

Moral: Mixing, as well as demixing, are something we *see* but that something has no effect on the MI of the system.

6.6 Reversible Assimilation Coupled with Expansion

In Section 6.2, we have seen that process II involves $\Delta S = 0$, and we concluded that from the thermodynamic point of view, this is a *non-process*. However, not every process for which $\Delta S = 0$, is a non-process.

Process V in Figure 6.7 looks deceptively simple. Indeed, no conspicuous process is observed in V, but underneath this apparent *non-process*, two processes are at work. One is expansion: each particle has changed its accessible volume from V to $2V$. In addition N indistinguishable particles and another N indistinguishable particles of the same kind are assimilated into into each other to form $2N$ indistinguishable particles. (Gibbs refers to this process as "mixing of gases of the same kind." I prefer to use the word "assimilation" for this kind of mixing, and reserve the term "mixing" for mixing different species.) Let us see quantitatively how much each of these processes contribute to the total change in the MI. The PF in the final and initial states are

$$Q_f = \frac{(2V)^{2N}}{(2N)!\Lambda^{3(2N)}}, \quad Q_i = \frac{V^N V^N}{N!N!\Lambda^{3(2N)}}. \tag{6.6.1}$$

For any finite N, we have

$$\Delta H_V = 2N \ln \frac{2V}{V} - \ln \frac{(2N)!}{(N!)^2} > 0. \tag{6.6.2}$$

Mathematically, this is always a positive quantity! This follows from the inequality

$$2^{2N} = (1+1)^{2N} = \sum_{i=1}^{2N} \binom{2N}{i} > \frac{(2N)!}{N!N!}. \tag{6.6.3}$$

Note that $\frac{(2N)!}{N!N!}$ is only one term in a sum of many positive numbers.

Thus, the MI always increases in process V. However, thermodynamically speaking, the difference between the two terms in (6.6.2) is insignificant and cannot be noticed experimentally. When N is very lage, ΔH_V in (6.6.2) is practically zero. This is easy to see if we use the Stirling approximation for the factorials $N!$ and $(2N)!$ in (6.6.2):

$$\Delta S_V = \lim_{N\to\infty} \Delta H_V = 2N \ln \frac{2V}{V} - 2N \ln 2 = 0. \tag{6.6.4}$$

We can now conclude that in process V, two things happen. First, we have an expansion that causes an increase in the MI exactly as in Section 6.1. In addition, there is also an assimilation process which contributes negatively to the MI. The two cancel each other for macroscopic systems (strictly for $N \to \infty$), so that the net result is a zero change in MI.

Exercise: Calculate ΔS_V for process V for the following two cases, and interpret the results in terms of change in the MI:

(i) $V_1 = V_2 = V$, and $N_1 \neq N_2$,
(ii) $V_1 \neq V_2$ and $N_1 = N_2 = N$.

Solutions: Using the partition functions (6.6.1) for the initial and final states, we get for the first case (equal volumes but unequal number of particles)

$$\Delta S_V \cong (N_1 + N_2)[\ln 2 + x_1 \ln x_1 + x_2 \ln x_2], \tag{6.6.5}$$

where $x_1 = N_1/(N_1 + N_2)$ and $x_2 = 1 - x_1$.

The first term is due to the increase of the accessible volume for each of the $N_1 + N_2$ particles, the same as in (6.6.4). The second and third terms are the contributions to ΔS due to assimilation, which is different from (6.6.4). This term has the form of the so-called "entropy of mixing," but clearly it has nothing to do with the mixing. For the special case, $x_1 = 1/2$, (6.6.5) reduces to (6.6.4).

For the second case (equal number of particles, but different initial volume) we have

$$\Delta S_V \cong -N \ln y_1 - N \ln y_2 - 2N \ln 2, \qquad (6.6.6)$$

where $y_1 = V_1/(V_1 + V_2)$ and $y_2 = 1 - y_1$. Here, the third term is due to assimilation. The change in MI due to volume change is different. N particles have changed the accessible volume from V_1 to V_1+V_2, and N particles have changed the accessible volume from V_2 to $V_1 + V_2$. When $V_1 = V_2 = V$, (6.6.6) reduces to (6.6.4).

6.7 Reflections on the Processes of Mixing and Assimilation

In this section, we reflect on the way we perceive the outcome of the two processes IV and V, and how our minds struggle to decide between two choices; to believe in what our eyes see, or to accept what the formal theoretical analysis tells us.

The two experiments IV and V are very simple and the theory we use to analyze them is very elementary, yet something very strange occurs when we compare what either our physical eyes or our "mental eyes" *see*, and what theory tells us.

As we have already seen in the previous sections, these issues run deep in our way of thinking and our way of visualizing the molecular processes. We shall separately discuss the two issues.

(i) *Does mixing increase the MI? It does not!*

The idea that mixing is an inherently irreversible process is not only written in textbooks of thermodynamics. It is not only the knowledge we acquired as students, learning thermodynamics. It is part of our daily experience. We see mixing occurring everywhere and everyday. Process IV is only one of many examples of mixing processes. A more familiar one; a drop of dark blue ink in a glass of water will mix with the water and color it light blue.

These are undeniable observations. We learn in all textbooks that in the mixing process, e.g., process IV, the MI or the entropy *increases*. We also learn that the MI increase is intimately associated with *increase* in disorder, and that seems very reasonable. It is therefore natural to associate *any* mixing process with decrease of order or increase of MI, and of course, irreversibility. The very usage of the term "mixed-up" in daily life implies *disordering*, whether it is two different gases or a thousand pairs of shoes — we mix them and we realize an increase of disorder. Thus, we intuitively, as well as intellectually, view mixing as one example of nature's manifestation of the arrow of time. Indeed, mixing is taught in the context of other examples of irreversible process — as flow of heat spontaneously from high to low temperature or flow of matter from high to low chemical potential.

We almost never observe a spontaneous demixing — as much as we never observe a spontaneous cooling of one part of a vessel of water and heating another part. So what could be more natural and appealing to our intuition than to accept that mixing is an irreversible process, that mixing causes disorder, and that mixing increases the MI in the system?

But alas, we have just learned that all of these are only an illusion — an illusion of our eyes. Mixing, by itself, is actually reversible, and both mixing and demixing, when properly coupled with expansion, can be rendered irreversible and if we eliminate the real thermodynamic "driving force" of expansion, we are left with bare mixing that has no effect on the thermodynamics in general, and on the MI, in particular.

It is admittedly shocking to learn that what we have seen everyday as mixing and what we thought we have fully understood is only an illusion. Nevertheless, we can quickly accommodate the new way of looking at the process of mixing. Mixing is something we do observe in our eyes, but that something has no effect on the MI of the system. We can easily subscribe to the new way of thinking — not to trust our eyes, but to trust our theoretical "microscopic" eyes; not to reach to a conclusion that what is conspicuous is necessarily the driving force, and that the real, genuine driving force might not be conspicuous. As Isaiah proclaimed: "*And he shall not judge after the sight of his eyes, neither reprove after the hearing of his ears.*"[13]

Once we are faced with the compelling argument as presented above, it is relatively easy to accept and to embrace the new conclusion. It is easy, since we know that both mixing and demixing can either be reversible or irreversible. It is easy, since we are offered a convincing alternative driving force, which facilitates the divorce from the notion that mixing is a driving force. We say *relatively* easy — easy with respect to what awaits us in the next section when discussing assimilation and deassimilation, which are anything but easy to accept and assimilate into our thinking.

But before plunging into the highly treacherous and slippery issue of assimilation, let us look at a fringe issue of semantics. I have often heard comments that the whole issue of "the entropy of mixing" is merely a matter of semantics. I adamantly disagree with that comment. Of course, there is *one* and only one aspect of the issue that has the flavor of being semantic. Of course, one can name the quantity $-k_B \sum x_i \ln x_i$ "entropy of mixing" or entropy of whatever, but one cannot dismiss the whole issue that mixing is "inherently irreversible" as an issue of semantics.[14] The latter is a profound issue, relevant to our understanding of the Second Law of Thermodynamics.

But even the admittedly semantic aspect of the issue is not fully justified. Naming a quantity "entropy of mixing" implies more than naming a person or a thing by an arbitrarily chosen name. A term like "entropy of mixing" is both inappropriate and potentially misleading. This term has misled so many of us to think of mixing as an irreversible process, as attested by most authors of textbooks on thermodynamics. It is fitting, therefore, to further elaborate on the "semantic" part of the issue. I will do that via a caricature rendition of a process.

Consider the process depicted in Figure 6.10, VIIa. N particles initially in a circle shape container (or sphere) of volume V are

---

[13] Isaiah 11:3. **ולא למראה עיניו ישפוט, ולא למשמע אוזניו יוכיח (ישעיה י"א ו)**.

[14] In this section, we discuss the common concept of "entropy of mixing." All the conclusions apply equally to the MI in the mixing process. The latter becomes equal to the former after multiplying by the Boltzmann constant.

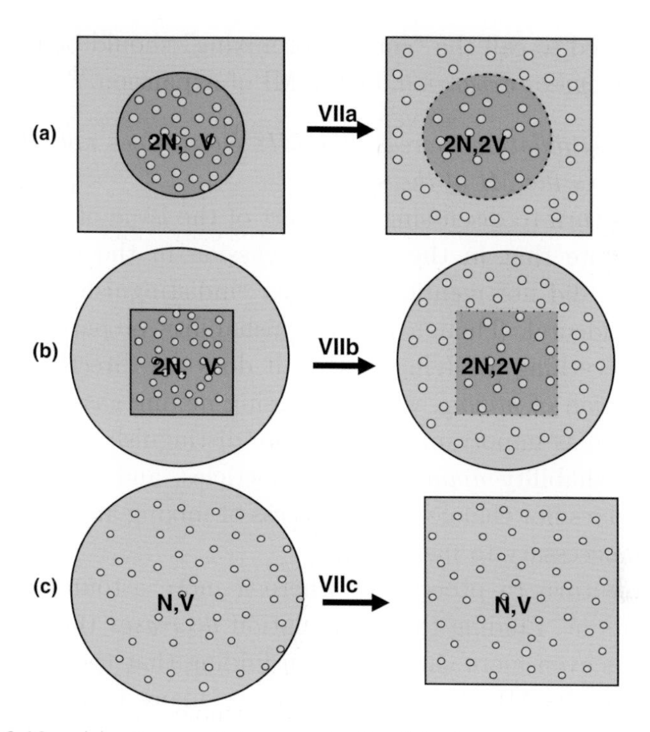

Figure 6.10. (a) "irreversible squaring", (b) "irreversible circling" and (c) "reversible squaring".

allowed to expand and fill a sphere (or a cube) of volume $2V$. The reverse of this process is shown in VIIb and a reversible version (in the sense that $\Delta S = 0$, see below) of the process is shown in VIIc. We can calculate the change in MI in process VIIa and find $\Delta S = 2N \ln 2$. The same is true for VIIb, but $\Delta S = 0$ for VIIc (assuming the system is macroscopic and neglecting surface effects).

Would anyone dare to refer to $\Delta S = 2N \ln 2$ in process VIIa as the "entropy of squaring," or in process VIIb, as the "entropy of circling?" Obviously, we all feel that though we *see* squaring in VIIa and we *see* circling in VIIb, the terms "entropy of squaring" and "entropy of circling" would not be appropriate. We all know and feel correctly that the "driving force" for both processes VIIa and VIIb is the expansion, and the quantity $2N \ln 2$ should appropriately be referred to as the entropy of expansion. In exactly the same sense, what we used to call the "entropy of mixing" should be referred to as the *entropy of expansion*, or the MI of expansion.[15]

(ii) *Does assimilation increase the MI? No, it does not! It actually decreases the MI of the system.*

We now turn to discussing the heart of the issue of assimilation. We first note that in the entire discussion in the previous subsection, we did not mention the word "indistinguishability." This is not accidental. Though indistinguishability of particles affects almost everything involving matter, it does not directly feature in the discussion of *mixing*, when discussing mixing gases of different kinds. What *is* important is that we *do* distinguish A from B. The indistinguishability *among* the N_A particles, and among the N_B particles does not change in the process of mixing and therefore we find it unnecessary to mention it.

We now turn to process V, where a more astounding revelation awaits us. Finding that assimilation *decreases* the MI (or the entropy) is even more shocking than finding that the mixing does not change the MI. After all, we do see many mixing processes occurring spontaneously. Our theoretical analysis in Section 6.3 tells us that assimilation processes *decrease* the MI. We must conclude that the reverse process of deassimilation should increase the MI. Yet the fact is that we never see around us processes of deassimilation occurring spontaneously. Therefore, it is hard to accept and harder to visualize why the deassimilation *increases* MI and assimilation *decreases* MI.

There is yet another, more profound reason why we have a hard time accepting and associating assimilation with the *decreasing* of MI. I venture to guess that even Gibbs himself would have been resentful of this idea. At this juncture, we can do no better than cite Gibbs himself[16]:

> *"If we should bring into contact two masses of the same kind of gas, they would also mix but there would be no increase in entropy.*

---

[15]The equivalence of process IV and two expansion processes was demonstrated in Figure 6.8.

[16]Gibbs (1906).

> *When we say that when two different gases mix by diffusion ... and the entropy receives a certain increase, we mean that the gases could be separated and brought to the same volume ... by means of certain changes in external bodies, for example, by the passage of a certain amount of heat from a warmer to a colder body. But when we say that when two gas masses of the same kind are mixed under similar circumstances, there is no change of energy or entropy, we do not mean that the gases which have been mixed can be separated without change to external bodies. On the contrary, the separation of the gases is entirely impossible."*

"*Entirely impossible*"! Why? Gibbs did not explain why. We feel, however, that his conclusion is obvious. Let us repeat what Gibbs says in plain words: though the mixing of *different* species, e.g., process IV *can* be reversed (by investing energy), the mixing of gases of the *same* species, e.g., process V, *cannot* be reversed.[17]

Gibbs' comment leaves us with an awesome feeling — something strange is going on in process V: the mixing in process IV, which we all observed as a spontaneous irreversible process, *can* be reversed. The "mixing" in process V, where we have observed *nothing*, there is no noticeable change in energy or in MI (or entropy). We are told that this process is *irreversible* in a more absolute sense: it *can never be reversed*, it is "*entirely impossible*." That sounds both obvious and paradoxical. It is obvious, since there exists no way of bringing each particle back to its original compartment. It is also paradoxical: how can it be that an *irreversible* mixing in process IV that causes an *increase* in the MI *can* be reversed, but the "mixing" in the *reversible* process V, where *no change* in the MI is observed, cannot be reversed — it is *entirely impossible*?

Is this another mysterious paradox borne out of Gibbs' writing on the entropy of mixing? To understand the troubling issues, let us analyze what Gibbs really meant by "entirely impossible."

There are at least three senses in which we can think of reversing process V. First, the easiest to conceive; let us call it the *mechanical sense.* We all know, and certainly Gibbs knew, that the equations of

[17] A very similar idea was expressed by Maxwell almost three decades before Gibbs. We shall comment on Maxwell's quotation at the end of this section.

motion are symmetrical with respect to time reversal. If we could, at least theoretically or mentally, reverse the motion of each particle, i.e., run the process V backward, we *could* in principle, reverse the process. However, in this sense, process IV can be reversed as well, i.e., this kind of reversal applies to both processes IV and V. It is clear that Gibbs did not mean "impossibility" in this sense. Hence, this sense is ruled out.

The second meaning, which is a little harder to conceive, can be referred to as the *fluctuation sense*. We all know, and probably Gibbs knew too, that the Second Law of Thermodynamics is not *absolute* — that a fluctuation, though very rare, can occur, in which the initial state *can* be observed. It can occur spontaneously by investing nothing. All we have to do is sit and watch. Watch for a very long time. How long? It depends on the size of the system. It might take many more years than the age of the Universe. But if we are willing to wait long enough, it *will* occur. For the present argument, it does not matter how long. It is clear that Gibbs could not have possibly meant "impossible" in this sense. True, the reversal is highly improbable but *possible*. However, reversal in this sense equally applies to both processes IV and V and Gibbs has singled out only the reversal of process V to be deemed "entirely impossible." Therefore, this sense is ruled out too.

The "impossibility" referred to by Gibbs is in the *thermodynamic sense*. Process IV, though spontaneous and irreversible, *can* be reversed by manipulating the system's thermodynamic variable. All we would need is to invest some work and we could get back to exactly the initial state of process IV. For instance, we could compress the system from $2V$ to V (investing energy by the amount $2N \ln 2$) then demix the two-component reversibly (reversing process II discussed in Section 6.2) and we are back *exactly* to the initial state.

Note that this reversal brings every A particle which originates from the left-hand compartment back to the left-hand compartment and similarly all B particles brought to the right-hand compartment. All we need is that all the A's are distinguishable from all the B's — the N_A particles are *ID* among themselves and the N_B, B particles are *ID* among themselves.

It is in this sense that Gibbs singled out process V and concluded that reversing the system to its initial state is "impossible," no matter how much energy we are willing to invest, and how many ingenious thermodynamic manipulations we are willing to devise to achieve that goal. We can never do that.

It is in this sense that we are left with a deep and compelling conviction that in process V something of a *stronger irreversibility* has occurred — something has been lost and can never be retrieved. Thus, even before knowing the connection between entropy and information, we already *feel* it in our bones that some information has been lost forever in process V. But what is that information which we have lost and because of which we are unable to bring back each particle to its original compartment? Not by reversing the equation of motion, not by waiting for an unimaginably long time but by experimental thermodynamic means? We understand clearly and vividly that something has been lost forever. We also know that the expansion part of the processes in both IV and V are the same. Therefore, the loss of that "something" in process V cannot be attributed to the increase of accessible volume for each particle. This will not single out process V. So what is left in process V which is not in process IV is the indistinguishability of *all* the particles in process V. It is at this juncture that our mental eyes betray us — including Gibbs. The deception is deeper than the deception we encountered in connection with the mixing process IV, where we saw mixing with our *real* eyes — irreversible mixing — only to be told by our microscopic eyes that the mixing did not have any effect on the MI.

Gibbs understood that process IV is irreversible. However, one can devise a contraption that separates A and B. All you need are partitions that are permeable only to A and only to B. He also understood that there can be no contraption that can separate between A particles that originate from the left compartment from A particles that originate from the right compartment. Recall that Gibbs' writings were at the end of the 19th century before quantum mechanics taught us that particles are *ID*. Had Gibbs thought from the outset of the particles as being *ID*, he would not have reached the conclusion that process V cannot be reversed. The question of whether or not process V could be reversed would not even be raised. In such a world, process V *can be reversed*, and ridiculously simply, with no investment of energy. Simply return back the partition into its original place and you are back *exactly* to the initial state, and the arrow in process V had been reversed.

Here is the apparent paradox. We were first convinced by Gibbs' argument that mixing of gases of the same kind cannot be reversed and that it is "entirely impossible." Lo and behold! We now find that not only it is not "entirely impossible," but it is so simply *possible*. So where did we go wrong?

The answer is this: our mental eyes have betrayed us. Every time we try to visualize process V we make a mental image of the system. In our mental eyes, we *label* the particles or assign them mental coordinates and mental trajectories. Then we are told, and we accept the fact, that particles are *ID*. At this point, we have to give up the labels or the coordinates of the particles. This results in a sense of loss of information. Therefore, we understand that process V cannot be reversed. But if we think from the outset that the particles are *ID*, then from the outset, we do not assign coordinates to the particles. Therefore, we do not have a sense of loss of information. *The reversal to the initial system becomes trivially possible.*

As far as we can tell from Gibbs' writings, he never discussed the reason for the entropy change in process IV. He realized though that while process IV is spontaneous and irreversible, it can be reversed by using some contraption that can distinguish between A and B, and by investing energy. He apparently did not realize that process IV, from the thermodynamic point of view, is just an expansion — nothing more.

Gibbs did contemplate the question of reversing process V. He concluded that such a reversal is "*entirely impossible.*" We cannot reconstruct his sequence of thoughts leading to this conclusion. However, when he thought of process V, he probably made a mental image of the system. In this mental image, particles have definite coordinates, or definite *labels*. This very labeling of the particles give a sense of possessing information, information about the location of each individual particle. Knowing, however, that the particles are *ID*, one has to relinquish this information. This results in a sense of informational loss — hence the conclusion that reversing process V is *entirely impossible.* But here our mental eyes fail us. By the very act of visualizing the individual particles as in process V, we already delude ourselves that we are in possession of information, information that we never had in the first place, and therefore we could not have lost.

Gibbs was well aware of the fact that in thermodynamics what matters is the *macroscopic* state of the system, not the detailed microscopic description of the states of the individual molecule.

In the same paragraph from which we took the quotation above, Gibbs writes (the italics are mine)

> "So when gases of *different kinds* are mixed, if we ask what changes in external bodies are necessary to bring the system to its original state, we do not mean a state in which each particle shall occupy more or less exactly the same position as at some previous epoch, but only a state which shall be *indistinguishable from* the previous one in its *sensible properties.* It is to *states* of *systems* thus *incompletely defined* that the problem of thermodynamics relate."

Thus, Gibbs clearly understood that it is the thermodynamic, or the macroscopic, states that can be reversed for a mixture of different kinds of molecules. Yet he failed to see that by placing the partition back after performing process V, one does recover the same *thermodynamic* state, which is indistinguishable from the initial state.

To summarize, in an expansion process I, we lose *real* information. We know more about the locations of the particles before the process than after the process. It is this *loss* of *information* that is responsible for the change in the MI in the expansion process. In the mixing process IV, we have again *real* loss of information, exactly as in the expansion process (only twice as large — one for the A's and one of the B's). In process V, again we lose *real* information due to the expansion — exactly as in the case of mixing I. But now we also *gained real* information by the assimilation process, which exactly cancels the loss of information due to the expansion. The net effect is no change in information (or in entropy).

I would like to end this long section with two pedagogical comments both involving the concept of indistinguishability of the particles. Quite a few times after I gave a talk on this topic, people who teach thermodynamics asked me, "how can you explain the entropy change in process IV within thermodynamics?" This is a very valid question. In a course in thermodynamics, you show students the two processes IV and V. It is clear that in IV there is a conspicuous mixing whereas in V nothing happens. Therefore, the only possible explanation for the increase in entropy is to assign it to the mixing process.

My answer is this: First, you cannot *explain* entropy change in any process within the realms of thermodynamics. That is true for the expansion of ideal gas, a spontaneous heat transfer from a hot to a cold body, or any other spontaneous process. Thermodynamics offers no explanation of any of these processes. The only way to understand entropy is within statistical mechanics. However, if one teaches only thermodynamics, one can explain that the entropy change in process IV is due only to expansion, nothing more. On the other hand, explaining the change in the MI (or in entropy) in either process V or the processes of assimilation is more subtle. There is no way but to admit that the particles are *ID*. That fact cannot be dealt with within classical thermodynamics. I believe, however, that the cyclic process depicted in Figure M.1 of Appendix M can help in understanding the role of labeling and unlabeling of particles as a source of the changes in information — hence of entropy — in any process involving changes in the numbers *ID* of particles.

The second comment is also related to the perception and understanding of the concept of *ID* particles.

James Clerk Maxwell, in a letter to J.W. Strutt in 1870 wrote (the italics are mine)[18]:

> "Moral. The Second Law of Thermodynamics has the same degree of truth as a statement that if you throw a tumblerful of water into the sea, you *cannot* get the *same* tumblerful of water out again."

[18]Quoted in Leff and Rex (1990) and Munowitz (2005).

If you have grasped the concept of *ID* of particles (here of water molecules), you will immediately realize the double fallacy of this statement. We realize that this statement was made 30 years before Gibbs analyzed the process of mixing and reached the conclusion quoted on pages 276–277. It is very clear that Maxwell, perhaps the greatest physicist of the 19th century, who was the first to introduce statistical arguments into scientific thought, did not, and perhaps could not, think in terms of *ID* particles. As we have discussed earlier — the Gibbs "entirely impossible" — we can also say in this case that we *can* get the same tumblerful of water from the sea[19]: simply fill-up the tumbler with sea water. This counter-intuitive conclusion can only be made if we rationally accept the idea that particles are *ID*. What we have in the tumbler is the *same* water, before and after we have thrown and retrieved it from the sea.[20] Note that initially the water in the tumbler is distinguishable from the water in the sea (see Appendix J). Throwing the water into the sea makes the water that originated from the tumbler indistinguishable from the water in the sea. After refilling the tumbler, the water in the tumbler is again distinguishable from the water in the sea. However, the water in the refilled tumbler is the same and ID from the water in the tumbler before it was emptied into the sea (except for a possible negligible difference in the number of water molecules in the two cases).

Although we feel it in our bones that this is not the *same* water, there is no way to prove it experimentally since this would require labeling the water molecules. Once we recognize the invalidity of the second part of the Maxwell statement, the invalidity of the first

[19]It is not clear what Maxwell had in mind when referring to the "sea," whether it is pure water or aqueous solution. For our purposes, it makes no difference. If we threw a tumblerful of water into the sea, we can easily get the *same* tumblerful of water by refilling it, from the sea. For simplicity, we shall assume that the water is pure, both in the tumbler and in the sea.

[20]It is understandable why Maxwell did err in this quoted statement. It is amusing that many authors make the same mistake even in the 21st century. For instance, Munowitz (2005) writes: "You can get another tumblerful of water out again, even one that looks the same superficially and macroscopically just like the original, but not *the* tumblerful you first tossed."

part follows suit, i.e., this statement is irrelevant to the Second Law of Thermodynamics.

Even thinking classically about the (labeled) particles, it is not true that we *cannot* retrieve (in an absolute sense) the *same* water molecules that were originally in the tumbler. It is not impossible but highly improbable! If the tumbler originally contained say 10^{23} water molecules, and the sea contains say $10^9 \times 10^{23} = 10^{32}$ water molecules, then filling the tumbler from the sea, we might retrieve the *same* water molecules. The probability of such an event is approximately $10^{-32\times 10^{23}}$.

This is a fantastically small number. However, thinking of water as made of indistinguishable molecules, the probability of retrieving the *same* molecules becomes nearly *one* (except that there might be a negligible change in the *number* of water molecules). It should be noted that though the order of magnitude of the probability of retrieving the *same* (classically thinking) water molecule is similar to the order of magnitude of the occurrence of a violation of the Second Law of Thermodynamics, the specific experiment referred to by Maxwell is irrelevant to the Second Law.

6.8 A Pure Spontaneous Deassimilation Process

We have seen that the assimilation process causes a decrease in MI. This leads us to conclude that the reverse of this process, the deassimilation process, should be associated with an increase in the MI. A purely assimilation process is one where V and T (or V and E) do not change, but only N changes. This is simply the reverse of process III described in Section 6.3, denoted process-III in Figure 6.11.

We all know that a spontaneous process causes an increase in MI. So, why do we never observe the reverse of process III occurring spontaneously? The answer is simple. Many processes can be described for which the MI change is positive but they do not actually occur as long as a constraint precludes its occurrence. The pure deassimilation in the reversed process III does not occur spontaneously as it is described in Figure 6.11a. However, we can simply do the deassimilation in III in two steps (Figure 6.11b). First, let

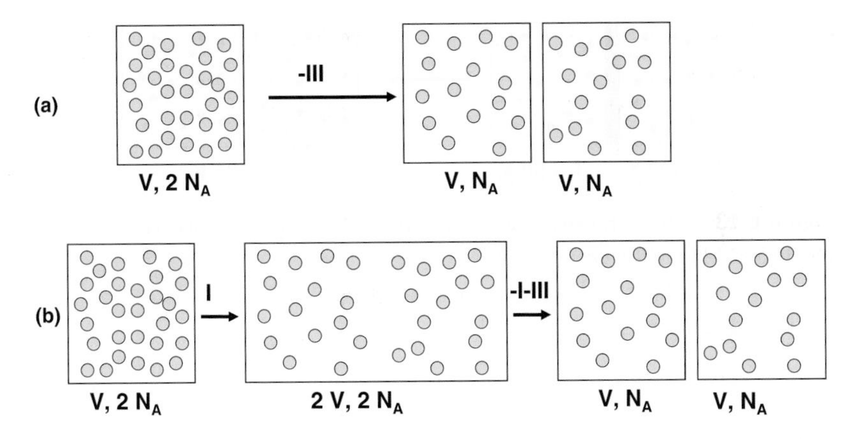

Figure 6.11. (a) Pure process of deassimilation, (b) The same process, as in (1), but in two steps; expansion followed by reduction of volume and deassimilation.

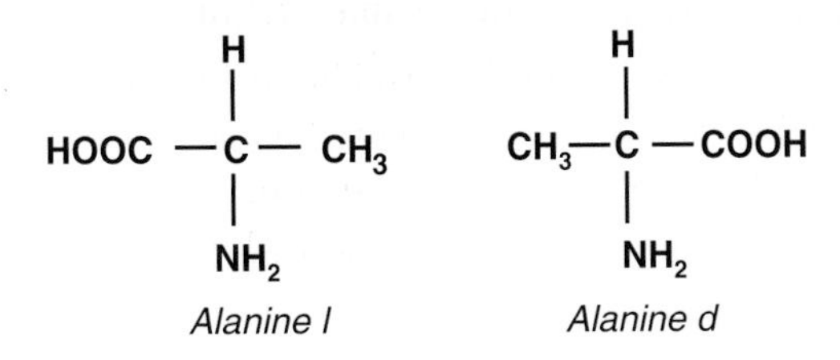

Figure 6.12. The two enantiomers of Alanine.

the system of $2N_A$ particles at V expand spontaneously into $2V$. Then divide the resulting volume into two halves. The net process is *pure* deassimilation. In this two-step process, we used the expansion (process I) part of the process as a "catalyst" for executing the net pure deassimilation.

Finally, we show that a *pure* deassimilation *can* occur spontaneously also in one step. Consider a system of $2N$ molecules, each of which contains one chiral center, say an alanine molecule (Figure 6.12). We initially prepare all the molecules in one form, say the d-form, in a volume V and temperature T. We next introduce a catalyst that facilitates the conversion of d- to l-form. Since l and d have exactly the same set of energy levels, hence, the same internal partition functions, the system will spontaneously evolve into a mixture of N molecules in the d-form and the N molecules in the

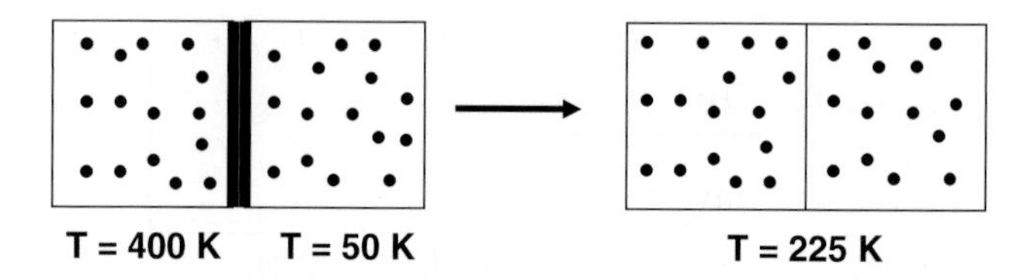

Figure 6.13. Heat transfer from a hot to a cold body. Initially, the two compartments are thermally insulated.

l-form. The net change in MI is

$$\Delta S = 2N \ln 2 > 0. \tag{6.8.1}$$

This is a *pure* deassimilation process — the system of $2N$ indistinguishable particles evolved into N particles of one kind and N particles of a second kind. The net effect is the same as in the reverse of process III where we achieve a split of the $2N$ particles into two groups — N in one box and another N in a second box. In both processes, the only change that took place is the deassimilation of the particles. The change in MI is positive. As we noted in Section 6.3, the decrease in MI in the assimilation process is counterintuitive; also, the increase in MI in the deassimilation process is counterintuitive. After all, we started with $2N$ particles which were initially indistinguishable, N of which acquired a new and different identity — and that certainly sounds like acquiring new information.

Another way of interpreting the process of racemization as described above is the following: If we start with say, pure d-molecules, each molecule has a set of energy levels. When the system changes to a mixture of d and l, we can view the system as composed of particles with the same energy levels, but now each energy level is doubly degenerated. This is the molecular reason for the increase in the MI.

It is interesting to note that some textbooks of physical organic chemistry refer to this process as "mixing" and the corresponding entropy change as "entropy of mixing" [see, for example, Eliel (1962), Jacques, Collet and Wilen (1981)]. Clearly, no mixing occurs in this process, only a spontaneous evolution of two components from a one-component system.

6.9 A Process Involving only Change in the Momentum Distribution

Up to this point, we have discussed changes in MI due to either the changes in location (changes in the volume V), or changes in the number of *ID* particles (changes in N). We now turn to discussing the simplest process where changes in momentum distribution is involved. This is also important from the historical point of view. It is one of the earliest processes for which the Second Law was formulated.

Consider the following process. We start with two systems of ideal gases, each containing N particles in a volume V, but the temperatures are different, say $T_1 = 50\,\mathrm{K}$ and $T_2 = 400\,\mathrm{K}$. Note also that T_1 and T_2 here are in Kelvin temperatures and the illustration in Figure 6.14 is for argon gas at these temperatures. We bring the two systems in contact (by placing a heat-conducting partition between them). Experimentally, we observe that the temperature of the hot gas will get lower, and the temperature of the cold gas will get higher. At equilibrium, we shall have a uniform temperature of $T = 225\,\mathrm{K}$ throughout the system.[21]

Clearly, heat or thermal energy is transferred from the hot to the cold gas. But how can we understand the change in the MI in this process? The qualitative discussion of this process has already been given in Section 1.2. Here, we present the more quantitative aspect of the change in MI in the process of heat transfer.

First, we note that temperature is associated with the distribution of molecular velocities. In Figure 6.14a, we illustrate the distribution of velocities for the two gases in the initial state. We see that the distribution is sharper for the lower temperature gas, and is more dispersed for the higher temperature gas. At thermal equilibrium, the distribution is somewhat intermediate between the two extremes, and is shown as a dashed curve in Figure 6.14a.

---

[21]The qualitative argument given here is valid for ideal gases. In general, when two different bodies at different temperatures are brought into thermal contact, the final temperature is somewhere between T_1 and T_2. However, to calculate the final temperature, we need to know the heat capacities of each system.

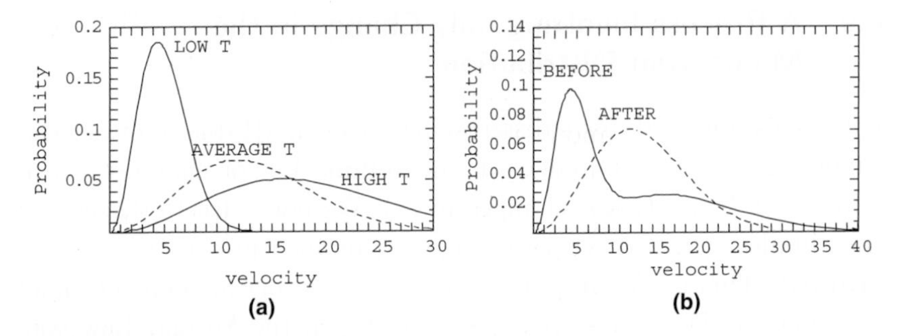

Figure 6.14. Velocity distributions (a) of the two systems before the thermal contact, (b) of the combined systems before and after the thermal contact.

What we observed experimentally is interpreted on a molecular level as the change in the distribution of molecular velocities. Some of the kinetic energies of the hotter gas is transferred to the colder gas so that a new, intermediary distribution is attained at equilibrium.

Within classical thermodynamics, we argue that since the total energy of the system is conserved in this process, and since in this particular example, the heat capacities of the two bodies are equal, the rise in the temperature of the cold gas must be equal to the decrease in the temperature of the hotter gas. If the final temperature is T, then

$$\begin{aligned} T &= T_1 + \Delta T, \\ T &= T_2 - \Delta T, \end{aligned} \tag{6.9.1}$$

and

$$T = \frac{T_1 + T_2}{2}. \tag{6.9.2}$$

The informational argument is different. We start with the velocity distributions of the two gases, $g_1(v)$ and $g_2(v)$ at temperatures T_1 and T_2, respectively. The variance of the velocity distribution of the combined initial (i) system is

$$\sigma_i^2 = \int_{-\infty}^{\infty} (v - \bar{v})^2 \left(\frac{g_1(v) + g_2(v)}{2} \right) dv = \frac{\sigma_1^2 + \sigma_2^2}{2}. \tag{6.9.3}$$

For simplicity, we can take v to be the x-component of the velocity. The three components of the velocities are independent — hence, their variances are equal.

We now apply Shannon's theorem for the continuous distribution (see Section 3.24). The theorem says that of all the distributions for which the variance is constant, the normal (or the Gaussian) distribution has a maximum MI. Since the total kinetic energy of the system before and after the process must be conserved, and since the average kinetic energy at equilibrium is proportional to the temperature, it follows that the final temperature T must be

$$T = \frac{T_1 + T_2}{2}. \tag{6.9.4}$$

The change in the MI in this process can easily be calculated from the Sackur–Tetrode equation (Section 5.4):

$$\begin{aligned}\Delta S &= S_f - S_i \\ &= \frac{3}{2}(2N)\ln T - \frac{3}{2}N\ln T_1 - \frac{3}{2}N\ln T_2 \\ &= \frac{3}{2}N\ln\frac{T}{T_1} + \frac{3}{2}N\ln\frac{T}{T_2},\end{aligned} \tag{6.9.5}$$

where the two terms correspond to the changes in the MI when the temperature changes from T_1 to T and from T_2 to T, respectively. Since T is the arithmetic average of T_1 and T_2, we have

$$\begin{aligned}\Delta S &= \frac{3}{2}N\ln\frac{T_1+T_2}{2T_1} + \frac{3}{2}N\ln\frac{T_1+T_2}{2T_2} \\ &= \frac{3}{2}N\ln\frac{(T_1+T_2)^2}{4T_1T_2} = \frac{3}{2}N\ln\frac{\left(\frac{T_1+T_2}{2}\right)^2}{T_1T_2} \\ &= 3N\ln\left[\frac{\frac{T_1+T_2}{2}}{\sqrt{T_1T_2}}\right] \geq 0.\end{aligned} \tag{6.9.6}$$

The last inequality follows from the inequality about the arithmetic and the geometric average (Appendix H):

$$\frac{T_1+T_2}{2} > \sqrt{T_1T_2}. \tag{6.9.7}$$

Thus, we have seen that in the spontaneous process as depicted in Figure 6.11, the MI has increased. This process is the simplest spontaneous process that involves an increase in the momentum MI. It should be noted, however, that the positive change in MI as calculated in (6.9.7) does not consist of an explanation of why the process proceeds spontaneously from the initial state to the final state. In both the initial and the final states, the momentum distribution is such that it maximizes the MI of the system. The Second Law of thermodynamics states that the MI in the final state must be larger than the MI in the initial state. To understand why the system proceeds spontaneously from the initial to the final states, we must appeal to a probabilistic argument. As will be shown in Section 6.12, the probability of the final distribution of the momenta is larger than the initial distribution.

6.10 A Process Involving Change in the Intermolecular Interaction Energy

Up to this point, we have discussed ideal gas systems, i.e., systems with no intermolecular interactions or negligible interactions. In Section 5.8, we derived the general expression for the MI due to the interactions among all the N particles. We discuss here a particularly simple process in which the intermolecular interaction changes. We assume that the density of the system is very low so that interaction among three or more particles can be neglected, but pair interactions are still operative. We have seen in Section 5.8 that the change in MI associated with turning on the (pairwise) interactions is

$$\Delta S = -\frac{N(N-1)}{2V^2} \iint g(\boldsymbol{R}_1, \boldsymbol{R}_2) \ln g(\boldsymbol{R}_1 \boldsymbol{R}_2) d\boldsymbol{R}_1 d\boldsymbol{R}_2. \quad (6.10.1)$$

We now design a process where the MI changes only due to changes in the interaction energy among the particles. We start with two boxes of equal volume V and equal number of particles N. The particles in the two boxes are different, but their intermolecular potential function $U(R)$ is very nearly the same, say two isotopes of argon. Figure 6.15 shows the form of the pair potential and the corresponding correlation function for a system at very low densities.

We now bring the two systems into one box of the same volume. Note that since the particles in the two boxes were different, we now

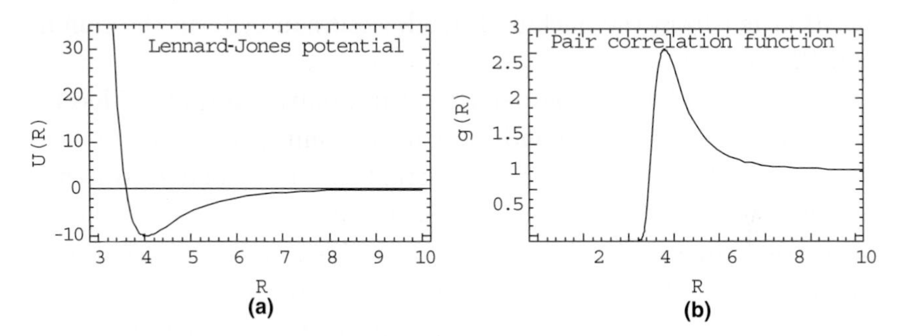

Figure 6.15. (a) Intermolecular pair potential, $U(R)$, (b) The corresponding pair correlation $g(R)$, at very low densities.

have a *mixture* of N particles of one kind, and N particles of the second kind. We have seen that this mixing process had no effect on the MI of the system. There is also no change in the volume of the system. If the process is carried out at constant temperature, then $g(R_1, R_2)$ is unchanged in the process.[22] The change in the MI in this process is thus

$$\begin{aligned}\Delta S &= S_f - S_i \\ &= -\left[\frac{2N(2N-1)}{2V^2} - 2\frac{N(N-1)}{2V^2}\right] \\ &\quad \times \iint g(\boldsymbol{R}_1, \boldsymbol{R}_2) \ln g(\boldsymbol{R}_1, \boldsymbol{R}_2) d\boldsymbol{R}_1 d\boldsymbol{R}_2. \end{aligned} \tag{6.10.2}$$

Note that we assume that, though the density has changed, $g(\boldsymbol{R}_1, \boldsymbol{R}_2)$ is unchanged. The only thing that has changed is the *total* number of *pairs* of interacting particles; from initially $N(N-1)/2$ pairs to $2N(2N-1)/2$ pairs in the final state.

It should be noted that in all of the examples discussed in the previous sections, the processes involved ideal gases, i.e., no interactions and no changes in internal degrees of freedom. In such a process, keeping the temperature constant is equivalent to keeping the total energy constant. Therefore, we could calculate the changes

[22] See Section 5.8 for the form of the function $g(R)$ at low densities. We discuss here only this case where the function $g(R)$ is independent of the density. Such systems can be studied experimentally, for example, gaseous argon.

in MI either from the Sackur–Tetrode equation or from the canonical PF.

In systems of interacting particles, the situation is quite different. The change in the MI depends on whether we carry out the process at constant temperature or at constant energy (isolated system). We demonstrate this difference with the simplest process of expansion. We assume again that the system is dilute enough so that only pairwise interactions are operative.

The mutual information due to intermolecular interactions in the system is given in (6.10.1). We now perform a spontaneous expansion, under two different conditions.

(a) At constant energy (isolated system)

In this process, the spontaneous expansion from volume V to $2V$ causes an increase in the locational MI as we calculated in Section 6.1. However, in contrast to Section 6.1 where we had ideal gas, here, the average intermolecular interaction energy will change in the process. Assuming that (6.10.1) is valid in this process, the number of *pairs* of interacting particles is unchanged in the process. However, the average interaction energy — hence the mutual information — changes in this process by the amount

$$\Delta S = -\frac{N(N-1)}{2(2V)^2}\iint g_f(\boldsymbol{R}_1,\boldsymbol{R}_2)\ln g_f(\boldsymbol{R}_1,\boldsymbol{R}_2)d\boldsymbol{R}_1 d\boldsymbol{R}_2 + \frac{N(N-1)}{2V^2}\iint g_i(\boldsymbol{R}_1,\boldsymbol{R}_2)\ln g_i(\boldsymbol{R}_1,\boldsymbol{R}_2)d\boldsymbol{R}_1 d\boldsymbol{R}_2. \quad (6.10.3)$$

There are two contributions to ΔS in (6.10.3). First, the density of particles changes from the initial density $\rho_i = \frac{N}{V}$ to the final density $\rho_f = \frac{N}{2V}$. Therefore, on average, the intermolecular interactions *decrease* (in absolute magnitude). To achieve that, energy must be invested.

Second, since the total energy is kept constant, the energy required to increase the average separation between the particles must come from the kinetic energy of the particles. Hence, the system's temperature must decrease. This will affect the pair correlation function from the initial value g_i to the final value g_f [see (5.8.19)]. The total change in MI will consist of three contributions. One is due to the changes in the locational MI, the second is due to the changes in the momentum MI, and the third is due to the change in the mutual information due to the interaction of the form (6.10.3).

The average interaction energy in a system of interacting particles is

$$\bar{U} = \frac{\rho^2}{2} \iint U(\boldsymbol{R}_1, \boldsymbol{R}_2) g(\boldsymbol{R}_1, \boldsymbol{R}_2) d\boldsymbol{R}_1 d\boldsymbol{R}_2, \tag{6.10.4}$$

where $U(\boldsymbol{R}_1, \boldsymbol{R}_2)$ in the interaction energy between two particles at $\boldsymbol{R}_1, \boldsymbol{R}_2$. Note that in general this quantity is negative.

(b) At constant temperature

If the system's temperature is maintained constant, then upon a spontaneous expansion, the changes in the locational MI will be the same as in the previous case. However, if we keep the temperature constant, then the pair correlation function will not change in the process. The change in the mutual information will be due only to changes in the density of the particles. Again, energy must be invested in order to increase the average separation between the pairs of particles. Since the temperature is kept constant, this energy must come from the heat reservoir (or from the thermostat). The net effect is that energy will flow into the system to maintain a constant temperature.

6.11 Some Baffling Experiments

We present here two "thought experiments" (which, in principle, could be carried out in practice if the right solutes and solvents, as specified below, could be found) that serve to demonstrate the moral of this chapter: what we observe with our eyes should not, in general, be trusted for reaching thermodynamic conclusions.

We recall the two processes IV and V as discussed in Section 6.7 (see Figures 6.6 and 6.7). In IV, we remove a partition that separates two different components of ideal gases. As is well known, Gibbs' energy change is[23]

$$\Delta G_{IV} = -2NT \ln 2, \tag{6.11.1}$$

[23]In this section, we use a constant temperature system and instead of the MI — we follow the changes in the Gibbs' energy of the system. One could carry out

where N is the number of molecules in each compartment. In process V, where we have the same components in the two compartments, the change in the Gibbs energy is

$$\Delta G_V = 0. \tag{6.11.2}$$

We now modify these experiments and show that in an apparently "no mixing" and "demixing" process, we obtain free energy changes that are *exactly* equal to (6.10.1).

To demonstrate these cases, consider again processes IV and V. Suppose that A and B are colored molecules, say A is blue and B is yellow. Assume also that $N_A = N_B = N$ and the systems are dilute enough so that the gases are ideal. We carry out two experiments in front of a class of students under a constant temperature. Removing the partitions in processes IV and V leads to the following conclusions:

(a) In process IV, we *observe mixing*. The entire system becomes green and we calculate $\Delta G_{IV} = -2NT \ln 2$.
(b) In process V, we *observe no mixing*. The entire system remains blue and we calculate $\Delta G_V = 0$.

After removing the partitions, we reach an equilibrium state. At equilibrium, we can introduce the partitions back into their original places and remove them with no discernible effect on the system and no change in the Gibbs energy.

Next, we place the two partitions back in their original places and modify the experimental situation as follows (Figure 6.16).

We fill the two compartments with two different liquids that are transparent to visible light. We also assume that we have chosen the solutes A and B in such a way that their colors are not affected by the presence of the solvents. We require also that the two liquids be completely immiscible with each other. (We may also use miscible solvents, but then we should require that partitions be permeable to A and B only and not to the solvents.)

---

the same process in an isolated system, but in that case, the temperature of the system would also change due to changes in the total interaction energy in the process. The reading of this section requires some familiarity with the theory of solutions. See, for example, Ben–Naim (2006).

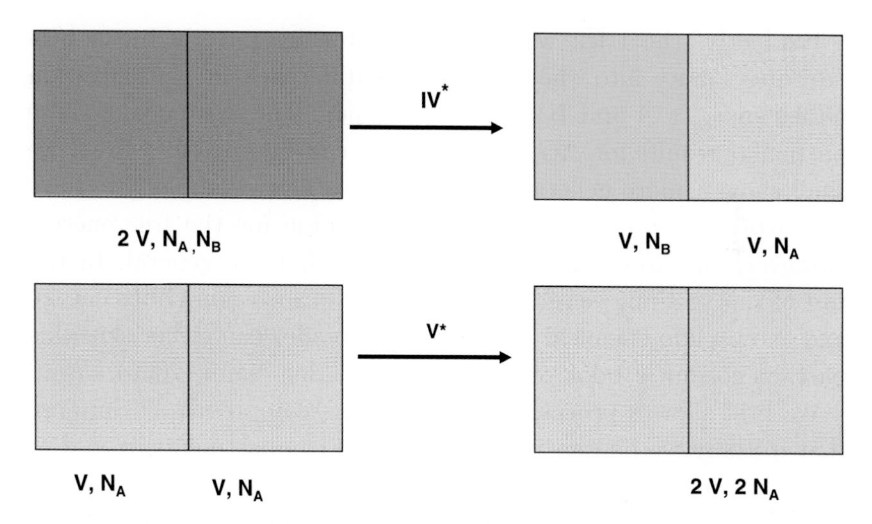

Figure 6.16. Two baffling experiments, described in Section 6.11. In these two processes, the removal of the partition between the two compartment, allow A and B to cross from one side to another, but not the solvents.

We now remove the partitions, the audience being unaware of the presence of the solvents in the two compartments. For the audience, the system looks exactly the same as in the original experiment. We claim that now the removal of the partition has caused the following results [compare with (a) and (b) above]:

(a*) In process IV*, we *observe demixing*. The entire green system separates into yellow and blue. We claim that $\Delta G_{IV*} = -2NT \ln 2$.

(b*) In process V*, we *observe no mixing*. The entire system remains blue, and we claim again that $\Delta G_{V*} = -2NT \ln 2$.

The results given above are correct. It is also true that removing the partitions causes only the flow of A or A and B between the two compartments, but not of the solvents (which are presumed to be immiscible).

Clearly, anyone who observes these experiments must be puzzled if he or she is not aware of the presence of the two solvents. For anyone in the audience, the compartments still contain the ideal gases, and therefore processes IV* and V* are expected to yield positive and zero Gibbs energy changes, respectively.

Evidently, the trick was made possible by introducing a new "driving force" into the system; the difference in the solvation Gibbs energies A and B in the two liquids. The existence of these particular results for ΔG_{IV*} and ΔG_{V*} can be proved.[24] Here, we shall show a more general result. In the above examples, we have endeavored to obtain a predetermined value for the free energy. However, the argument can be made much more general. In the rest of this section, we rely on the concept of solvation Gibbs energy and the pseudo-chemical potential. The reader can either skip this part, or consult a book on solvation, e.g., Ben-Naim (2006).

We first show a process of spontaneous "compression." Suppose that initially we had an ideal gas in both compartments, α and γ, such that initially we have $N_A^{\alpha i} = N_A^{\gamma i} = 1/2N_T$, $V^\alpha = V^\gamma = V$, where N_T is the total number of molecules in the system, $N_T = N_A^\alpha + N_A^\gamma$, $\rho_T = N_T/2V$.

Let us now replace one phase, say α by a transparent solvent. At equilibrium, we have

$$\mu_A^{\alpha f} = \mu_A^{*\alpha} + T\ln\rho_A^{\alpha f} = \mu_A^{*\gamma} + T\ln\rho_A^{\gamma f}, \tag{6.11.3}$$

where $\mu_A^{*\gamma}$ is the pseudo-chemical potential of A in an ideal gas phase, and the solvation Gibbs energy of A in α is, by definition[25]

$$\begin{aligned}\Delta G_A^{*\alpha} &= \mu_A^{*\alpha} - \mu_A^{*\gamma} = T\ln(\rho_A^{\gamma f}/\rho_A^{\alpha f})_{eq}\\ &= T\ln\frac{1-x}{x},\end{aligned} \tag{6.11.4}$$

where $x = N_A^{\alpha f}/N_T = \rho_A^{\alpha f}/2\rho_T$ is the mole fraction of A in the phase α at equilibrium. Clearly, the more negative the solvation Gibbs energy of A in the phase α is (i.e., $\Delta G_A^{*\alpha} < 0$) the smaller the ratio $(1-x)/x$, or the closer x is to unity. By choosing a solvent that strongly solvates A, we can make x as close to unity as we wish. The audience, who is not aware of the presence of a solvent in α (it is a transparent liquid), would be astonished to observe that the gas A that was initially distributed evenly in the two compartments is now concentrated in one compartment α. Clearly, the

[24] Ben-Naim (1987).

[25] See Ben-Naim (2006).

"driving force" that produced this spontaneous "condensation" is the solvation Gibbs energy of A in α.

We can calculate the Gibbs energy change in this process carried out at constant temperature

$$\Delta G = G^f - G^i = [N_A^{\alpha f}\mu_A^{\alpha f} + N_A^{\gamma f}\mu_A^{\gamma f}] - [N_A^{\alpha i}\mu_A^{\alpha i} + N_A^{\gamma i}\mu_A^{\gamma i}],$$

$$\frac{\Delta G}{N_T} = x[\mu_A^{*\alpha} + T\ln\rho_A^{\alpha f}] + (1-x)[\mu_A^{*ig} + T\ln\rho_A^{\gamma f}]$$
$$- \left[\frac{1}{2}(\mu_A^{*ig} + T\ln\rho) + \frac{1}{2}(\mu_A^{*ig} + T\ln\rho)\right], \qquad (6.11.5)$$

where we $\mu_A^{*\gamma} = \mu_A^{*ig}$ and $\rho_A^{\alpha i} = \rho_A^{\gamma i} = \rho_T = N_T/2V$.
Hence, from (6.11.4) and (6.11.5):

$$\frac{\Delta G}{N_T} = x[\mu_A^{*\alpha} - \mu_A^{*ig}] + Tx\ln 2x + T(1-x)\ln 2(1-x)$$
$$= Tx\ln\frac{1-x}{x} + Tx\ln x + T(1-x)\ln(1-x) + T\ln 2$$
$$= T\ln 2(1-x). \qquad (6.11.6)$$

Thus, for each chosen x (determined by choosing ΔG_A^*), we can calculate the corresponding change in Gibbs energy, or $\Delta G/TN_T$. Figure 6.17a shows the plot of $\Delta G/TN_T$ as a function of x. For $x = 0$, we start with the value of ln 2. As expected, the larger x, the larger and negative is $\Delta G/TN_T$. Thus, we see that we can achieve a value of x as close to one as we wish.

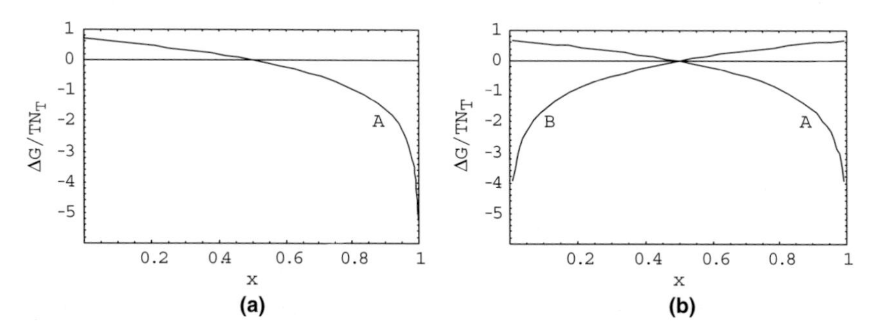

Figure 6.17. (a) The Gibbs energy change for the "compression" process, (6.11.6), (b) The same function for the A and the B components in the de-mixing experiment.

The same argument can be applied to performing a spontaneous demixing process. Simply start with two types of molecules, say A and B, in the two compartments. Then fill one compartment with a transparent solvent that strongly solvates the A molecules, as in the example discussed above. The second compartment can be filled with a liquid that strongly solvates B molecules. In this case, removing the partition between the two compartments will result in "compression" of nearly all A into one compartment, and "compression" of nearly all B in the other compartment. This is effectively a spontaneous demixing of A and B — driven by the solvation Gibbs energy. The change in the Gibbs enery for the two processes of "compression" are shown in Figure 6.17b.

6.12 The Second Law of Thermodynamics

The Second Law of Thermodynamics was originally formulated in terms of some specific experimental observations such as the maximal efficiency of a heat engine operating between two temperatures, the spontaneous heat flow from a hot to a cold body, or the spontaneous expansion of an ideal gas. Clausius' formulation of the Second Law in terms of the entropy is much more general and refers to any spontaneous process occurring in isolated systems. It states that there exists a function S, which Clausius called entropy (but here referred to as MI), that is a state function, and that in an isolated system never decreases.

In the classical non-atomistic formulation[26] of the Second Law, entropy *always* increases in a spontaneous process in an isolated system, and it stays constant *forever* once it reaches the equilibrium state. The law is stated in absolute terms.

The atomistic, or Boltzmann's, formulation of the Second Law, is essentially the same except for the "absoluteness" of its validity. Instead of the terms "*always*" and "*forever*," we must use the softer terms "almost always" and "almost forever."

Both formulations of the Second Law, the non-atomistic as well as the atomistic, consist of statements about *what* is the thing that

[26]Here, classical refers to the original non-atomistic formulation. See, for example, Denbigh (1966).

changes, and in which direction that thing changes in a spontaneous process. None offers any *explanation* to the question of *why* the entropy changes in one direction; it makes no difference whether one refers to entropy as a measure of disorder or a measure of MI, all these are only possible descriptions of the *thing* that changes. It is not true, as claimed in some textbooks, that a system left unattained, or undisturbed, such as a child's bedroom or the books in a library, will tend to a state of greater disorder.

The only convincing answer to the question "why" is probabilistic. It is this probabilistic answer that also reduces the Second Law to a matter of plain common sense.[27] Basically, it says that the system will go from a relatively low probability state to a high probability state. If we accept the frequency interpretation of probability, then the last statement is tantamount to saying that a single system will spend a larger fraction of the time in states that have larger probability, i.e., states having larger frequency of occurence.

As it stands, the Boltzmann equation does not contain the word nor the concept of probability explicitly. It connects between the MI (or the entropy), and the total number of states. We also know that the MI is *defined* for any distribution $(p_1, \ldots, p_n)$. It is an average quantity $-\sum p_i \ln p_i$. The principle of maximum MI states that there exists a distribution $(p_1^*, \ldots, p_n^*)$ that maximizes the MI, i.e., $S(p_1, \ldots, p_n)$ has a maximum at the specific distribution $(p_1^*, \ldots, p_n^*)$. This statement, although formulated in terms of the probability distribution, does not consist of a probabilistic interpretation of the Second Law. The latter is provided by a probability distribution that is *defined on* the probability distribution $(p_1, \ldots, p_n)$. We shall denote the latter probability by **Pr**. It is this probability that attains a maximum at the equilibrium state, which is characterized by the distribution $(p_1^*, \ldots, p_n^*)$, and we shall denote it by $\mathbf{Pr}(p_1^*, \ldots, p_n^*)$. The existence of these "two levels" of distributions can be potentially confusing. Therefore, we shall refer to $(p_1, \ldots, p_n)$ as the *state* distribution, and to **Pr** as the *super* probability.

We next describe a few examples which will clarify the connection between the state distribution $(p_1, \ldots, p_n)$, which describes the

[27] See Ben-Naim (2007).

state of the system or of the ensemble, on the one hand, and the quantity MI, which is defined *on* the distribution $(p_1, \ldots, p_n)$, and provides an answer to the question of "what" is the thing that is changing in a spontaneous process on the other hand. On the same distribution $(p_1, \ldots, p_n)$, we define the super probability function, denoted $\mathbf{Pr}(p_1, \ldots, p_n)$, which harbors the answer to the question of "*why*," and hence provides a probabilistic explanation of the Second Law of Thermodynamics.

(i) *Ideal gas in two compartments*

We start with a classical system of N non-interacting particles (ideal gas) in a volume $2V$ at constant energy E. We divide the volume of the system into two parts, each of volume V (Figure 6.18). We define the *microscopic* state of the system when we are given $E, 2V, N$, and in addition we know which specific particles are in compartment R, and which specific particles are in compartment L (Figure 6.18a). The *macroscopic* description of the same system is $(E, 2V, N; n)$, where n is the *number* of particles in the compartment L (Figure 6.18b). Thus, in the microscopic description, we are given a specific configuration of the system as if the particles were labeled $1, 2, \ldots, N$, and we are given the specific information about which particle is in which compartment. In the macroscopic description, only the number n is specified.

Clearly, if we know only that n particles are in L and $N - n$ particles are in R, we have

$$W(n) = \frac{N!}{n!(N-n)!}, \tag{6.12.1}$$

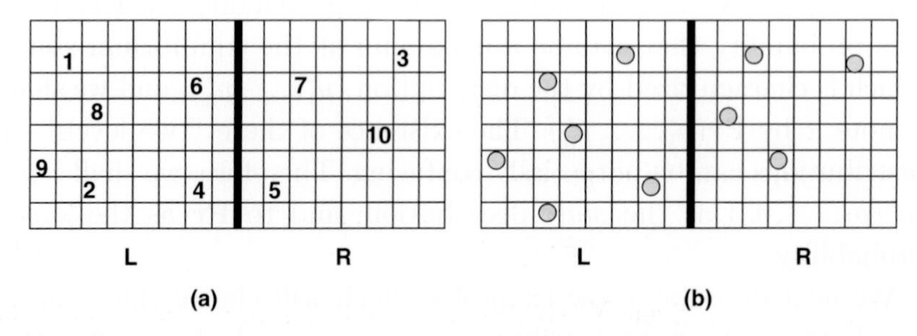

Figure 6.18. (a) A specific or microscopic description of the state of the system, (b) A generic, or a macroscopic description of the system.

specific configurations that are consistent with the requirement that there are n particles in L.

The first postulate of statistical mechanics states that all *specific* configurations of the system are equally probable. Clearly, the total number of specific configurations is

$$W_T = \sum_{n=0}^{N} W(n) = \sum_{n=0}^{N} \frac{N!}{n!(N-n)!} = 2^N. \tag{6.12.2}$$

Using the *classical* definition of probability (see Section 2.3), we can write the probability of finding n particles in L as

$$P_N(n) = \frac{W(n)}{W_T} = \left(\frac{1}{2}\right)^N \frac{N!}{n!(N-n)!}. \tag{6.12.3}$$

It is easy to show that $W(n)$, or $P_N(n)$ has a maximum as a function of n. The condition for the maximum is

$$\begin{aligned} \frac{\partial \ln W(n)}{\partial n} &= \frac{1}{W(n)} \frac{\partial W(n)}{\partial n} \\ &= -\ln n^* + \ln(N - n^*) = 0 \end{aligned} \tag{6.12.4}$$

or equivalently

$$n^* = \frac{N}{2}. \tag{6.12.5}$$

This is a maximum since

$$\frac{\partial^2 \ln W(n)}{\partial n^2} = \frac{-1}{n} - \frac{1}{N-n} = \frac{-N}{(N-n)n} < 0. \tag{6.12.6}$$

Thus, the function $W(n)$ or $P_N(n)$ has a maximum with respect to n (keeping $E, 2V, N$ constant). The value of the maximum of $W(n)$ is

$$W(n^*) = \frac{N!}{[(N/2)!]^2}, \tag{6.12.7}$$

and the corresponding probability is

$$P_N(n^*) = \frac{W(n^*)}{2^N}. \tag{6.12.8}$$

Thus, for any given N, there exists a maximum of the probability $P_N(n)$. Therefore, if we prepare a system with any initial distribution of particles n and $N - n$ in the two compartments, and let the system evolve, the system's *macroscopic* state (here described by the parameter n) will change from a state of lower probability to a state of higher probability. We can perform the process in a quasi-static way by opening and closing a small window between the two compartments. [For a detailed description of the transition probabilities at cach stage of the process, see Ben-Naim (2007).]As can be seen in Figure 6.19, the maximal value of the probability $P_N(n^*)$ is a *decreasing* function of N. Thus, for any starting value of n, the system will evolve towards n^* [this tendency will be very strong for large deviations from n^*, and weaker as we reach closer to n^*; for a detailed example, see Ben-Naim (2007)]. However, as N increases, the value of the maximum number of configurations $W(n^*)$ *increases* with N, but the value of the maximal probability $P_N(n^*)$ *decreases* with N.[28]

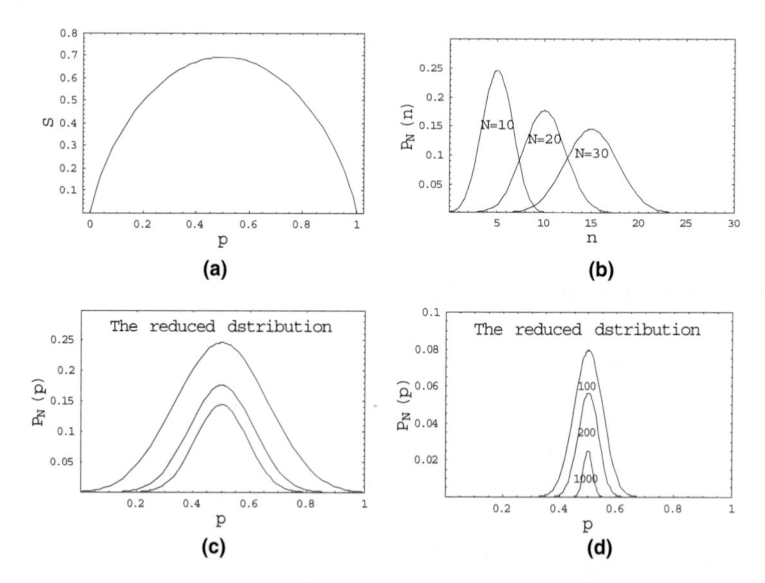

Figure 6.19. (a) The function $S(p, 1-p)$, (b) The distribution (6.12.3) for different values of N, (c) The same distribution as in (b), but as a function of the reduced variable $p = n/N$ and (d) The same distributions as in (c), but for $N = 100$, 200 and 1000.

[28]This fact has caused some confusion in the literature. Though it is true that as N becomes larger, $W(n^*)$ *increases* with N, the denominator of (6.12.8) 2^N increases with N faster than $W(n^*)$.

To appreciate the significance of this fact, consider a few cases:

(a) $N = 2$
Suppose we have a total of $N = 2$ particles. In this case, we have the following possible macroscopic descriptions and the corresponding probabilities:

$$\begin{array}{ccc} n = 0, & n = 1, & n = 2, \\ P_N(0) = \frac{1}{4}, & P_N(1) = \frac{1}{2}, & P_N(2) = \frac{1}{4}. \end{array} \tag{6.12.9}$$

This means that on average, we can expect to find the configuration $n = 1$ (i.e., one particle in each compartment) about half of the time, but each of the configurations $n = 0$ and $n = 2$ only a quarter of the time.

(b) $N = 4$
For the case $N = 4$, we have a distribution with a maximal probability $P_N(2) = 6/16$, which is *smaller* than $\frac{1}{2}$. In this case, the system will spend only 3/8 of the time in the maximal state $n^* = 2$.

(c) $N = 10$
For $N = 10$, we can calculate the maximum at $n^* = 5$, which is

$$P_{10}(n^* = 5) = 0.246.$$

Thus, as N increases, $W(n^*)$ increases, but $P_N(n^*)$ decreases. For instance, for $N = 1000$, the maximal probability is only $P_{1000}(n^*) = 0.0252$. See Figure 2.13.

To examine how $P_N(n^*)$ changes with N, we use the Stirling approximation (Appendix E) in the form

$$n! \cong \left(\frac{n}{e}\right)^n \sqrt{2\pi n} \tag{6.12.10}$$

and get

$$P_N\left(n^* = \frac{N}{2}\right) \approx \sqrt{\frac{2}{\pi N}}. \tag{6.12.11}$$

Thus, as N increases, the maximal probability decreases as $N^{-1/2}$. In practice, we know that when the system reaches the state of equilibrium, it stays there "forever." The reason is that the macroscopic state of equilibrium is not the same as the exact

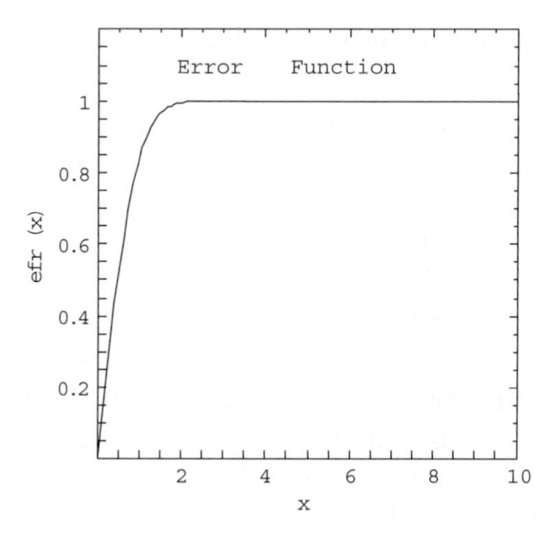

Figure 6.20. The function *erf*(x).

state for which $n^* = \frac{N}{2}$, but this state along with a small neighborhood of n^*, say $n^* - \delta N \leq n \leq n^* + \delta N$, where δ is small enough such that no experimental measurement can detect the difference between n^* and $n^* \pm \delta N$. (Note that δN means $\delta \times N$ and not a small increment in N.)To calculate the probability of finding the system in the neighborhood of n^*, we use the approximation.

$$
\begin{aligned}
&P_N(n^* - \delta N \leq n \leq n^* + \delta N) \\
&\quad = \sum_{n=n^*-\delta N}^{n^*+\delta N} P_N(n) \\
&\quad \approx \int_{n=n^*-\delta N}^{n=n^*+\delta N} \frac{1}{\sqrt{\pi N/2}} \exp\left(\frac{-\left(n - \frac{N}{2}\right)^2}{N/2}\right) dn. \qquad (6.12.12)
\end{aligned}
$$

This is the error function *erf*$(\sqrt{2\delta N})$ (see Section 2.10). Figure 6.20 shows that the function *erf*(x) is almost unity when N is of the order of 10. This means that when N is of the order of 10^{23}, we can allow deviations of $\delta \approx 0.001$, or even smaller, yet the probability of finding the system at, or near, n^* will be almost one. It is for this reason that when the system reaches the neighborhood of n^*, it will stay at n^* or near n^* for most of the time. For N of the order of 10^{23}, "most of the time" means *always*.

The above-mentioned specific example provides an explanation for the fact that the system will "always" change in one direction, and *always* stay at the equilibrium state once that state is reached. The tendency towards a state of larger probability is equivalent to the statement that events that are supposed to occur more frequently will occur more frequently. This is plain common sense. The fact that we do not observe deviations from either the monotonic change of n towards n^*, nor deviations from n^*, is a result of our inability to detect very small changes in n.

In the above example, we have given an *explanation* of the Second Law in terms of probabilities. We shall next turn to the relationship between these probabilities and the MI. Before doing that, let us reformulate the problem in such a way that it will be easy to generalize it. Instead of n and $N-n$, we define the fractions p and q:

$$p = \frac{n}{N}, \quad q = \frac{N-n}{N}, \tag{6.12.13}$$

i.e., p is the fraction of particles in the L compartment and q, the fraction in the R compartment. Clearly, (p, q) with $p + q = 1$ is a probability distribution.[29] We now proceed to calculate the super probability defined on this probability distribution.

The number of specific configurations for which there are n particles in L is

$$W(p, q) = \binom{N}{pN}, \tag{6.12.14}$$

and the corresponding super probability is

$$\mathbf{Pr}(p, q) = \left(\frac{1}{2}\right)^N \binom{N}{pN}. \tag{6.12.15}$$

Note that $\mathbf{Pr}(p, q)$ is the same distribution as in (6.12.3), but here only expressed in term of the distribution (p, q). For each distribution (p, q), we can define the measure of MI as

$$H(p, q) = -p \ln p - q \ln q. \tag{6.12.16}$$

[29]Note that at equilibrium the probability of finding a *specific* particle in L is $1/2$. However, if we prepare the system with n particles in L, then the probability of finding a specific particle in L is n/N.

When $N \to \infty$, we can use the Stirling approximation to rewrite the right-hand side of (6.12.15) as

$$\ln \binom{N}{pN} \xrightarrow{N\to\infty} -N[p\ln p + q\ln q] - \ln\sqrt{2\pi Npq}$$
$$= NH(p,q) - \frac{1}{2}\ln(2\pi N pq). \qquad (6.12.17)$$

Hence, in this approximation

$$\mathbf{Pr}(p,q) \cong \left(\frac{1}{2}\right)^N \frac{\exp[NH(p,q)]}{\sqrt{2\pi Npq}} \quad \text{(for large } N\text{)}. \qquad (6.12.18)$$

This is the relationship between the MI defined on the distribution (p,q), and the super probability $\mathbf{Pr}(p,q)$ defined on the same distribution. Clearly, these two functions are very different (see Figure 6.19). $H(p,q)$ has a maximum at the point $p^* = q^* = 1/2$. This maximum is independent of N. On the other hand, $\mathbf{Pr}(p,q)$ has a sharp maximum at the point $p^* = q^* = 1/2$. The value of the maximum $\mathbf{Pr}(\frac{1}{2},\frac{1}{2})$, decreases with N.

We stress again that H is defined for any arbitrary state distribution $(p,\ q)$. To relate changes in H to changes in the thermodynamic entropy, we have to perform the transition from the initial state $(n, N-n)$ [or the distribution (p,q)], to the final equilibrium state $(N/2, N/2)$[or the distribution $(1/2, 1/2)$], in a quasi-static process. Suppose we open a small window between the two compartments, and allow for an infinitesimal change dn, we have from (6.12.16), after substituting (6.12.13), $dH = \frac{\partial H}{\partial n}dn = \frac{dn}{N}\ln\frac{N-n}{n}$. When $N-n > n$, $dn > 0$ and $dH > 0$; also when $N-n < n$, then $dn < 0$ and $dH > 0$. This is exactly the same entropy change calculated from either the partition function or the Sackur–Tetrode equation for an ideal gas system. The result for the change in entropy per particle is: $dS = \frac{dn}{N}\ln\frac{N-n}{n}$.

Thus, the change in the MI is the same as the change in entropy calculated from the Sackur–Tetrode equation. This identification is rendered possible only when we make an infinitesimal change in n, so that the process, from the initial to the final states, may be viewed as sequence of equilibrium states. We note also that at the final equilibrium state, we place the partition between the two compartments. The net change in H, or in S, is due only to the change in the number of indistinguishable particles.

We can now summarize our findings for this specific example as follows. For any state distribution (p, q) with $p + q = 1$, we can define the MI by (6.12.16). When N is macroscopically large, changes in MI turn into changes in the thermodynamic MI. This function has a maximum at $p^* = q^* = 1/2$ — hence the statement that the macroscopic system will tend towards the maximum value of S (consistent with the fixed values of E, V, N). This provides an explanation of "what" is the thing that changes in this process. On the other hand, the super probability $\mathbf{Pr}(p, q)$ in (6.12.15), also has a sharp maximum at $p^* = q^* = 1/2$. This fact provides the explanation "why" the system will move towards the state distribution $p^* = q^* = 1/2$. Thus, for any initially prepared distribution (p, q), the system will evolve towards the distribution (p^*, q^*) for which the super probability $\mathbf{Pr}(p, q)$ has a maximum. This is also the distribution for which the MI, or the entropy, is maximum, for a system characterized by the variables E,V,N. For very large N, the relation between the two functions is (6.12.18).

Once the distribution reaches the state for which $p^* = q^* = 1/2$, or nearly this state, it will practically stay in this neighborhood forever. There will be fluctuations away from (p^*, q^*). Small fluctuations will be frequent, but unobservable and un-measurable. Large fluctuations are in principle measurable, but are extremely improbable — hence almost never observable.

(ii) *Ideal gas in r compartments*

We briefly discuss here a generalization of the previous case. Instead of two compartments, we now have r compartments each of volume V. The whole system contains N particles and has a fixed energy E. This case does not add any new idea. However, it facilitates the generalization discussed in subsections (iii).

Clearly, we can *prepare* a system with any prescribed distribution $N_1, N_2, \ldots, N_r$, $(\sum N_i = N)$, where N_i is the number of particles in the compartment i (Figure 6.21). We also know that once we remove the partitions between the compartments, the system will

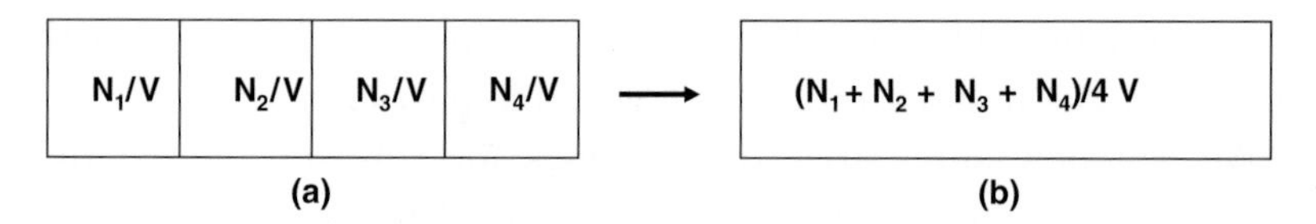

Figure 6.21. (a) An initial prepared distribution (different densities in each volume) and (b) The final equilibrium uniform density.

evolve from the initial distribution to the final uniform distribution $N_1 = N_2 = \cdots = N_r = N/r$.

Following similar reasoning as in the previous example, we define the state distribution $p_1, \ldots, p_r$, such that $p_i = \frac{N_i}{N}$ is the fraction of particles in the ith compartment $(\sum p_i = 1)$.

The corresponding MI is

$$H(p_1, \ldots, p_r) = -\sum p_i \ln p_i. \tag{6.12.19}$$

This quantity is defined for any distribution $p_1, \ldots, p_r$. At the moment of removal of the partitions between the r compartments, the quantity p_i is the probability of finding a specific particle in the ith compartment. This probability distribution changes with time towards the distribution $(p_1^*, \ldots, p_r^*)$ for which the super probability function

$$\mathbf{Pr}(p_1, \cdots p_r) = \left(\frac{1}{r}\right)^N \frac{N!}{\prod_{i=1}^r N_i!} \tag{6.12.20}$$

has a maximum. This probability provides the "driving force"[30] for the evolution of the system from the initial distribution $p_1, \ldots, p_r$, having lower probability, to the final distribution $(p_1^*, \ldots, p_r^*)$. It turns out that the distribution $p_1^* = p_2^* = p_r^* = \frac{1}{r}$, which maximizes $\mathbf{Pr}$, is also the distribution that maximizes the MI in (6.12.19); the connection between the two functions for $N \to \infty$ is

$$\mathbf{Pr}(p_1, \ldots, p_r) \cong \left(\frac{1}{r}\right)^N \frac{\exp[NH(p_1, \ldots, p_r)]}{\sqrt{(2\pi N)^{r-1} p_1 \times p_2 \times \cdots \times p_r}}. \tag{6.12.21}$$

[30]In this section, we use the phrase "driving force" in its colloquial sense. This is not a *force* in the physical sense.

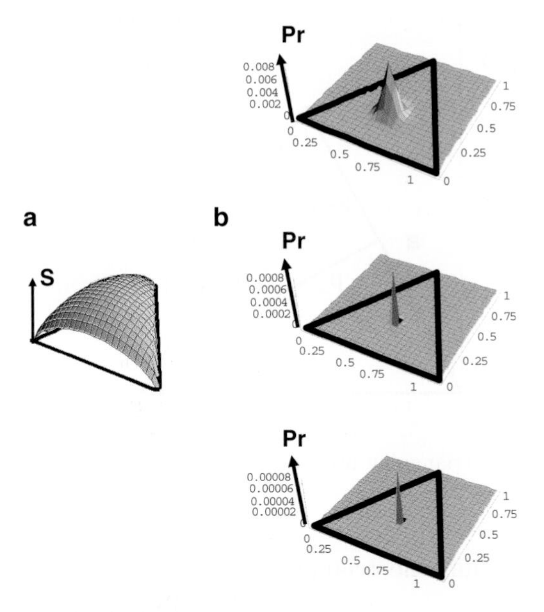

Figure 6.22. (a) The MI (6.12.19), for the case $r = 3$ and (b) The probability distribution, (6.12.20), for $r = 3$ and three values of N:100, 1000 and 10000. The horizontal axes are as in Figure 6.23

At the point $p_1^* = p_2^* = \cdots = p_r^* = \frac{1}{r}$, (6.12.21) reduces to

$$\mathbf{Pr}\left(\frac{1}{r}, \ldots, \frac{1}{r}\right) = \left(\frac{1}{r}\right)^N \frac{r^N}{\sqrt{(2\pi N)^{r-1}(1/r)^r}}$$

$$= \sqrt{\frac{r^r}{(2\pi N)^{r-1}}}. \tag{6.12.22}$$

This quantity sharply decreases with N. However, if we take a neighborhood of the point $p_1^* = p_r^* = \cdots = p_r = \frac{1}{r}$, the probability of finding the system at, or near, the point $p_1^* = p_r^* = \cdots = p_r = \frac{1}{r}$ tends to *unity* as N increases and becomes macroscopically large. In Figure 6.22, we show the function $\mathbf{Pr}(p_1, p_2, p_3)$ with $p_1 + p_2 + p_3 = 1$, defined in (6.12.20) and the function $S(p_1, p_2, p_3)$ defined in (6.12.19) for the case $r = 3$. The two functions are plotted over the equilateral triangle region whose edge is a $a = 2/\sqrt{3}$. With this choice, any point within the triangle represents a triplet of numbers p_1, p_2, p_3, such as $\sum p_i = 1$. The distances between

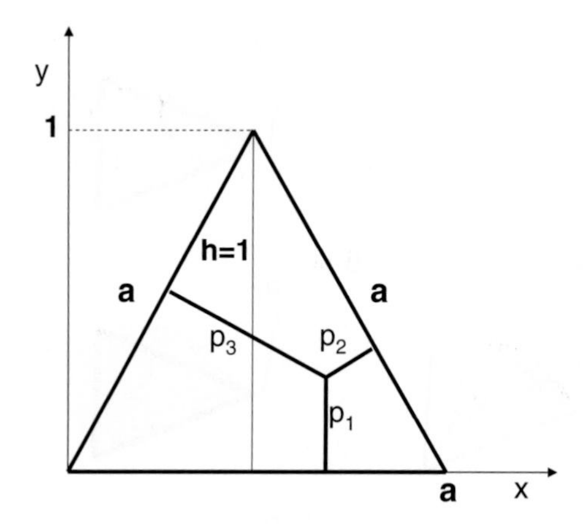

Figure 6.23. The triangular area over which the S and Pr are plotted In Fig. 6.22.

any point p within the triangle and the three edges is shown in Figure 6.23.

The transformation between p_1, p_2 and p_3, and the Cartesian coordinates is[31]

$$\begin{aligned} p_1 &= y, \\ p_2 &= \left(a - x - \frac{y}{\sqrt{3}}\right)\frac{\sqrt{3}}{2}, \\ p_3 &= 1 - p_1 - p_2. \end{aligned} \tag{6.12.23}$$

The range of the variations of x and y is

$$\begin{gathered} 0 \leq y \leq 1, \\ \frac{y}{\sqrt{3}} \leq x \leq a - \frac{y}{\sqrt{3}}. \end{gathered} \tag{6.12.24}$$

As in the case of $r = 2$, here also we see that as N increases the distribution becomes sharper, and the value of the maximum probability becomes smaller (note the scale of the probability axis in Figure 6.22).

[31] These relationships are based on a theorem of geometry. The sum of the three perpendicular lines p_1, p_2 and p_3 in Figure 6.23 is equal to h (here $h = 1$).

We note again that the changes of H as defined in (6.12.19) for a quasi-static process is

$$dH = \sum_{i=1}^{r} \frac{\partial H}{\partial N_i} dN_i = \frac{-1}{N} \sum_{i=1}^{r} dN_i \ln N_i.$$

This is the same change in the entropy of the system calculated from either the partition function or from the Sackur–Tetrode equation (applied for each compartment). The result is

$$dS = \frac{-1}{N} \sum_{i=1}^{r} dN_i \ln N_i.$$

Note that if at each step of the process, we close the windows between the compartments, the net change in the MI or the entropy will be due to changes in the number of indistinguishable particles only.

(iii) *The more general case*
In the previous examples, we had a classical system described by the locations and momenta of all the particles. We also restricted ourselves to changes in the locational MI only. In the more general case, the distribution of momenta might also changes (e.g., in heat transfer from a hot to a cold body). Yet, in a more general case, the system is described in terms of its quantum mechanical states.

Suppose we have again a system of N particles contained in a volume V and with a fixed total energy E. The second postulate of statistical mechanics states that all the microstates states $i(i = 1, 2, \ldots, W_T)$ of a system characterized by E, V, N at equilibrium have equal probability $1/W_T$. For this equilibrium state, the Boltzmann relation is

$$S = \ln W_T. \qquad (6.12.25)$$

In the previous examples, we have *prepared* a single thermodynamic system at some arbitrary distribution of particles in the various compartments. We then removed the constraints (i.e., removed the partitions between the compartments), and followed the evolution of the system. For a general system of interacting particles, it is impossible to prepare a *single* system with an initial arbitrary distribution $(p_1, \ldots, p_{W_T})$. Instead, we can imagine an ensemble of

M systems all characterized by the same thermodynamic variables E, V, N. In this ensemble, we can imagine that we have *prepared* the ensemble in such a way that M_1 systems are in the quantum mechanical state 1, M_2 in state 2, and so on. Thus, the total number of systems in the ensemble is[32]

$$M = \sum_{i=1}^{W_T} M_i. \tag{6.12.26}$$

Clearly, such an ensemble does not represent an equilibrium system characterized by E, V, N. Note that this ensemble is not the micro-canonical ensemble discussed in Chapter 5. Here, each system is in a single quantum mechanical state. This is a highly hypothetical ensemble. Recall that even the micro-canonical ensemble is highly hypothetical. There exists no absolute isolated system with exactly fixed energy E. For any initially *prepared* ensemble with distribution $(p_1, \ldots, p_{W_T})$, where $p_i = \frac{M_i}{M}$ is the fraction of systems in the ensemble being in state i,[33] we can write the super probability distribution is

$$\begin{aligned} \mathbf{Pr}(p_1, \ldots, p_{W_T}) &= \frac{W(M_1, \ldots, M_{W_T})}{\sum W(M_1, \ldots, M_{W_T})} \\ &= \left(\frac{1}{W_T}\right)^M \frac{M!}{\prod_{i=1}^{W_T} M_i!}. \end{aligned} \tag{6.12.27}$$

This super probability follows from the second postulate of statistical mechanics and from the classical definition of probability.

The MI defined on the same distribution $p_1, \ldots, p_{W_T}$ is

$$H(p_1, \ldots, p_{W_T}) = -\sum_{i=1}^{W_T} p_i \ln p_i. \tag{6.12.28}$$

[32]This hypothetical ensemble can be constructed from the micro-canonical ensemble as follows: We make a measurement on each of the systems, and determine its quantum mechanical state, and then we choose M_1 systems in state 1, M_2 systems in state 2, and so on. In this way, we get the new hypothetical ensemble.

[33]p_i is the probability of finding a system, picked at random from the ensemble, in state i.

It should be noted that there is a fundamental difference between this case and the previous two cases. In the previous cases, it was possible to prepare a system at any distribution $N_1, \ldots, N_r$ which is at equilibrium. We can also perform a quasi-static precess of removing the constraint on the initial distribution towards the final distribution $(1/r, 1/r, \ldots, 1/r)$. For example, we can open and close small windows between the compartments so that at each stage of the process, each compartment is in an internal equilibrium state. Hence, the thermodynamic entropy is defined at each stage of the quasi-static process.

In the general case, each system in the ensemble is in one specific quantum mechanical state. This is clearly not an equilibrium state. Hence, the thermodynamic entropy is not defined for the initially prepared system.

Nevertheless, the measure of MI is definable for such an ensemble since this is definable for any distribution $p_1, \ldots, p_r$. As we shall see below, the relation between the MI and the super probability is the same as in the previous cases. However, the MI is equal to the thermodynamic entropy, neither in the initial state nor along the way towards the final equilibrium state. It is only at the final equilibrium state that the MI, denoted H, becomes identical to the thermodynamic entropy S.

The argument regarding the evolution of the ensemble's distribution towards equilibrium is essentially the same as in the previous examples. The number of particles is now replaced by the number of systems in the ensemble.

If we apply a small perturbation to each of the systems, the proposed ensemble will evolve forwards the distribution $p_1^* = p_2^* = \cdots = p_{W_T}^*$ that maximises both the super distribution and the MI. The maximal value of H is the thermodynamic MI, which we denote by S and is given by

$$\begin{aligned} S = H\left(\frac{1}{W_T}, \ldots, \frac{1}{W_T}\right) &= -\sum_{i=1}^{W_T} \frac{1}{W_T} \ln \frac{1}{W_T} \\ &= \ln W_T, \end{aligned} \tag{6.12.29}$$

which is the Boltzmann relation (6.12.25). Note that only the maximal value of H is identified with the thermodynamic MI.

So far we have made a statement about what the *thing* that changes and attains a maximum value at equilibrium is. The "driving force" for this change is the super probability $\mathbf{Pr}(p_1, \ldots, p_{W_T})$. This function has a sharp maximum for the same distribution $p_i^*, \ldots, p_{W_T}^*$ that maximizes the MI. The value of the maximal probability is

$$\mathbf{Pr}(p_i^*, \ldots, p_{W_T}^*) = \left(\frac{1}{W_T}\right)^M \frac{M!}{\prod_{i=1}^{W_T}(p_i^* M)!}. \tag{6.12.30}$$

The connection between the super probability function $\mathbf{Pr}(p_1, \ldots, p_{W_T})$ and the MI function $H(p_1, \ldots, p_{W_T})$ is formally the same as in previous sections. We take the limit $M \to \infty$ and use the Stirling approximation for all the $M_i!$ to obtain

$$\mathbf{Pr}(p_1, \ldots, p_{W_T}) = \left(\frac{1}{W_T}\right)^M \frac{\exp[MH(p_1, \ldots, p_{W_T})]}{\sqrt{(2\pi M)^{W_T - 1} \prod_{i=1}^{W_T} p_i}}. \tag{6.12.31}$$

Again, we conclude that the MI $MH(p_1, \ldots, p_{W_T})$ is defined for any arbitrary distribution $(p_1, \ldots, p_{W_T})$, with $\sum p_i = 1$. It attains the thermodynamic value of the MI $MH(p_1^*, \ldots, p_{W_T}^*)$ for the specific distribution $p_i^* = W_T^{-1}$. The "driving force" for the evolution of the ensemble's distribution towards the equilibrium distribution $p_1^*, \ldots, p_{W_T}^*$ is the super probability $\mathbf{Pr}(p_1, \ldots, p_{W_T})$. This has a sharp maximum at $p_1^*, \ldots, p_{W_T}^*$. The value of the maximal probability for $M \to \infty$, is

$$\mathbf{Pr}(p_1^*, \ldots, p_{W_T}^*) = \sqrt{\frac{W_T^{W_T}}{(2\pi M)^{W_T - 1}}}. \tag{6.12.32}$$

This is the general relation between the super probability at equilibrium and the total number of states of a single system. The latter is related to the MI by (6.12.29).

To summarize this section, we note that (6.12.31) can be viewed in two ways, or in two directions. One can formulate the Second Law by stating that the entropy (or the MI) increases in a spontaneous process in an isolated system. Such a formulation does not provide any explanation of the Second Law. It reaches a dead end, and it does not matter whether you refer to S as entropy, disorder or MI.

The second view is to start with a description of the ensemble and the corresponding super probability as in (6.12.27). We then interpret the probability of an event as the fraction of time the system will spend at that event. Therefore, the system will spend more time at events (or states) of higher probability. This, in itself, provides an explanation of the "driving force" for the evolution of the system; from a state of low to a state of high probability. Equation (6.12.31) translates the last statement from the language of probability into the language of the entropy (or the MI). In this view, we have not only an interpretation for the quantity S as MI, but also an answer to the question of "why" the system evolves in the specific direction of increasing the entropy (or the MI). In this view, the statement of the Second Law is reduced to a statement of common sense, nothing more [for an elementary discussion of this topic, see Ben-Naim (2007)].

A very common way of explaining the Second Law is to use the example of the scattered pages of a novel, say Tolstoy's "War and Peace." It is true that there are many more ways of collecting the scattered pages of the book in the wrong order than in the right order. This is also a good example of an extremely improbable event (collecting the pages in the exact order of the book). However, this example might be misleading if used to explain the Second Law of Thermodynamics. Clearly, the "right order" of the pages of the book depends on the content written on each of the pages. A person who does not know the language in which the book is written might not distinguish between the right and the wrong "order" of the pages. From such considerations, it is easy to reach the conclusion that entropy or MI is a subjective quantity.

Note, however, that the "original order" of the pages of any book is meaningful even when the pages are all blank, or contain the same content. Since the pages of a book are classical objects, they are distinguishable. Therefore, it is highly improbable to collect the scattered pages of a book into the "original order", just as it is highly improbable to retrieve the tumblerful of water from the sea (see page 282). If the pages of the book were indistinguishable (see Appendix J) then the probability of collating the pages in the "original order" becomes one!

Finally, one should realize that the Second Law is crucially dependent on the atomic constituency of matter. If matter were not made-up of a huge number of atoms, the Second Law would not have existed, and the entropy would not have been defined. The same is true for the concepts of temperature and heat. Hence, also the Zeroth and the First Law of Thermodynamics would not have been formulated.

The atomic constitution of matter is perhaps the most important discovery of science. On this matter, I cannot do any better then citing Feynman (1996):

> "*If, in some cataclysm, all scientific knowledge were to be destroyed, and only one sentence passed on to the next generations of creatures, what system would contain the most information in the fewest words?*
>
> *I believe it is the atomic hypothesis (or atomic fact, or whatever you wish to call it) that all things are made of atoms.*"

Appendices

Appendix A Newton's binomial theorem and some useful identities involving binomial coefficients

The binomial coefficient is defined by

$$\binom{n}{i} = \frac{n!}{i!(n-i)!}. \tag{A.1}$$

The Newton bionomial theorem is

$$(x+y)^n = \sum_{i=0}^{n} \binom{n}{i} x^i y^{n-i}. \tag{A.2}$$

The proof, by mathematical induction on n, is quite straightforward. For $n = 1$ and $n = 2$, we have the familiar equalities

$$(x+y)^1 = x^1 y^0 + x^0 y^1 = x + y, \tag{A.3}$$

$$(x+y)^2 = x^2 + 2xy + y^2. \tag{A.4}$$

Assuming that (A.1) is true for n, it is easy to show that it is also true for $n+1$:

$$(x+y)^{n+1}$$

$$= (x+y)^n(x+y) = (x+y)\sum_{i=0}^{n} \binom{n}{i} x^i y^{n-i}$$

$$
\begin{aligned}
&= \sum_{i=0}^{n} \binom{n}{i} x^{i+1} y^{n-i} + \sum_{i=0}^{n} \binom{n}{i} x^{i} y^{n+1-i} \\
&= \sum_{l=1}^{n+1} \binom{n}{l-1} x^{l} y^{n-l+1} + \sum_{i=0}^{n} \binom{n}{i} x^{i} y^{n+1-i} \\
&= \sum_{i=1}^{n+1} \binom{n}{i-1} x^{i} y^{n-i+1} + \sum_{i=0}^{n} \binom{n}{i} x^{i} y^{n+1-i} \\
&= \sum_{i=1}^{n} \binom{n}{i-1} x^{i} y^{n-i+1} + \sum_{i=1}^{n} \binom{n}{i} x^{i} y^{n+1-i} + x^{n+1} + y^{n+1} \\
&= \sum_{i=1}^{n} \left[\binom{n}{i-1} + \binom{n}{i} \right] x^{i} y^{n-i+1} + x^{n+1} + y^{n+1} \\
&= \sum_{i=0}^{n+1} \binom{n+1}{i} x^{i} y^{n-i+1} = (x+y)^{(n+1)} \qquad \text{(A.5)}
\end{aligned}
$$

Note that in the last step, we used identity (A.8) below.

The generalization of (A.1) to m variables is called the multinational theorem. Denote

$$
\binom{n}{i_1, i_2, i_3, \ldots, i_m} = \frac{n!}{\prod_{k=1}^{m} i_k!}. \qquad \text{(A.6)}
$$

The multinomial theorem is

$$
\left(\sum_{i=1}^{m} x_i \right)^n = \sum_{i_1, i_2, \ldots, i_m} \binom{n}{i_1, i_2, \ldots, i_m} \prod_{k=1}^{m} x_k^{i_k}, \qquad \text{(A.7)}
$$

where the sum is over all sequences $i_1, i_2, \ldots, i_m$ such that $\sum_{k=1}^{m} i_k = n$.

Some useful identities involving binomial coefficients are:

$$
\binom{n}{i-1} + \binom{n}{i} = \binom{n+1}{i} \qquad \text{(A.8)}
$$

or equivalently

$$
\binom{n}{i} = \binom{n-1}{i} + \binom{n-1}{i-1}, \qquad \text{(A.9)}
$$

$$
\binom{n}{i} = \binom{n}{n-i}, \qquad \text{(A.10)}
$$

$$k\binom{n}{k} = n\binom{n-1}{k-1}, \tag{A.11}$$

$$\binom{n}{m} = \frac{\binom{n}{i}\binom{n-i}{m-i}}{\binom{m}{i}}, \tag{A.12}$$

$$\binom{2n}{n} = \sum_{i=0}^{n}\binom{n}{i}^2 = \frac{4n-2}{n}\binom{2n-2}{n-1}, \tag{A.13}$$

$$\sum_{i=0}^{n}(-1)^i\binom{n}{i} = 0, \tag{A.14}$$

$$\sum_{i=0}^{n} i\binom{n}{i} = n2^{n-1}, \tag{A.15}$$

$$\sum_{i=0}^{n}\binom{n}{i} = 2^n, \tag{A.16}$$

$$\sum_{i=0}^{n} i\binom{n}{i}^2 = \binom{2n}{n}\frac{n}{2}. \tag{A.17}$$

Appendix B The total number of states in the Fermi–Dirac and the Bose–Eistein statistics

The results of this appendix are well known and can be found in any textbook on statistical mechanics.

In the Fermi–Dirac (FD) statistics, we are given N indistinguishable particles to be placed in M boxes (with $M \geq N$) with the condition that no more than one particle can occupy a given box.

In this case, the counting is straightforward. For N distinguishable particles, the total number of ways of placing the N particles in M boxes is simply $M(M-1)(M-2)\cdots(M-N+1)$. Next, we divide by $N!$ to correct for indistinguishability of the particles. The result is

$$W_{FD}^{(N)} = \frac{M(M-1)\cdots(M-N+1)}{N!} = \frac{M!}{N!(M-N)!} = \binom{M}{N}. \tag{B.1}$$

For BE particles, we do not impose any restriction on the number of particles in each box. To calculate the number of arrangements, we place N particles and $M - 1$ sticks on a line. Clearly, $M - 1$ sticks define M boxes; an example of an arrangement for $N = 18$ and $M = 7$ is

$$\circ\circ\circ|\circ\circ|\circ\circ\circ|\circ|\circ\circ|\quad\circ\circ\circ\circ|\circ\circ\circ$$

Since there are no restrictions on the number of particles in any single box, the total number of arrangements is exactly the total number of permutations of the $N + M - 1$ objects (particles and sticks). But in this count, we counted too many arrangements; we must divide by $(M - 1)!N!$ to account for over-counting configurations that are indistinguishable. Thus, the required number of indistinguishable configurations is

$$W_{BE}^{(N)} = \frac{(M+N-1)!}{N!(M-1)!} = \binom{N+M-1}{N}. \tag{B.2}$$

To obtain the Maxwell–Boltzmann (MB) statistics, we can either take the limit of either (B.1) or (B.2) as $M \gg N$, or do the calculation directly.

If $M \gg N$, the limit of (B.1), using the Stirling approximation (see Appendix E), is

$$\begin{aligned}\ln W_{FD}^{(N)} &\cong M\ln M - N\ln N - (M-N)\ln(M-N)\\ &= -M\ln\left(1-\frac{N}{M}\right) + N\ln\left(\frac{M}{N}-1\right)\\ &\cong N\ln\frac{M}{N} + N = \ln\frac{M^N e^N}{N^N} \cong \ln\frac{M^N}{N!}.\end{aligned} \tag{B.3}$$

Hence,

$$W_{FD}^{(N)} \to \frac{M^N}{N!}. \tag{B.4}$$

The same limit can also be obtained from

$$W_{BE}^{(N)} \to \frac{M^N}{N!}. \tag{B.5}$$

Figure 6.5 shows W_{FD}, W_{BE} and W_{MB} for fixed M and varying N, and for a fixed $N = 100$ and varying $100 \leq M \leq 2000$.

Exercise: Calculate all the possible configurations for the FD and BE statistics for $M = 4$ and $N = 3$.

Appendix C Pair and triplet independence between events

We present here two examples where dependence or independence between two events does not imply dependence or independence between three events, and *vice versa.*

(i) *Pairwise independence does not imply triplewise independence*

Consider a board of total area S being hit by a ball. The probability of hitting a certain region is assumed to be proportional to its area. On this board, we draw three regions A–C (Figure C.1). If the area of the entire board is chosen as unity, and the areas of A–C are 1/10 of S, then we have the probabilities

$$P(S) = 1, \quad P(A) = P(B) = P(C) = \frac{1}{10}. \tag{C.1}$$

In this example, we have chosen the regions A–C in such a way that the area of the intersection (region I) is 1/100 of S. Hence, in this case we have

$$\begin{aligned} P(A \cdot B) = P(B \cdot C) = P(A \cdot C) = \frac{1}{100} &= P(A)P(B) \\ = P(A)P(C) = P(B)P(C). \end{aligned} \tag{C.2}$$

Thus, we have pairwise independence, e.g., the probability of hitting A and B is the product of the probabilities $P(A)$ and $P(B)$. However, in this system

$$P(A \cdot B \cdot C) = \frac{1}{100} \neq P(A)P(B)P(C) = \frac{1}{1000}. \tag{C.3}$$

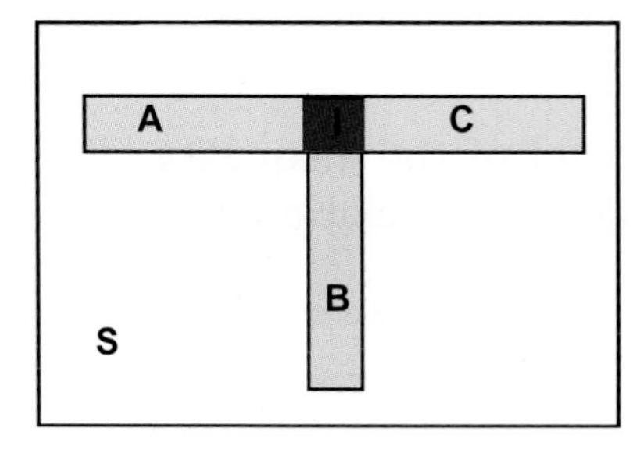

Figure C.1. Three events A, B and C and the intersection I. The areas are given in equations (C.1) and (C.2)

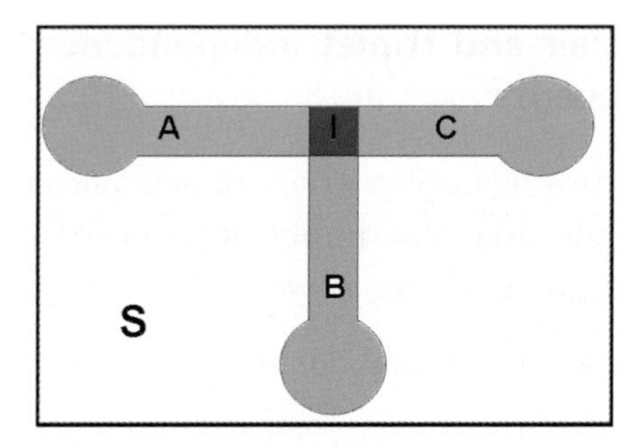

Figure C.2. Three events A, B and C and the intersection I. The areas are given in (C.1) and (C.4).

The probability of hitting A–C is not the product of the probabilities $P(A), P(B)$–$P(C)$, i.e., there is no triplewise independence.

(ii) *Triplewise independence does not imply pairwise independence*

In this example (Figure C.2), the total area of S is unity. Also, the area of each region A–C is again 1/10 of S. But now the intersection I of the three regions has an area 1/1000 of S. In this case we have

$$\frac{1}{1000} = P(A \cdot B \cdot C) = P(A)P(B)P(C). \tag{C.4}$$

The probability of the event $(A \cdot B \cdot C)$ is the product of the three probabilities $P(A), P(B)$ and $P(C)$, On the other hand,

$$\frac{1}{1000} = P(A \cdot B) \neq P(A)P(B) = \frac{1}{100}, \tag{C.5}$$

and similarly for $P(A \cdot C)$ and $P(B \cdot C)$. Hence, in this system there is triplewise independence, but not pairwise independence.

Appendix D Proof of the inequality $|R(X, Y)| \leq 1$ for the correlation coefficient

The correlation coefficient between two random variables X and Y is defined by

$$R(X, Y) = \frac{Cov(X, Y)}{\sigma_x \sigma_y} = \frac{E(X \cdot Y) - E(X)E(Y)}{\sqrt{Var(X)}\sqrt{Var(y)}}, \tag{D.1}$$

where it is assumed that $Var(X) \neq 0$ and $Var(Y) \neq 0$ and the positive square root is taken in the denominator of (D.1).

We have seen in Section 2.7 that

$$\begin{aligned} Var(X+Y) &= Var(X) + Var(Y) + 2Cov(X,Y) \\ &= Var(X) + Var(Y) + 2R(X,Y)\sqrt{Var(X)\,Var(Y)}. \end{aligned} \tag{D.2}$$

We have to show that

$$-1 \leq R(X,Y) \leq 1. \tag{D.3}$$

First, we show that $R(X,Y)$ does not change if we transform to new random variables, defined by

$$X' = \alpha X + a, \quad Y' = \beta Y + b \tag{D.4}$$

with $\alpha > 0, \beta > 0$. This follows from the properties

$$E(\alpha X + a) = \alpha E(X) + a, \tag{D.5}$$

$$E(\beta Y + b) = \beta E(X) + b, \tag{D.6}$$

$$E[(\alpha X + a)(\beta X + b)] = \alpha\beta E(X \cdot Y) + \alpha b E(X) + a\beta E(Y) + ab. \tag{D.7}$$

Hence,

$$\begin{aligned} Cov(X',Y') &= E(X' \cdot Y') - E(X')E(Y') \\ &= \alpha\beta[E(X \cdot Y) - E(X)E(Y)], \end{aligned} \tag{D.8}$$

$$Var(X') = Var(\alpha X + a) = \alpha^2\, Var(X), \tag{D.9}$$

$$Var(Y') = Var(\beta Y + b) = \beta^2\, Var(Y), \tag{D.10}$$

$$R(X',Y') = \frac{\alpha\beta[E(X \cdot Y) - E(X)E(Y)]}{\alpha\sqrt{Var(X)}\,\beta\sqrt{Var(Y)}} = R(X,Y). \tag{D.11}$$

Because of the invariance property (D.11), we can assume that X and Y have average equal zero and variance equal to one. If this is not the case, we can transform into a new *rv* as follows:

$$X' = \frac{X - E(X)}{\sqrt{Var(X)}}, \quad Y' = \frac{Y - E(Y)}{\sqrt{Var(Y)}} \tag{D.12}$$

so that X' and Y' have these properties.

For such "normalized" random variables, we have [note that the variance is always positive $Var(X) \geq 0$]

$$\begin{aligned} Var(X+Y) &= Var(X) + Var(Y) + 2R(Y,Y)\sqrt{Var(X)\,Var(Y)} \\ &= 1 + 1 + 2R(X,Y) \geq 0. \end{aligned} \tag{D.13}$$

Hence,

$$R(X,Y) \geq -1. \tag{D.14}$$

Similarly, we can write

$$\begin{aligned} Var(X-Y) &= Var(X) + Var(Y) - 2R(X,Y) \\ &= 1 + 1 - 2R(X,Y) \geq 0. \end{aligned} \tag{D.15}$$

Hence,

$$R(X,Y) \leq 1. \tag{D.16}$$

The last two results can be written as

$$|R(X,Y)| \leq 1. \tag{D.17}$$

Note that the two limits $+1$ and -1 are attainable. For $X = Y$, we have $R(X,X) = 1$ and for $Y = -X, R(X,-X) = -1$. Thus, positive and negative correlations correspond to positive and negative values $R(X,Y)$, respectively.

Exercise: Calculate $R(X,Y)$ for the two *rv*: X being the outcomes of throwing a fair dice, and $Y = -X$.

Solution:

$$E(X) = 3.5, E(Y) = -3.5, E(X \cdot Y) = \frac{-91}{6}, \tag{D.18}$$

$$Var(X) = \frac{35}{12}, \quad Var(Y) = (-1)^2\, Var(X) = \frac{35}{12}, \tag{D.19}$$

$$R(X,Y) = -1. \tag{D.20}$$

Note that the values of the two *rv* do not necessarily have to be of opposite signs to get a negative correlation. The following example shows that X and Y attain only positive values but the correlation is negative.

Exercise: X is, as before, the *rv* associated with throwing a fair die (with values 1, 2, 3, 4, 5, 6) and Y is defined on the same sample space, but

$$X(i) = i \quad \text{for } i = 1, 2, \ldots, 6,$$
$$Y(i) = |7 - i|, \tag{D.21}$$

i.e., Y gets the value of the "opposite face" of i. These are:

$$Y(1) = 6, Y(2) = 5, Y(3) = 4, Y(4) = 3, Y(5) = 2, Y(6) = 1, \tag{D.22}$$

$$E(X) = E(Y) = 3.5, \; Var(X) = Var(Y) = \frac{35}{12}, \tag{D.23}$$

$$E(X \cdot Y) = \frac{1}{6}(1 \times 6 + 2 \times 5 + 3 \times 4 + 4 \times 3 + 5 \times 2 + 6 \times 1)$$
$$= \frac{28}{3}, \tag{D.24}$$

and

$$R(X, Y) = \frac{-35}{35} = -1. \tag{D.25}$$

Thus, we have the largest negative correlation coefficient although the values of X and Y are all positive. What makes the correlation negative is that a positive deviation from the average of one *rv* is correlated with a negative correlation from the average of the second *rv*. We can see this clearly from the explicit form of the $Cov(X, Y)$:

$$(6 - 3.5)(1 - 3.5) + (5 - 3.5)(2 - 3.5) + (4 - 3.5)(3 - 3.5)$$
$$+ (3 - 3.5)(4 - 3.5) + (2 - 3.5)(5 - 3.5) + (1 - 3.5)(6 - 3.5).$$

Each of the terms in this sum is negative; hence, the correlation is negative too.

A special case of a correlation function is for two characteristic functions. Let A and B be two events, and let I_A and I_B be the corresponding characteristic functions:

$$I_A(\omega) = \begin{cases} 1 & \text{if } \omega \in A, \\ 0 & \text{if } \omega \notin A, \end{cases}$$

$$I_B(\omega) = \begin{cases} 1 & \text{if } \omega \in B, \\ 0 & \text{if } \omega \notin B. \end{cases} \tag{D.26}$$

The corresponding averages are

$$\begin{aligned}
E(I_A) &= P(A), \quad E(I_B) = P(B), \\
E(I_A \cdot I_B) &= P(A \cap B), \\
Var(I_A) &= P(A)P(\bar{A}) = P(A)(1 - P(A)), \\
Var(I_B) &= P(A)P(\bar{B}) = P(B)(1 - P(B)), \\
R(I_A, I_B) &= \frac{P(A \cap B) - P(A)P(B)}{\sqrt{P(A)P(\bar{A})P(B)P(\bar{B})}} = R(A, B).
\end{aligned} \tag{D.27}$$

If $A = B$, then there is maximum positive correlation and $R(A, B) = 1$. When $B = \bar{A}$, there is maximum negative correlation and $R(A, B) = -1$. Note also that in this case, independent events (i.e., $P(A \cap B) = 0$) is equivalent to uncorrelated events $R(A, B) = 0$.

Appendix E The Stirling approximation

A very useful approximation for $n!$ is the following:

$$n! \approx \sqrt{2\pi n}\, n^n e^{-n}. \tag{E.1}$$

We usually use this approximation for $\ln(n!)$ and for very large n, say of the order of 10^{23}. In this case

$$\ln(n!) \cong n \ln n - n + \frac{1}{2} \ln(2\pi n). \tag{E.2}$$

When n is very large, one can neglect the last term on the right-hand side of (E.2). A detailed proof of this approximation can be found in Feller (1957). The general idea is as follows. We can write

$$\ln(n!) = \sum_{i=1}^{n} \ln i = \sum_{i=1}^{n} \Delta i \ln i, \tag{E.3}$$

where $\Delta i = (i + 1) - i = 1$. For very large n, this sum can be approximated by the integral

$$\sum_{i=1}^{n} \Delta i \ln i \approx \int_1^n \ln(i) di = n \ln n - n - 1 \approx n \ln n - n. \tag{E.4}$$

Exercise: The binomial distribution is

$$P_N(n) = \binom{N}{n} p^n (1 - p)^{N-n}. \tag{E.5}$$

Use the Stirling approximation to show that the value of n for which $P_N(n)$ has a maximum is the same as the average value of n.

Solution:

$$\frac{\partial \ln P_N(n)}{\partial n} = -\ln n + \ln(N-n) + \ln p - \ln(1-p) = 0. \quad \text{(E.6)}$$

Hence, the maximum is attained for $n = n^*$, which fulfills the equation

$$\frac{n^*}{N-n^*} = \frac{p}{1-p} \quad \text{(E.7)}$$

or equvalently

$$n^* = pN. \quad \text{(E.8)}$$

It is easy to see that this is a maximum. We have already seen in Section 2.8 that n^* is also the average value of n.

Appendix F Proof of the form of the function H

Here, we shall assume that the function H fulfills the three requirements as stated in Section 3.2 and show how one gets the form of this function [there are other proofs that are based on different set of requirements; see Khinchin (1957) and Katz (1967)].

We have an experiment with n outcomes $A_1, \ldots, A_n$. These are assumed to be mutually exclusive. We group them into r sets, each of which contains m_k elements ($k = 1, 2, \ldots, r$), and $\sum_{k=1}^{r} m_k = n$.

We denote the new events $A'_1, A'_2, \ldots, A'_r$. These are defined in terms of original outcomes[1]:

$$\begin{aligned}
A'_1 &= \{A_1 + A_2 + A_3 + \cdots + A_{m_1}\} \\
A'_2 &= \{A_{m_1+1} + A_{m_1+2} + \cdots + A_{m_1+m_2}\} \\
A'_3 &= \{A_{m_1+m_2+1} + A_{m_1+m_2+2} + \cdots + A_{m_1+m_2+m_3}\} \\
&\vdots \\
A'_r &= \left\{A_{\sum_{k=1}^{r-1} m_k+1} + \cdots + A_{n=\sum_{k=1}^{r} m_k}\right\}.
\end{aligned} \quad \text{(F.1)}$$

[1]The plus sign in (F.1) means union of events. The original outcomes are not necessarily elementary events.

Since the original events $\{A_i\}$ are mutually exclusive, the probabilities of the new events are simply the sum of the probabilities of the original events that are included in the new event, thus

$$p'_1 = P(A'_1) = \sum_{i=1}^{m_1} p_i,$$

$$p'_2 = P(A'_2) = \sum_{i=m_1+1}^{m_1+m_2} p_i,$$

$$\vdots$$

$$p'_r = P(A'_r) = \sum_{i=\sum_{k=1}^{r-1} m_k+1}^{n} p_i. \tag{F.2}$$

It is convenient to denote the events included in the kth group as $A_1^k \cdots A_{m_k}^k$, and the corresponding probabilities as $p_1^k \cdots p_{m_k}^k$ (the superscript is the index of the kth group, and the subscript is the index of the event within the kth group).

In this notation, the requirement of consistency for the function H is written as:

$$H(p_1, \ldots, p_n) = H(p'_1, \ldots, p'_r) + \sum_{k=1}^{r} p'_k H\left(\frac{p_1^k}{p'_k}, \ldots, \frac{p_{m_k}^k}{p'_k}\right). \tag{F.3}$$

Thus, the total MI of the original set of events should be equal to the MI of the new set of r compound events, plus an average MI within the groups of events $k = 1, 2, \ldots, r$.

We now assume that the original probabilities are all rational numbers, i.e., they can be written as

$$p_i = \frac{M_i}{\sum_{j=1}^{n} M_j}, \tag{F.4}$$

where M_i are non-negative integers. The assumption of rational probabilities in (F.4) is not really serious, since in any event we know the probabilities up to a given accuracy, and we can always cast this into a rational number.

We next "expand" our set of outcomes. Instead of the original n outcomes $A_1 \cdots A_n$, we build a new set of outcomes in such a way that A_1 consists of M_1 elements of equal probability $\frac{1}{M}$, A_2 consists

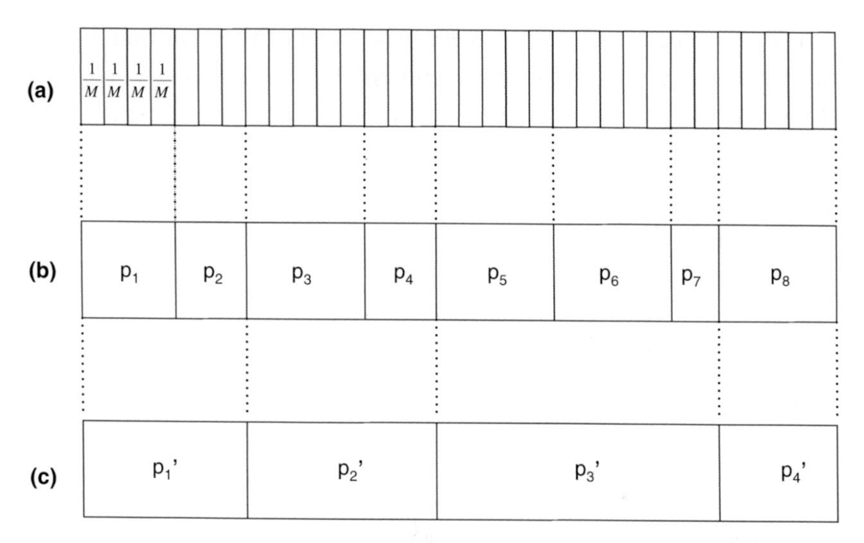

Figure F.1. The three "levels" of describing the set of events, as described in Appendix F.

of M_2 elements of equal probability $\frac{1}{M}$, etc., where $M = \sum_{i=1}^{n} M_i$ (see Figure F.1).

With this expansion into M events, all of the equal probability, we can rewrite the requirement of consistency on the expanded set of events as

$$H\left(\frac{1}{M}, \ldots, \frac{1}{M}\right) = H(p_1, \ldots, p_n) + \sum_{i=1}^{n} p_i H\left(\frac{1/M}{M_i/M}, \ldots, \frac{1/M}{M_i/M}\right). \tag{F.5}$$

Note that in (F.5), we have on the left-hand side the MI of M events of equal probability. These are regrouped in such a way that the kth group is exactly the event A_k of the original set of events in which we have started.

We now define the function

$$F(M) = H\left(\frac{1}{M}, \ldots, \frac{1}{M}\right) \tag{F.6}$$

and rewrite (F.5) as

$$F(M) = H(p_1, \ldots, p_n) + \sum_{i=1}^{n} p_i F(M_i). \tag{F.7}$$

It is now clear that if we can find the form of the function $F(M)$, we can use (F.7) to obtain the required function $H(p_1, \ldots, p_n)$ for any distribution $p_1, \ldots, p_n$.

We now digress to the mathematical problem of finding the function $F(M)$. We choose a particular case when all M_i are equal, i.e.,

$$M_i = m, \quad \sum_{i=1}^{n} M_i = nm = M. \tag{F.8}$$

In this particular case, the probabilities are

$$p_i = \frac{M_i}{M} = \frac{m}{M} = \frac{1}{n}. \tag{F.9}$$

Therefore, for this particular case, we have

$$H(p_1, \ldots, p_n) = H\left(\frac{1}{n}, \ldots, \frac{1}{n}\right) = F(n) \tag{F.10}$$

and

$$\sum_{i=1}^{n} p_i F(M_i) = F(m). \tag{F.11}$$

Hence, (F.7) reduces, for this case, to

$$F(M) = F(n) + F(m) \tag{F.12}$$

or equivalently

$$F(n \times m) = F(n) + F(m). \tag{F.13}$$

One can easily prove that the only function that has this property is the logarithm function [for details of the proof, see Khinchin (1957) or Katz (1967)], i.e.,

$$\log(n \times m) = \log(n) + \log(m), \tag{F.14}$$

where we left the base of the logarithm unspecified. Thus, we found that

$$F(M) = \log M. \tag{F.15}$$

Having found the form of the function $F(M)$, we can now go back to (F.7), and write for the general case where $p_1, \ldots, p_n$ are unequal:

$$\begin{aligned} H(p_1, \ldots, p_n) &= F(M) - \sum_{i=1}^{n} p_i F(M_i) \\ &= \log M - \sum_{i=1}^{n} p_i \log M_i \\ &= -\sum p_i \log \frac{M_i}{M} = -\sum_{i=1}^{n} p_i \log p_i. \end{aligned} \tag{F.16}$$

Thus, we found the general form of the function H:

$$H(p_1, \ldots, p_n) = -\sum_{i=1}^{n} p_i \log p_i \tag{F.17}$$

for any distribution $p_1, \ldots, p_n$. Because of the continuity requirement of H, we can approximate any distribution by rational numbers of the form (F.4). Therefore, this is not a real limitation on the validity of the proof of the function H.

Appendix G The method of Lagrange undetermined multipliers

Let $F(x)$ be a function of a single independent variable x. The condition for an extremum is

$$\frac{dF}{dx} = 0 \tag{G.1}$$

and for a maximum, we also need the condition

$$\frac{d^2F}{dx^2} < 0. \tag{G.2}$$

For a function $F(x_1, \ldots, x_n)$ of n independent variables, the condition for an extremum is

$$dF = \sum_{i=1}^{n} \frac{\partial F}{\partial x_i} dx_i = 0, \tag{G.3}$$

where $\partial F/\partial x_i$ is the partial derivative of F with respect to the variable x_i, keeping all other variable constant. Since (G.3) must hold for any variation in the independent variables, it follows from (G.3) that[2]

$$\frac{\partial F}{\partial x_i} = 0 \quad \text{for each } i = 1, \ldots, n. \tag{G.4}$$

The derivatives should be evaluated at the point of the extremum, say $x_1^0, \ldots, x_n^0$.

Now, suppose we need to find an extremum of a function $F(x_1, \ldots, x_n)$, but with the constraint that the variables $x_1, \ldots, x_n$ are not independent. The simplest case is that the variables are to satisfy a condition that can be written as

$$G(x_1, \ldots, x_n) = c, \tag{G.5}$$

where c is a constant. Clearly, if (G.5) is to be satisfied, the variations in $x_1, \ldots, x_n$ are not independent, since

$$dG = \sum_{i=1}^{n} \frac{\partial G}{\partial x_i} = 0. \tag{G.6}$$

If the function $G(x_1, \ldots, x_n)$ is given explicitly, one can eliminate one of the variables, say x_1, express it in terms of the others, say $x_1 = f(x_2, \ldots, x_n)$, and substitute in $F(x_1, \ldots, x_n)$ to obtain a function of $n - 1$ independent variables.

The procedure due to Lagrange is much more simple and elegant. One first defines the auxiliary function L, by subtracting λ times the constraint (G.5) from F to obtain

$$L \equiv F(x_1, \ldots, x_n) - \lambda[G(x_1, \ldots, x_n) - c], \tag{G.7}$$

where λ is any constant, independent of $x_1, \ldots, x_n$. We now write the total difference of L as

$$dL = \sum_{i=1}^{n} \left(\frac{\partial F}{\partial x_i} - \lambda \frac{\partial G}{\partial x_i} \right) dx_i. \tag{G.8}$$

[2]For instance, one could take $dx_1 \neq 0$ and $dx_i = 0$ for all $i \neq 1$ and obtain the condition $\frac{\partial F}{\partial x_1} = 0$.

Since λ can be chosen at will, we can choose λ in such a way that one of the terms in (G.8) is zero,[3] e.g.,

$$\frac{\partial F}{\partial x_1} - \lambda \frac{\partial G}{\partial x_1} = 0 \quad \text{or} \quad \lambda = \frac{\partial F/\partial x_1}{\partial G/\partial x_1}. \tag{G.9}$$

From (G.8) and (G.9), it follows that all the coefficients in (G.8) must be zero. Thus, we have the conditions

$$\frac{\partial F}{\partial x_i} - \lambda \frac{\partial G}{\partial x_i} = 0 \quad \text{for} \quad i = 2, 3, \ldots, n, \tag{G.10}$$

where λ is determined by (G.9). Thus, we have altogether n conditions in (G.9) and (G.10) that can be solved to find the point of extremum. The same procedure can be generalized to cases when there are more than one constraint.

Exercise: Find the maximum of the function $f(x, y) = xy$, subject to the condition $x^2 + y^2 = 1$.

Solution: Define $F(x, y) = xy - \lambda(1 - x^2 - y^2)$. Take the derivatives with respect to x and with respect to y, and the condition of maximum is

$$\frac{\partial F}{\partial x} = y - 2\lambda x = 0, \quad \frac{\partial F}{\partial y} = x - 2\lambda y = 0. \tag{G.11}$$

From these two equations, we get

$$\lambda = \frac{y}{2x}, \quad \lambda = \frac{x}{2y}. \tag{G.12}$$

Hence, the maximum is attained at

$$x^0 = y^0 = \pm\sqrt{1/2}, \tag{G.13}$$

and the value of the maximum f is

$$f_{\max} = 1/2. \tag{G.14}$$

[3]Note that since all the derivatives in (G.3) as well as in (G.8) are evaluated at the point of extremum, say $x_1^0, \ldots, x_n^0$, the parameter λ depends on this point.

Appendix H Some inequalities for concave functions

In this appendix, we shall present some important inequalities that are used in connection with information theory. Proofs can be found in Yaglom and Yaglom.[4]

Definition: A function is said to be concave[5] in some region (a, b) if for any two points x' and x'' such that $a \leq x' \leq b$, $a \leq x'' \leq b$, the entire straight line connecting the two points $f(x')$ and $f(x'')$ is *below* the function $f(x)$ (Figure H.1).

An important property of a concave function is the following: a function $f(x)$ is concave in (a, b) if and only if, the second derivative is negative in (a, b).

Examples of concave functions are (Figure H.2):

(i)
$$f(x) = \log x. \tag{H.1}$$

This function is defined for $x > 0$. The first two derivatives are:

$$f'(x) = \frac{1}{x}, \quad f''(x) = \frac{-1}{x^2} < 0 \quad \text{for } x > 0. \tag{H.2}$$

(ii)
$$f(x) = -x \log x. \tag{H.3}$$

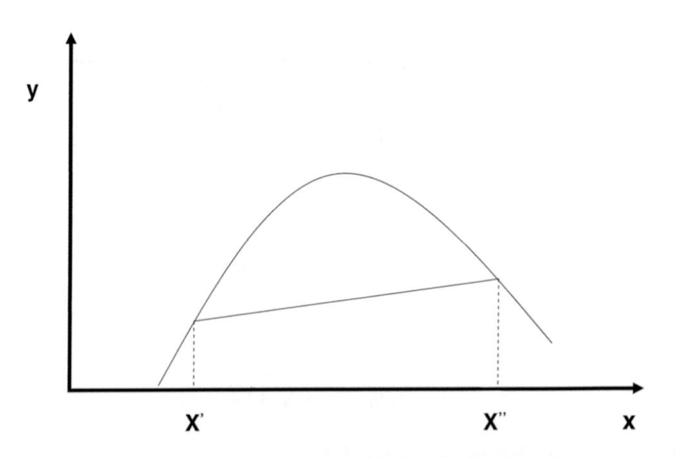

Figure H.1. A convex downward function.

---

[4] Yaglom, A.M. and Yaglom, I.M. (1983).

[5] Some authors refer to what we call a "concave function" as a "convex function." We use the definition of "concave function" as described above and illustrated in Figure H.1.

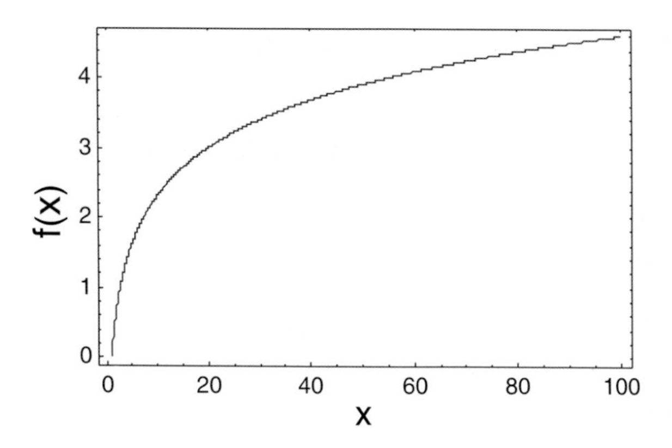

Figure H.2. The function $f(x) = \log(x)$.

This function is defined for $x \geq 0$ and its first two derivatives are:

$$f'(x) = -1 - \log x, \quad f''(x) = \frac{-1}{x} < 0 \quad \text{for } x \geq 0. \tag{H.4}$$

(iii) The function

$$-x \log x - (1 - x) \log(1 - x) \quad \text{for } 0 \leq x \leq 1 \tag{H.5}$$

is concave in $0 \leq x \leq 1$, and its first two derivatives are:

$$f'(x) = -\log x + \log(1-x), \quad f''(x) = \frac{-1}{x(1-x)} < 0 \quad \text{for } 0 < x < 1. \tag{H.6}$$

We now present a few inequalities for concave functions.

Inequality I

If $f(x)$ is concave in (a, b), and if $a \leq x_1 \leq x_2 \leq b$, then

$$\frac{f(x_1) + f(x_2)}{2} \leq f\left(\frac{x_1 + x_2}{2}\right), \tag{H.7}$$

i.e., the value of the function f at the point $\frac{x_1+x_2}{2}$ is always larger than the point $(f(x_1)+f(x_2))/2$ on the straight line connecting the points $f(x_1)$ and $f(x_2)$. This property follows immediately from the definition, and it is clear from Figure H.1. A more general statement of this property is:

Inequality II

$$\lambda f(x_1) + (1-\lambda) f(x_2) \\ \leq f(\lambda x_1 + (1-\lambda) x_2) \quad \text{for any } \lambda \text{, such } 0 \leq \lambda \leq 1. \tag{H.8}$$

As an example, applying inequality (H.7) to the function (H.1), we get

$$\frac{\log x_1 + \log x_2}{2} \leq \log\left(\frac{x_1 + x_2}{2}\right) \tag{H.9}$$

or equivalently

$$(x_1 x_2)^{1/2} \leq \frac{x_1 + x_2}{2}. \tag{H.10}$$

This is a well-known inequality, i.e., the geometrical average of two different and positive numbers x_1, x_2 is always smaller than their arithmetic average.

Applying the second inequality (II) to this function, we obtain

$$\lambda \log x_1 + (1-\lambda) \log x_2 \leq \log(\lambda x_1 + (1-\lambda) x_2) \quad \text{for } 0 \leq \lambda \leq 1 \tag{H.11}$$

or equivalently

$$x_1^{\lambda} x_2^{(1-\lambda)} \leq \lambda x_1 + (1-\lambda) x_2, \tag{H.12}$$

which is a generalization of (H.10).

Inequality III

Another generalization of the inequality I is:

$$\frac{f(x_1) + f(x_2) + \cdots + f(x_m)}{m} \leq f\left(\frac{x_1 + x_2 + \cdots + x_m}{m}\right). \tag{H.13}$$

Or more generally:

Inequality IV

$$\lambda_1 f(x_1) + \lambda_2 f(x_2) + \cdots + \lambda_m f(x_m) \\ \leqslant f(\lambda_1 x_1 + \lambda_2 x_2 + \cdots + \lambda_m x_m) \quad \text{for } \lambda_i \geq 0 \text{ and } \sum_{i=1}^{m} \lambda_i = 1. \tag{H.14}$$

The inequalities (III) and (IV) can easily be proven by mathematical induction over n. Applying (IV) for the function (H.1), we have

$$\sum_{i=1}^{m} \lambda_i \log x_i \leq \log \left(\sum_{i=1}^{m} \lambda_i x_i \right) \tag{H.15}$$

or equivalently

$$\prod_{i=1}^{m} x_i^{\lambda_i} \leq \sum_{i=1}^{m} \lambda_i x_i. \tag{H.16}$$

An important inequality that follows from (H.15) is the following

Let $p_1, \ldots, p_m$ and $q_1, \ldots, q_m$ be any of the two distributions, such that $\sum_{i=1}^{m} p_i = \sum_{i=1}^{m} q_i = 1$, $p_i > 0$ and $q_i > 0$. We choose $\lambda_i = p_i$ and $x_i = \frac{q_i}{p_i}$. Since $\sum \lambda_i = 1$, and since the function $\log x$ is concave for any $x > 0$, then application of (H.15) for this choice gives

$$\sum_{i=1}^{m} p_i \log \frac{q_i}{p_i} \leq \log \left(\sum_{i=1}^{m} p_i \frac{q_i}{p_i} \right) = \log \sum q_i = 0. \tag{H.17}$$

Hence, it follows from (H.17) that

$$\sum_{i=1}^{m} p_i \log q_i \leq \sum_{i=1}^{m} p_i \log p_i. \tag{H.18}$$

This inequality was used in Section 3.2.2. In Section 3.2.2, we proved this inequality from the inequality $\ln x \leq x - 1$ for any $x > 0$. The latter inequality follows from the following considerations: define the function $f(x) = \ln x - (x - 1)$, the derivative of which is

$$f'(x) = \frac{1}{x} - 1 = \frac{1 - x}{x}. \tag{H.19}$$

For $0 < x < 1$, this derivative is positive. Hence, in this region, $f(x)$ is an increasing function of x. Since $f(1) = 0$, it follows that $f(x) \leq 0$, or equivalently $\ln x \leq x - 1$ for $x < 1$.

For $x > 1$, $f'(x)$is negative. Hence, $f(x)$ is a decreasing function of x. Since $f(1) = 0$, it follows that $f(x) \leq 0$, or equivalently $\ln x \leq x - 1$ for $x > 1$. The equality sign holds for $x = 1$. Figure H.3 shows the function $\ln(x)$ and the function $x - 1$. The geometrical meaning of the inequality $\ln x \leq x - 1$ is that the straight line $y = x - 1$ is always above the line $y = \ln x$, for any $x > 0$.

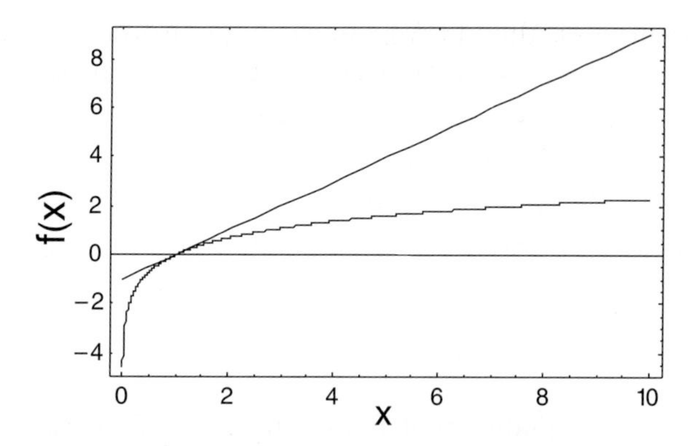

Figure H.3. The logarithm and the linear functions.

Another application of (H.14), for the function (H.3) is

$$-\sum_{i=1}^{m} \lambda_i x_i \log x_i \leq -\left(\sum_{i=1}^{m} \lambda_i x_i\right) \log \left(\sum_{i=1}^{m} \lambda_i x_i\right)$$

$$\text{for } \lambda_i \geq 0, \sum \lambda_i = 1. \tag{H.20}$$

Two important conclusions result from the concavity of the function $f(x) = -x \log x$.

(i) The missing information (MI) of an experiment having m outcomes cannot exceed the MI of an experiment with equally probable outcomes. Thus, if we choose $x_i = p_i$, the probabilities of the outcomes of an experiment we have from (H.13)

$$-\frac{\sum x_i \log x_i}{m} \leq \left(\frac{\sum x_i}{m}\right) \log \left(\frac{\sum x_i}{m}\right). \tag{H.21}$$

Multiplying by m and choosing $x_i = p_i$, $\sum x_i = 1$, we get

$$-\sum_{i=1}^{m} p_i \log p_i \leq -\log \frac{1}{m} = \log m, \tag{H.22}$$

i.e., the MI of an experiment with m equally probable events is always larger than the MI of m events with unequal probability. The equality holds if and if only all $p_i = p = 1/m$. (We have seen this Chapter 3, proven by the method of Lagrange undetermined multipliers.)

(ii) The second result that follows from the concavity of the function $f(x) = -x \log x$ is the following. Suppose we have two random variables, or two experiments, X and Y. Let p_i be the probabilities of the events $x_i = (i = 1, \ldots, n)$.[6] Let us denote by $P(y_j/x_i)$, the probability of y_j, given the event x_i. Substituting $\lambda_i \to p_i$ and $x_i \to P(y_j/x_i)$ in (H.20), we obtain

$$-\sum_{i=1}^{n} p_i P(y_j/x_i) \log P(y_j/x_i) \leq \left[-\sum_{i=1}^{n} p_i P(y_j/x_i)\right] \log \left[\sum p_i P(y_j/x_i)\right]. \tag{H.23}$$

We now apply the theorem of total probability (see Section 2.6), which for this case, is

$$\sum_{i=1}^{n} p_i P(y_j/x_i) = P(y_j). \tag{H.24}$$

Hence, (H.23) can be rewritten as

$$-\sum_{i=1}^{n} p_i P(y_j/x_i) \log P(y_j/x_i) \leq -P(y_j) \log P(y_j). \tag{H.25}$$

The last inequality is valid for any $j(j = 1, 2, \ldots, m)$. Summing over all $j(j = 1, \ldots, m)$, we obtain

$$-\sum_{i=1}^{n} p_i \sum_{j=1}^{m} P(y_j/x_i) \log P(y_j/x_i) \leq -\sum_{j=1}^{m} P(y_j) \log P(y_j), \tag{H.26}$$

which in the notation of Section 3.2.2 is

$$\sum_{i=1}^{n} p_i H(Y/x_i) \leq H(Y), \tag{H.27}$$

or equivalently

$$H(Y/X) \leq H(Y). \tag{H.28}$$

---

[6]The "event x_i" is a shorthand notation for the event $\{X = x_i\}$, and similarly the event y_i, is a shorthand for $\{Y = y_i\}$.

Note that X and Y are the two random variables or two *experiments*, the outcome of which are $x_1, \ldots, x_n$ and $y_1, \ldots, y_m$, respectively. The equality in (H.28) holds when the two experiments are independent, which we have already seen in Section 3.2.

Appendix I The MI for the continuous case

In this appendix, we discuss the passage to the limit of a continuous distribution. We assume that the random variable X can attain any value within the interval (a, b), and that there exists a probability density $f(x)$, such that

$$\Pr(x_1 \leq X \leq x_2) = \int_{x_1}^{x_2} f(x)dx \tag{I.1}$$

and

$$\int_a^b f(x)dx = 1. \tag{I.2}$$

We now divide the interval (a, b) into n intervals, each of size $\delta = (b - a)/n$. We denote the points.

$$x_1 = a, \quad x_i = a + (i - 1)\delta, \quad x_{n+1} = a + n\delta = b. \tag{I.3}$$

Thus, the probability

$$P(i, n) = \int_{x_i}^{x_{i+1}} f(x)dx \tag{I.4}$$

is the probability of finding the value between x_i and x_{i+1}, for a given subdivision into n intervals.

The MI associated with the probability distribution $P(i, n)$, for a fixed n, is

$$H(n) = -\sum_{i=1}^{n} P(i, n) \log P(i, n). \tag{I.5}$$

Clearly, since $H(n)$ is defined for a finite value of n, there is no problem in using (I.5) for any fixed n.

Substituting (I.4) into (I.5), we have [note that $\delta = (b-a)/n$]

$$\begin{aligned} H(n) &= -\sum_{i=1}^{n}\left[\int_{x_i}^{x_{i+1}} f(x)dx\right]\log\left[\int_{x_i}^{x_{i+1}} f(x)dx\right] \\ &= -\sum_{i=1}^{n}[\bar{f}(i,n)\delta]\log[\bar{f}(i,n)\delta] \\ &= -\sum_{i=1}^{n}\left[\bar{f}(i,n)\frac{(b-a)}{n}\right]\log[\bar{f}(i,n)] \\ &\quad -\left[\sum_{i=1}^{n}\bar{f}(i,n)\frac{(b-a)}{n}\right]\left[\log\left(\frac{(b-a)}{n}\right)\right], \end{aligned} \tag{I.6}$$

where $\bar{f}(i,n)$ is some value of the function $f(x)$ between $f(x_i)$ and $f(x_{i+1})$, for a specific value of n. When $n \to \infty$, we have

$$\lim_{n\to\infty}\sum_{i=1}^{n}\bar{f}(i,n)\frac{(b-a)}{n} = \int_a^b f(x)dx = 1, \tag{I.7}$$

$$\lim_{n\to\infty}\sum_{i=1}^{n}\bar{f}(i,n)\log[\bar{f}(i,n)]\frac{(b-a)}{n} = \int_a^b f(x)\log f(x)dx. \tag{I.8}$$

The two limits in (I.7) and (I.8) are basically the definition of the Rieman integral. These are presumed to be finite [note, however, that the quantity in (I.8) might either be positive or negative]. Hence, in this limit we have

$$H = \lim_{n\to\infty} H(n) = -\int_a^b f(x)\log f(x)dx - \lim_{n\to\infty}\log\left[\frac{b-a}{n}\right]. \tag{I.9}$$

Clearly, the second term on the right-hand side of (I.9) diverges when $n \to \infty$. The reason for this divergence is clear. The larger the n, the larger the number of intervals, and the more information is needed to locate a point on the segment (a, b). Note, however, that this divergent term does not depend on the distribution density $f(x)$. It only depends on how we choose to divide the segment (a, b). Therefore, when we calculate the *differences* in H for different distributions, say $f(x)$ and $g(x)$, we can take the limit $n \to \infty$ *after* the formation of the difference, i.e.,

$$\Delta H = \lim_{n\to\infty} \Delta H(n) = -\int_a^b f(x)\log f(x)dx + \int_a^b g(x)\log g(x)dx. \tag{I.10}$$

Here the divergent part does not appear and the quantity ΔH is finite in the limit $n \to \infty$ [assuming the two integrals in (I.10) exist]. It should also be noted that the quantity H always depends on how accurately we are interested to locate a particle in a segment (a, b) (or in a volume V in the 3-dimensional case); in other words, H depends on δ. It is only the difference ΔH that is independent of δ. In practice, we always have a limited accuracy for any measurable quantity; therefore, the strict mathematical limit of $n \to \infty$ is never used in practice (see also Sections 3.24 and 4.4).

Thus, for a continuous *rv* we shall always use, as Shannon did, the definition of the MI as

$$H = -\int_a^b f(x)\log f(x)dx. \tag{I.11}$$

One should be careful, however, in interpreting this quantity. This MI does not have all the properties of H as for the finite case. First, because H in (I.11) does not need to be positive, as in the finite case. Second, H in (I.11) is not zero when we specify the outcome of the experiment. In the finite case, if we know the outcome, then H is zero. Here, knowing the answer means that $f(x)$ becomes a Dirac delta function and H in (I.11) diverges.[7] We shall use the definition (I.11) for H only when we are interested in the *differences* in the MI, in which case no difficulties arise even in the limit $n \to \infty$.

The extension of the inequality (3.2.19) to the continuous case is as follows[8]: Let $f(x)$ and $g(x)$ be any two probability density functions. If the two integrals $-\int_{-\infty}^{\infty} f(x)\log g(x)dx$ and $-\int_{-\infty}^{\infty} f(x)\log f(x)dx$ exist, then the following inequality holds:

$$-\int_{-\infty}^{\infty} f(x)\log\frac{f(x)}{g(x)}dx \le 0, \tag{I.12}$$

with equality if and only if $f(x) = g(x)$ for almost all x.

[7]There is another problem with the invariance of the MI to a linear transformation of the random variable. We shall not be concerned with this aspect of H.
[8]See Ash (1965).

It should be noted that the quantity $I(X;Y)$, the mutual information for X and Y, is defined as the difference between two uncertainties; therefore, there exists no problem in extending this quantity to the case of continuous random variables. Thus, for two random variables with densities $f(x)$ and $g(y)$, we define

$$I(X;Y) = \int_{-\infty}^{\infty}\int_{-\infty}^{\infty} F(x,y)\log\left[\frac{F(x,y)}{f(x)g(y)}\right] dxdy, \qquad \text{(I.13)}$$

where $F(x,y)$ is the density distribution for the joint distribution of X and Y, and $f(x)$ and $g(y)$ are the marginal distributions:

$$f(x) = \int F(x,y)dy, \qquad \text{(I.14)}$$

$$g(y) = \int F(x,y)dx. \qquad \text{(I.15)}$$

Appendix J Identical and indistinguishable (ID) particles

In daily life, we use the two concepts of identical and indistinguishable (ID) as synonyms. In physics, we distinguish between the two terms. Two particles are said to be identical if we cannot tell the difference between them. Two identical particles are said to be distinguishable if they can be given a *label*, in the sense that at each point of time, we can tell that this particle has this label, and that particle has that label. It should be said from the outset that by "label" we do not mean a tag, a number or any other kind of label that one *attached* to the particles. Clearly, such a label will make the identical particles un-identical, and hence distinguishable. What we need is a label that we can *assign* to the particles without affecting their identity. Or put better: two identical particles are said to be ID if we cannot find a label that distinguishes between the two, yet does not affect their being identical. Quite often, it is said that, "In quantum mechanics, identical particles are indistinguishable in principle."[9] This does not consist of a definition of what makes particles ID. Clearly, even in classical mechanics, one can say that "identical particles are indistinguishable in principle." Unless one defines indistinguishability, this sentence is merely a tautology. The

[9] Kubo (1965).

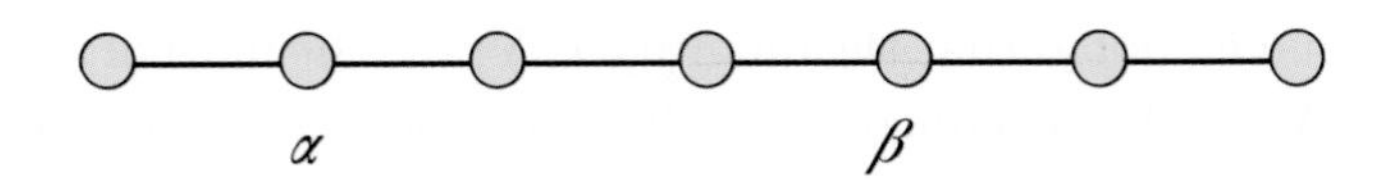

Figure J.1. Two identical, but distinguishable particles on two lattice points.

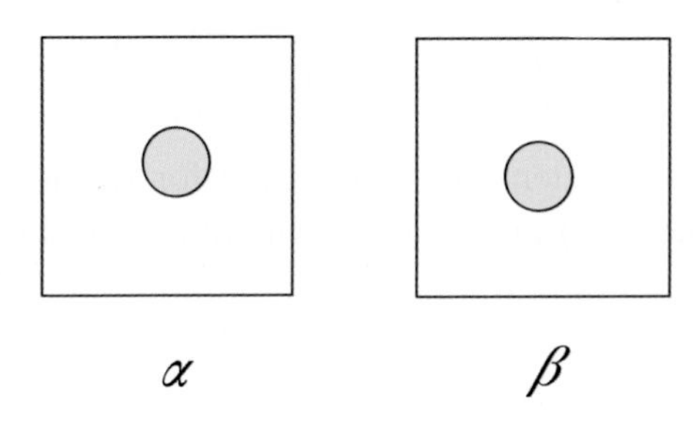

Figure J.2. Two identical, but distinguishable particles in two different boxes.

difference between classical and quantum mechanical views is that in the former, one can use the *trajectory* of a particle to make it distinguishable from another identical particle. This assignment of trajectories as labels is, *in principle*, not allowed in quantum mechanics.

Two kinds of "labels" are frequently encountered.

(i) Two identical particles at two different lattice sites, say α and β (Figure J.1).
(ii) Two identical particles in different boxes, say boxes α and β (Figure J.2).

In these two cases, we can always say that one particle is in site α (or box α), and the other is in site β (or box β). Clearly, these labels, though making the particles distinguishable, do not affect the identity of the particles. The situation is different when the two particles are in the same box. In this case, the two particles are *identical* and *indistinguishable* (ID). The reason is that these particles cannot be labeled. This distinction originates from the different ways particles are viewed in classical and in quantum mechanics; this brings us to the third way of labeling, which is "permitted" in a classical, but not in a quantum mechanical, world.

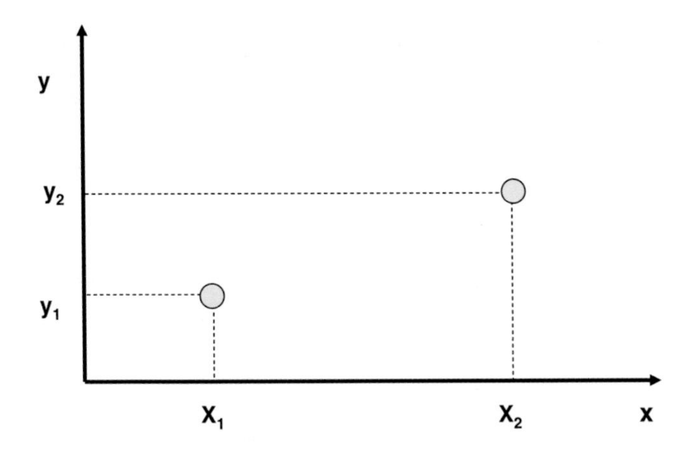

Figure J.3. Two particles that can be labeled by their coordinates.

In classical mechanics, the particles can be, in principle, labeled. For instance, the two particles shown in Figure J.3 can be assigned coordinates, say $\mathbf{R}_1 = (x_1, y_1, z_1)$, and $\mathbf{R}_2 = (x_2, y_2, z_2)$ at some time $t = 0$. When we write the equations of motion of N particles, we label the particles $1, 2, \ldots, N$. We solve the equations and, in principle, we can follow the trajectories of each particle having a specific label. That means that at each time t, we can tell which particle had the "label" $\mathbf{R}_1$ at time $t = 0$, the "label" $\mathbf{R}_2$ at time $t = 0$, etc. The trajectoy of each particle is unique and can serve as a label. The particles are thus distinguishable, though identical, if these labels are maintained.

In quantum mechanics, this labeling of microscopic particles is, in principle, impossible.[10] Even if we start with two particles with an *assigned* label at time $t = 0$, after some time, we cannot tell which particle originates from $\mathbf{R}_1$, which particle from $\mathbf{R}_2$, and so on.[11]

---

[10]Rushbrooke (1949) expressed a different opinion on this matter: "...there is nothing quantum mechanical about this result. The division by $N!$ is an essentially classical necessity produced by the indistinguishability of the systems."

[11]It is sometimes said that the labels are "erased," or the information on the labels is "lost." These statements could be potentially misleading since the *assigned* information is not really there. When two particles in the same box collide, the Heisenberg uncertainty principle prevents us from telling which outgoing particle can be identified with which incoming particle.

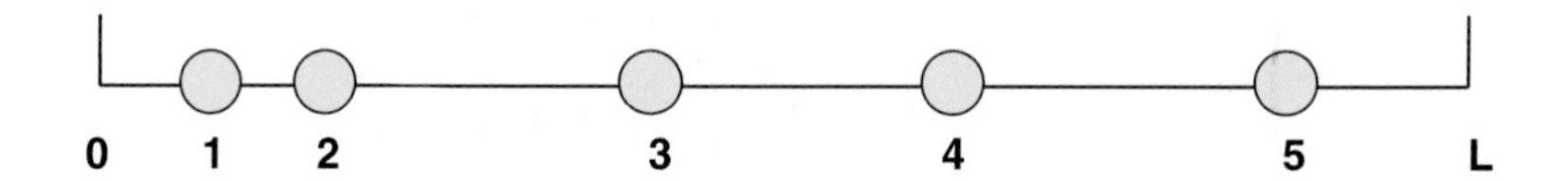

Figure J.4. A one-dimensional fluid.

It should be noted that in both examples (i) and (ii) as given above, the particles can either vibrate about the lattice points, or wander in the entire volume V of the box. The particles can reach an arbitrary close distance from each other. Yet, if we can always say that this particle belongs to site α (or to box α), and that that particle belongs to site β (or to box β), then the two particles have different labels and therefore are considered to be distinguishable (though identical). The labels, in this case, are α and β.

Another case of labeling of particles is a one-dimensional liquid (Figure J.4). If particles are assumed to be impenetrable (i.e., the intra-molecular potential is infinitely repulsive at very short distances), then we can label the particles as first, second, third, etc. In this case, the particles are considered to be distinguishable.

In the earlier literature, a distinction used to be made between *localized* and non-localized particles,[12] instead of distinguishable and indistinguishable. "Localized" normally refers to a particle at a lattice point, but it can also refer to particles in this or that box, i.e., particles that belong to site α, or to box α, can be said to be localized at α, or labeled by the index α.

We can generalize the concept of ID to many particles as follows: suppose that we have two boxes (Figure J.5), N_1 particles in α, N_2 particles in β, all particles being identical. We treat all the N_1 particles as ID among themselves, and the N_2 particles as ID among themselves, but the particles in α are *distinguishable* from the particles in β. It is very often said that if we *exchange* any two ID particles we get an indistinguishable state. That is true for identical particles as well. In the example of N_1 particles in one box and N_2 particles in another box it is true that exchanging one particle from α with one particle in β leaves the *state* of the

[12] See, for example, Rushbrooke (1949).

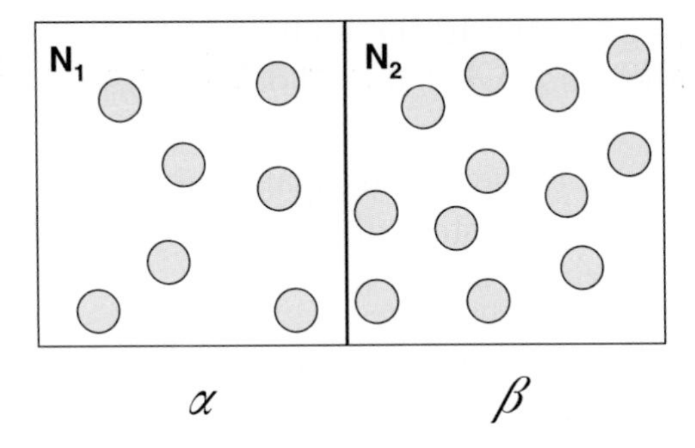

Figure J.5. Identical particles in two different boxes. The particles in one box are distinguishable from particles in the second box.

entire system unchanged, yet the two particles are not considered as ID. The reason is that the particles in α can be labeled as α, and particles in β can be labeled as β, and these labels are retained as long as the walls of the boxes are impenetrable. Thus, particles in α are *identical* but distinguishable from particles in β. Had we considered particles in α to be indistinguishable from particles in β, we would have gotten the following absurd result: removing the partition between the two compartments in Figure J.5 causes a change in free energy

$$\begin{aligned}\Delta A &= -T \ln \frac{Q_{final}}{Q_{initial}} = -T \ln \frac{(2V)^{N_1+N_2}(N_1+N_2)!}{(N_1+N_2)!V^{N_1}V^{N_2}} \\ &= -T \ln 2^{(N_1+N_2)} = -T(N_1+N_2)\ln 2 < 0. \end{aligned} \tag{J.1}$$

Clearly, there should be no change in the free energy if the volumes of α and β are the same and $N_1 = N_2 = N$. The error in the above reasoning is that we have assumed that the $(N_1 + N_2)$ particles in *both* boxes are ID. Instead, we should view the initial state as having N_1 ID particles in α, and N_2 ID particles in β, but in the final state, we have $(N_1 + N_2)$ ID particles in the combined volume $2V$. In this view, the change in free energy is

$$\begin{aligned}\Delta A &= -T \ln \frac{Q_{final}}{Q_{initial}} = -T \ln \frac{(2V)^{N_1+N_2}N_1!N_2!}{(N_1+N_2)!V^{N_1}V^{N_2}} \\ &= -T \ln 2^{(N_1+N_2)} - T \ln \frac{N_1!N_2!}{(N_1+N_2)!}. \end{aligned} \tag{J.2}$$

If $N_1 = N_2 = N$, then in the limit of large N, these two terms cancel each other and $\Delta A = 0$ as should be (see also Sections 6.6 and 6.7).

A special case is a one-dimensional liquid. The configurational partition function of a system of N identical particles in a box of length L can be written in two forms:

$$\begin{aligned} Q &= \frac{1}{N!}\int_0^L \cdots \int_0^L dX_1 \cdots dX_N \exp[-\beta U_N(X_1,\ldots,X_N)] \\ &= \int_{X_{N-1}}^L dX_N \int_{X_{N-2}}^{X_N} dX_{N-1} \cdots \\ &\quad \times \int_0^{X_2} dX_1 \exp[-\beta U_N(X_1,\ldots,X_N)]. \end{aligned} \tag{J.3}$$

In the first expression on the right-hand side of (J.3), we allow all the configurations $0 \le X_i \le L$ for each particle. In this case, the particle are ID. Hence, in performing the integration for each particle between 0 and L, we over-count configurations; thus, we need to divide by $N!$ to correct for over-counting. In the second expression, we impose *order* on the particles, i.e., we distinguish between the first, the second, the third ... the Nth particle — hence, no correction of $N!$ is required. In this case, the particles have *labels*; first, second, third, etc., and hence, though they are identical, they are distinguishable. In the example shown in Figure J.5, the individual particles in each compartment cannot be labeled. However, N_1 particles in one group have a common label (say α), and N_2 particles have a common label as well (say β). Once we remove the partition, these labels are no longer in effect, and all the $(N_1 + N_2)$ particles become indistinguishable.

An important difference between the concepts of *identity* and ID is the following.

Classically, we can conceive of a process that changes some attribute of the particles (say billiard balls) continuously so that they are transformed from being *different* to being *identical.* For instance, we can change the color, the tag or the form of particles (Figure J.6) continuously to make them identical, but not ID. One cannot change continuously from *identical* to indistinguishable, i.e., particles can either be labeled or not. The indistinguishability of

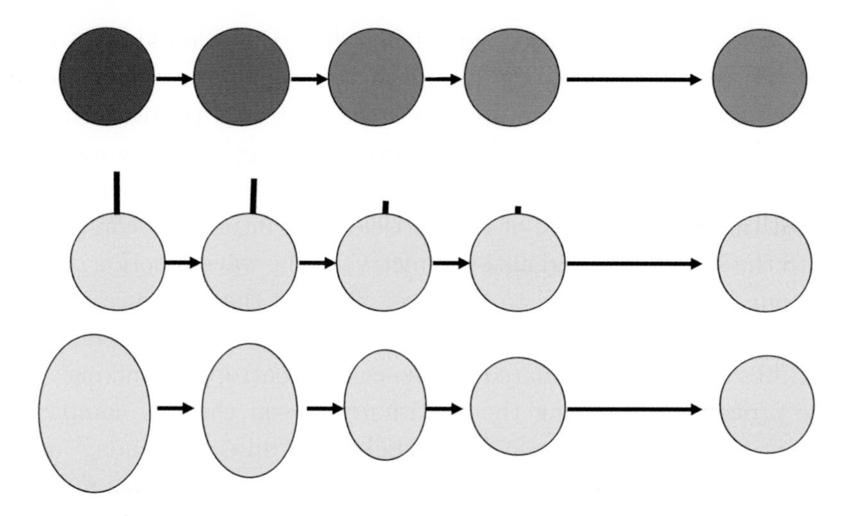

Figure J.6. Continuous change of colors, labels or shapes.

the particles is not a property that we can see or measure. If particles can exchange spontaneously in such a way that we cannot tell which particle is which, then the two identical particles are also ID. Particles localized on lattice points, or particles in different boxes cannot exchange spontaneously and therefore are considered to be label-able — hence, distinguishable. Thus, particles are either ID or not ID; there is no continuous transition from distinguishable to ID. As Lesk (1980) wrote: "*Indistinguishability is necessarily an all-or-nothing phenomenon.*" Not recognizing this discontinuity of the property of ID leads to the so-called Gibbs paradox. This is discussed in Appendix O.

It should be noted that the concept of impenetrable walls or impenetrable particles (in one dimension) is an idealization. In practice, whenever the barrier energy is so high that the probability of penetration is negligibly small, we say that the wall is impenetrable. Gibbs was the first to introduce the correction of $N!$ to the classical partition function. When writing the partition function for the combined systems as in Figure J.5 or for mixtures, Gibbs introduced $N_1!$ for one box and $N_2!$ for the second box. He reasoned that a configuration in which two particles are exchanged are considered as the same configuration. Hence, one must divide by $N_1!$ (or $N_2!$) for over-counting configurations in one box, or in the second box — but not $(N_1 + N_2)!$ since, by definition, we do not count configurations for which particles in one box are exchanged with particles in the second box. Gibbs' correction for over-counting of configurations was introduced before quantum mechanics. It was only later recognized that this property of the particles was intimately connected with the symmetry (and anti-symmetry) of the wave function of the system. This in turn led to the recognition of the existence of two types of indistinguishable particles, the Bosons and the Fermions.

Gibbs correctly calculated the so-called "entropy of mixing" in the process of removing the partition between the two compartments in Figure J.5, once for particles of "different kinds," and once for particles of the "same kind." However, Gibbs was apparently thinking of the particles as being in principle label-able, not recognizing the full implication of the indistinguishability of particles of the same kind. We further discuss this aspect of Gibbs' thoughts in Section 6.6 and Appendix O. In Section 6.8, we showed that there exists a spontaneous process (referred to as deassimilation) in which a group of particles can acquire a collective label spontaneously. This spontaneous process involves an increase in the MI of the system.

Appendix K The equivalence of the Boltzmann's and Jaynes' procedures to obtain the fundamental distribution of the canonical ensemble

In Chapter 5, we discussed the application of the maximum MI principle to derive the canonical distribution. Here, we show the equivalence between the traditional method and the maximum MI method.

The traditional method of obtaining the Boltzmann distribution is well known.[13] We construct an ensemble of M isolated systems $(M \to \infty)$, each of which has the same thermodynamical characterization E, V, N. We next bring the system to thermal contact. The volume and the number of particles remain fixed, but the energy

[13]See, for example, Hill (1960), Rushbrook (1965).

of each system can now fluctuate. We assume, for simplicity that each system can be in any one of the finite number of energy levels $E_i (i = 1, 2, \ldots, r)$. Let M_i be the number of systems being in a state having energy E_i. Clearly, the two conditions must be satisfied:

$$\sum_{i=1}^{r} M_i = M, \tag{K.1}$$

$$\sum_{i=1}^{r} M_i E_i = E_T = M\bar{E}, \tag{K.2}$$

where $\bar{E}$ is the average energy of a system in the ensemble.

The number of quantum mechanical states of the whole ensemble with a specific distribution $\mathbf{M} = (M_1, M_2, \ldots, M_r)$ is given by

$$W(\mathbf{M}) = \frac{M!}{\prod_{i=1}^{r} M_i!}, \tag{K.3}$$

and the total number of states of the ensemble as a whole is

$$W_T = \sum_{\mathbf{M}} W(\mathbf{M}), \tag{K.4}$$

where the sum is over all possible $\mathbf{M}$ subject to the constraints (K.1) and (K.2).

The Boltzmann procedure is to find the maximum of (K.3) subject to the two constraints (K.1) and (K.2). The result is

$$x_i^* = \frac{M_i^*}{M} \quad (i = 1, 2, \ldots, r), \tag{K.5}$$

where x_i^* is the most probable fraction of systems in the ensemble being in a state of energy E_i. Using the Lagrange method of undetermined multipliers, one obtains the well-known Boltzmann distribution

$$x_i^* = \frac{M_i^*}{M} = \frac{\exp[-\beta E_i]}{\sum_{j=1}^{r} \exp[-\beta E_j]}. \tag{K.6}$$

Jaynes' procedure can be obtained as follows. We write the logarithm of $W(\mathbf{M})$ in (K.3) and assuming the Stirling approximation for all $M_i!$, we get, in the limit $M \to \infty$

$$\begin{aligned}\ln W(\mathbf{M}) &\cong M \ln M - \sum_{i=1}^{r} M_i \ln M_i \\ &= M\left[-\sum_{i=1}^{r} p_i \ln p_i\right] \\ &= MH(p_1, p_2, \ldots, p_r),\end{aligned} \tag{K.7}$$

where $p_i = M_i/M$.

Thus, instead of maximizing $W(\mathbf{M})$ subject to the conditions (K.1) and (K.2), we can maximize $H(p_1, \ldots, p_r)$ subject to the conditions $\sum p_i = 1$ and $\sum p_i E_i = \bar{E} = E_T/M$. The two methods are equivalent when $M \to \infty$.

Clearly, the interpretation of H as a measure of *missing information* does not make the Jaynes' approach more subjective than the Boltzmann procedure, where the word "information" is not mentioned. Both of these approaches deal with physically objective quantities.

Appendix L An alternative derivation of the Sackur–Tetrode equation

Consider a point particle confined to a rectangular box of edges L_x, L_y, L_z. We assume that the walls of the box are such that they provide an infinitely high potential barrier so that the probability of crossing the walls is null. For such a system, the Schrödinger equation is

$$\frac{-h^2}{8\pi^2 m}\left(\frac{\partial^2}{\partial x^2} + \frac{\partial^2}{\partial y^2} + \frac{\partial^2}{\partial z^2}\right)\psi = E\psi, \tag{L.1}$$

where h is the Planck constant. With the boundary conditions imposed by the walls, the solutions of the Schrödinger equation have the form

$$\psi(x, y, z) = \sin\left(\frac{\pi x}{L_x} i\right) \sin\left(\frac{\pi y}{L_y} j\right) \sin\left(\frac{\pi z}{L_z} k\right), \tag{L.2}$$

and the corresponding energies are

$$E(i, j, k) = \frac{h^2}{8m}\left(\frac{i^2}{L_x^2} + \frac{j^2}{L_y^2} + \frac{k^2}{L_z^2}\right). \tag{L.3}$$

The solutions (L.2) are referred to as the eigenfunctions, and the energies (L.3) as the eigenvalues of the Schrödinger equation for this system, and i, j, k are integers.

For simplicity, we assume that the box is a cube of length L, in which case (L.3) reduces to

$$E(i,j,k) = \frac{h^2}{8mL^2}(i^2 + j^2 + k^2). \tag{L.4}$$

Note that the same energy is obtained for any triplet of integers, having the same sum $(i^2 + j^2 + k^2)$.

Next, we calculate the number of stationary states having energy between zero to ε. We also assume that i, j and k are large and can be treated as continuous variables. Thus, the number of stationary states having energy between zero and ε is equal to the volume of the sphere having radius r, where r is given by

$$r = \sqrt{i^2 + j^2 + k^2} = \sqrt{\frac{8mL^2\varepsilon}{h^2}}. \tag{L.5}$$

The volume of the sphere of radius r is $\frac{4\pi r^3}{3}$, but since we are interested only in triplets of positive numbers $(i^2 + j^2 + k^2)$, we need only 1/8 of this sphere which is $\pi r^3/6$. The total number of states within this "volume" that has energy between zero to ε is

$$W(\varepsilon) = \frac{\pi r^3}{6} = \frac{\pi}{6}\left(\frac{8mL^2\varepsilon}{h^2}\right)^{3/2} = \frac{\pi}{6}\left(\frac{8m\varepsilon}{h^2}\right)^{3/2} V. \tag{L.6}$$

The number of states with energy between ε and $\varepsilon + d\varepsilon$ is thus

$$W(\varepsilon + d\varepsilon) - W(\varepsilon) \approx \frac{\partial W}{\partial \varepsilon} d\varepsilon = \frac{\pi}{6}\left(\frac{8m}{h^2}\right)^{3/2} V \frac{3}{2}\varepsilon^{1/2} d\varepsilon = \varpi(\varepsilon) d\varepsilon. \tag{L.7}$$

The partition function for one molecule is thus

$$q = \int_0^\infty \varpi(\varepsilon)\exp(-\beta\varepsilon) d\varepsilon. \tag{L.8}$$

Changing variables $x = \beta\varepsilon$, the integral in (L.8) becomes a definite integral and of the form

$$\int_0^\infty \sqrt{x}\exp(-x)dx = \sqrt{\frac{\pi}{4}}. \tag{L.9}$$

Hence, from (L.7)–(L.9), we obtain

$$q = \frac{V}{\Lambda^3}, \tag{L.10}$$

where

$$\Lambda^3 = \left(\frac{h}{\sqrt{2\pi mT}}\right)^3. \tag{L.11}$$

The partition function for a system of N such particles is

$$Q(T, V, N) = \frac{V^N}{N!\Lambda^{3N}}. \tag{L.12}$$

Note that this form is valid for the case of the Boltzmann statistics, i.e., when the number of states is very large compared with N, i.e., when

$$\left(\frac{2\pi mT}{h^2}\right)^{3/2} V \gg N, \tag{L.13}$$

or equivalently, when

$$\frac{N}{V}\Lambda^3 \ll 1. \tag{L.14}$$

The quantity Λ^3 is referred to as either the momentum partition function, or as the thermal de Broglie wavelength. Indeed, for a particle with the velocity $v \approx \sqrt{T/m}$, using the relation $\Lambda = \frac{h}{p} = \frac{h}{mv}$, we get

$$\Lambda \approx \frac{h}{\sqrt{2\pi mT}}. \tag{L.15}$$

The condition $\rho\Lambda^3 \ll 1$ essentially states that the thermal wavelength should be much smaller than the average distance between the particles.

From the canonical partition function (L.12), we can get all the thermodynamic quantities for an ideal gas (see Section 5.3). In particular, we have for the MI of an ideal gas

$$\begin{aligned} S &= \left(\frac{-\partial A}{\partial T}\right)_{V,N} = \left(\frac{\partial(T \ln Q)}{\partial T}\right)_{V,N} \\ &= N \ln\left[\frac{V}{N\Lambda^3}\right] + \frac{5}{2}N, \end{aligned} \tag{L.16}$$

which is the Sackur–Tetrode equation. (See Sections 4.3 and 5.4.)

Appendix M Labeling and un-labeling of particles

In Section 6.3, we discussed the pure assimilation process. It is *pure* in the sense that *only* the number of indistinguishable particles is changed. The volume accessible to each particle does not change, nor does the temperature. Hence, there is no change in the MI associated with either the locational distribution or the velocity distribution of the particles. As was pointed out in Sections 6.6 and 6.7, there is a conceptual difficulty in interpreting the kind of informational *gain* in the assimilation process (or the loss of information in the deassimilation process). This difficulty is probably the reason for Gibbs' erroneous conclusion that the reversing of process V of Section 6.6 is "entirely impossible."

The interpretation offered here applies to all the three cases discussed in Section 6.3, but for simplicity, we assume that $M \gg N$; hence the change in the MI is always

$$\Delta H_{qm} = -\ln N < 0. \tag{M.1}$$

To interpret this result, suppose we could do the same process V, as in Section 6.3, in a purely *classical* world. By classical world, I do not mean the MB statistics but a world where the particles are *distinguishable*, i.e., particles can be labeled. Hence, the mutual information among the particles is zero (see Section 4.3).

Now we perform the same process in the classical (cl) world and in a quantum mechanical (qm) world, as shown in Figure M.1.

In the quantum world, the change in the MI, ΔH_{qm} is as in (M.1). However, in the classical world, the same process does not involve any change in the MI. Hence, we have

$$\Delta H_{cl} = 0. \tag{M.2}$$

The transfer of each box from the quantum world to the classical world involves the change of MI due to labeling the particles (i.e., changing from indistinguishable to distinguishable particles). These are

$$\Delta H_a = 0, \tag{M.3}$$

i.e., the MI of a system with *one* particle is unchanged by labeling it.

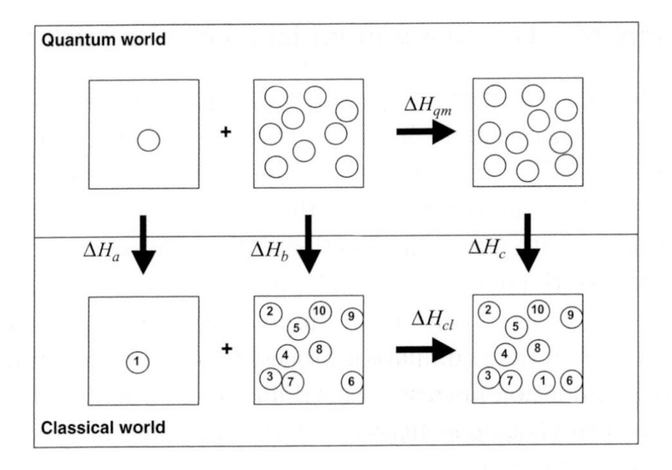

Figure M.1. The cycilc process of assimilation as described in Appendix M.

On the other hand,

$$\Delta H_b = \ln(N-1)! \tag{M.4}$$

and

$$\Delta H_c = \ln N!, \tag{M.5}$$

i.e., labeling involves increase in the MI (or un-labeling involves decrease in the MI). Thus, instead of carrying out the process in the quantum world, we first transform the two boxes into the classical world, perform the process there and then transfer the resulting system back to the quantum world. The balance of information is

$$\begin{aligned} \Delta H_{qm} &= \Delta H_a + \Delta H_b - \Delta H_c + \Delta H_{cl} \\ &= -\ln N!/(N-1)! = -\ln N. \end{aligned} \tag{M.6}$$

Thus, the reduction of the MI in the assimilation process is a result of the *difference* in the MI associated with un-labeling N and $(N-1)$ particles, respectively.

Appendix N Replacing a sum by its maximal term

In statistical thermodynamics, we frequently use the approximation of replacing a sum of a huge number of terms by only the maximal term. Intuitively, this sounds a very unreasonable approximation, in particular when the sum consists of many positive terms.

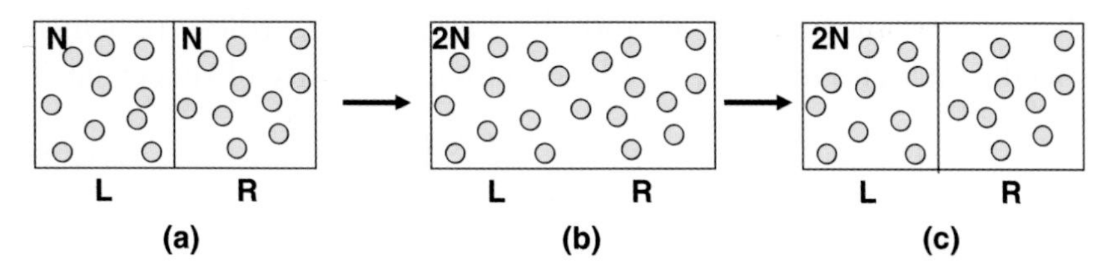

Figure N.1. Removing and reinserting a partition between two compartments.

We demonstrate the general principle with a simple example. Suppose we have two compartments R and L of equal volumes and containing N particles each (Figure N.1a).

We start with the identity which is a particular case of the Binomial theorem:

$$(1+1)^N = \sum_{n=0}^{N} \binom{N}{n} = \sum_{n=0}^{N} \frac{N!}{n!(N-n)!}. \tag{N.1}$$

Neither this identity, nor the approximation which we shall describe soon, has anything to do with any physical system. However, it is helpful to think of it in terms of configurations of the particles in the two boxes. We define a *specific* configuration of the system as a list of the *specific* particles that are in compartment R and in compartment L. We assume that the probability of any specific particle to be found in R is $1/2$. Likewise, the probability to be found in L is $1/2$.

We also assume that the events "particle i is in R" are independent events (for $i = 1, 2, \ldots, N$). Therefore, the probability of finding any specific set of particles, say $1, 2, 3, \ldots, k$ in R, and the remaining particles, $k+1, k+2, \ldots, N$ in L is

$$\Pr(\textit{specific}) = \left(\frac{1}{2}\right)^k \left(\frac{1}{2}\right)^{N-k} = \left(\frac{1}{2}\right)^N. \tag{N.2}$$

Thus, all of the 2^N specific configurations have equal probability 2^{-N}.

Next, we define the generic configurations, as the event that compartment R contains n particles, and the compartment L contains $N-n$ particles. Clearly, since all the specific configurations are disjoint events, the probability of the generic event "n particles in R" is the sum of the probabilities of all the specific events for which there are n particles in R and $N-n$ in L, regardless of which specific particles are in R and in L. This probability of this event is

$$P_N(n) = \frac{1}{2^N}\binom{N}{n}. \tag{N.3}$$

This probability follows from the classical definition of probability (Section 2.3). There are $\binom{N}{n}$ specific events that conform with the generic event "n particle in R and $N-n$ in L," and the total number of specific events is 2^N.

Recall that the Stirling approximation is quite good for n of the order of 100, and it is excellent for numbers of the order of Avogadro numbers 10^{23} (Appendix E). We use the approximation in the form

$$n! \approx \left(\frac{n}{e}\right)^n \sqrt{2\pi n}. \tag{N.4}$$

Since all the terms in (N.1) are positive, we have the inequality

$$2^N = \sum_{n=0}^{N} \frac{N!}{n!(N-h)!} \geq \frac{N!}{[(N/2)!]^2}. \tag{N.5}$$

Recall that at $n = N/2$, the probability (N.3) has a maximum. Thus, on the right-hand side of (N.5), we have the maximal terms of the sum on the left-hand side. We now apply the Stirling approximation (N.4) to the right-hand side of (N.5) to obtain

$$2^N \geq \frac{N!}{[(N/2)!]^2} \approx 2^N \sqrt{\frac{2}{\pi N}}, \tag{N.6}$$

or equivalently

$$\ln \sum_{n=0}^{N} \binom{N}{n} - \ln \binom{N}{N/2} \geq \ln \sqrt{\frac{2}{\pi N}}. \tag{N.7}$$

In thermodynamics we are dealing with the logarithm of the sums (N.1) in which case we have

$$\ln \sum_{n=0}^{N} \binom{N}{n} = N \ln 2$$

$$> \ln \binom{N}{N/2} \approx N \ln 2 + \ln \sqrt{\frac{2}{\pi N}}. \tag{N.8}$$

Therefore, when dealing with the logarithm of the sum in (N.1), we see that as N increases, the difference between the logarithm of the sum and the logarithm of the maximal term is of the order of $\ln N$, whereas the sum itself is of the order of N. As an example, for $N = 10^{23}$, we have

$$\ln \sum_{n=0}^{N} \binom{N}{n} \cong 10^{23} \ln 2 \tag{N.9}$$

$$\ln \binom{N}{N/2} \cong 10^{23} \ln 2 + \ln \sqrt{\frac{2}{\pi 10^{23}}}. \tag{N.10}$$

We conclude that for a thermodynamic system, when N is very large, we can use the Stirling approximation in the form

$$n! \approx \left(\frac{n}{e}\right)^n. \tag{N.11}$$

Applying (N.11) to (N.9) and (N.10), we get an approximate equality, i.e.,

$$\ln \sum_{n=0}^{N} \binom{N}{n} \approx \ln \binom{N}{N/2}. \tag{N.12}$$

This "equality" should be understood in the sense that the difference of the two quantities in (N.12) is negligible compared with the value of each quantity.

One should be careful not to apply the approximation (N.11) to the probability in (N.3) which will lead to the absurd result. Thus, for $n = N/2$ in (N.3), we get

$$P_N(n = N/2) \approx \frac{2^N}{2^N} = 1. \tag{N.13}$$

Instead, one must use the approximation (N.4) to obtain

$$P_N(n = N/2) = \frac{2^N}{2^N} \sqrt{\frac{2}{\pi N}}. \tag{N.14}$$

Thus, the probability of finding n particles in R and $(N - n)$ particles in L has a maximum at $n = N/2$. The *value* of the maximum probability *decrease* with N as $N^{-1/2}$.

We conclude this appendix with a comment on the experimental significance of the approximation (N.12). Suppose we start with the two compartments having initially exactly $N/2$ in each compartment. We remove the partition and let the particles "mix" (Figure N.1b). In the new system, all of the 2^N specific configurations have the same probability 2^{-N}. We next place the partition in exactly the original position. The new system will now differ from the initial system for two reasons (see Figure N.1c):

(i) Different specific particles will occupy R and L.
(ii) The number of particles in R can be any number between zero to N.

In spite of this huge number of possibilities for the final state, we can claim that thermodynamically speaking, the final state will be "equal" to the initial state. The reasons are: first, the particles are indistinguishable; therefore, there are only $(N + 1)$ possible generic configurations, i.e., we cannot distinguish between all the 2^N specific configurations. Second, the number of particles in R will most likely be different from $N/2$, but the difference will be very small compared with $N/2$ itself. Therefore, we can conclude that the final state (after placing the partition) will be, thermodynamically speaking, identical to the initial state.

Appendix O The Gibbs paradox (GP)

There are several versions of the so-called Gibbs paradox (GP). None is really a paradox, and none was viewed as a paradox by Gibbs himself. We shall discuss here two versions of the GP and show that one arises from treating the particle classically, not recognizing the indistinguishability (ID) of the particles, the other involves the fallacious idea that the ID of the particles is a property that can be changed continuously.

The first "paradox" arises when we use the (purely) classical partition function. As we saw in Section 5.3, the classical PF gives the correct equation of state, the correct heat capacity and some other properties of an ideal gas. It fails to give the correct MI or the chemical potential. More specifically, the MI (or the entropy) of a system, derived from the classical partition function, does not have the additive property, or more generally, S_{Class} is not an extensive function of the variable E, V, N.

As we have also seen in Section 5.3, by correcting the classical PF for the ID of the particles, the thermodynamic MI becomes an extensive function.

Clearly, the fact that a wrong result is obtained from the classical PF does not consist of a paradox. In the history of science, there are abundant examples of incorrect results obtained from an inappropriate theory. For instance, lattice models of the liquid states yielded some correct results for mixtures, but failed to predict the entropy of liquids. This discrepancy was never considered to be a paradox; it was "fixed" in an *ad hoc* manner by adding the so-called *communal entropy*.[14] However, this remedy was abandoned later when it was recognized that a lattice model is inherently inappropriate to describe the liquid state.

The second version of the "paradox" is associated with the so-called "entropy of mixing." In Chapter 6, we discussed two processes that were analyzed by Gibbs. As we saw in Sections 6.6 and 6.7, the MI (or the entropy) change in processes IV and V of Figures 6.6 and 6.7 are

$$\Delta S_{IV} = 2N \ln 2, \tag{O.1}$$

$$\Delta S_{V} = 0. \tag{O.2}$$

The paradox in this case is often stated as follows.[15] Suppose we could have changed the "extent of ID" of the particle continuously, say linearly from the distinguishable particles to ID. Similar to the processes depicted in Figure J.6. If we do that, we should have expected that the value of ΔS should also change continuously from ΔS_{IV}, when the two components A and B are distinguishable, to ΔS_V when they become ID. The fact is that one never observes any intermediary value between ΔS_{IV} and ΔS_V. The fact that

[14]See Ben-Naim (2006).

[15]Note that if one uses the purely classical PF for ideal gases, one does not get the results (O.1) and (O.2) which are consistent with experiments. We assume here that (O.1) and (O.2) were derived from the classical limit of the quantum mechanical PF (see also Section 5.3), i.e., after introducing the correction due to the ID of the particles.

ΔS changes *discontinuously* as one changes the ID *continuously* is viewed as a paradox. However, there is no paradox here, and there was no allusion to any paradox in Gibbs' writings. The fact that the density of water changes discontinuously when the temperature changes continuously, say between 90°C to 110°C, is not viewed as a paradox. Furthermore, the presumed continuous change in the "extent of ID" of the particles is now recognized as, in principle, invalid. Particles are either distinguishable or ID — there are no intermediate values of indistinguishability.

In process IV, the two components A and B are different, whereas in process V, the particles are all identical. In the initial state, the particles in each compartment are ID among themselves, but the particles in one compartment are distinguishable from the particles that are in the second compartment.

Upon removal of the partition in process IV, the particles A and B remain distinguishable. On the other hand, removal of the partition in process V makes all the particles in the system indistinguishable.

As we noted in Section 6.4, Gibbs did notice the remarkable fact that the "entropy of mixing" is independent of the degree of similarity between the particles. This fact seems to be more puzzling than the fact that ΔS collapses discontinuously to zero when the particles become identical. However, the independence of the entropy of mixing on the *kind* of particles is only puzzling when we view the *mixing* itself as the cause of $\Delta S > 0$. Once we recognize that it is the expansion (i.e., the increase of the accessible volume for each particle), not the mixing, that is responsible for the positive change in the MI, the puzzlement evaporates.

The real paradox that arises from Gibbs' writings and that seemed to elude the attention of scientists is the following. In analyzing the two processes IV and V, Gibbs correctly obtained the results (O.1) and (O.2). Today, we would say that process IV, for which $\Delta S > 0$, is irreversible — in the sense that it cannot be reversed spontaneously. On the other hand, process V in which $\Delta S = 0$ is said to be reversible.

However, Gibbs concluded (see Section 6.7) that process IV *can* be *reversed* — in the sense that the system can be brought to its initial state. This reversal of process IV would require investing energy. However, for process V, Gibbs concluded that its reversal is "entirely impossible." Here is a paradox: How can a process, which is deemed to be *reversible* (process V), be at the same time "*entirely impossible*" to reverse? As I have discussed in Section 6.7, this apparent paradox is only an illusion. It is an illusion arising from our mental imaging of process V, in which particles are assigned mental coordinates, and mental trajectories.

Appendix P The solution to the three prisoner's problem

Formulation of the mathematical problem

We present here the solution to the three prisoners' problem from Section 2.6.2. Define the following three events (Figure P.1):

$$
\begin{aligned}
Af &= \{\text{Prisoner } A \text{ is going to be freed}\}, \quad P(Af) = 1/3,\\
Bf &= \{\text{Prisoner } B \text{ is going to be freed}\}, \quad P(Bf) = 1/3,\\
Cf &= \{\text{Prisoner } C \text{ is going to be freed}\}, \quad P(Cf) = 1/3, \qquad \text{(P.1)}
\end{aligned}
$$

and the following three events:

$$
\begin{aligned}
Ad &= \{\text{Prisoner } A \text{ is going to die}\}, \quad P(Ad) = 2/3,\\
Bd &= \{\text{Prisoner } B \text{ is going to die}\}, \quad P(Bd) = 2/3,\\
Cd &= \{\text{Prisoner } C \text{ is going to die}\}, \quad P(Cd) = 2/3 \qquad \text{(P.2)}
\end{aligned}
$$

We also denote by $W(B)$ the following event:

$$
\begin{aligned}
&W(B)\\
&\quad = \{\text{The warden points at } B \text{ and says that he will be executed}\}. \qquad \text{(P.3)}
\end{aligned}
$$

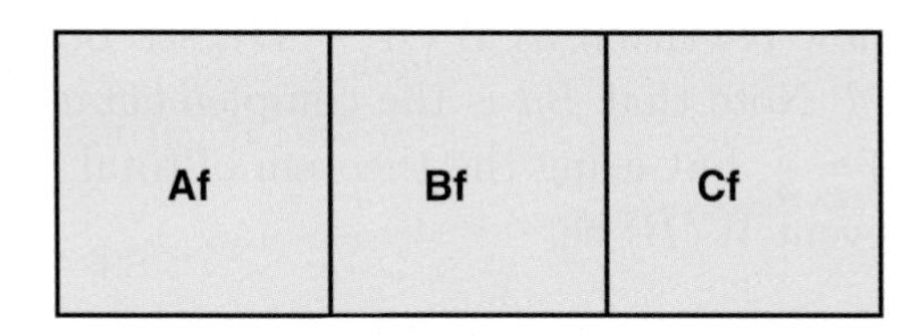

Figure P.1. The Venn diagram for the three events Af, Bf, and Cf.

First "solution"

A asks the question. He *knows* that B will die. Therefore, the probability that A will be freed, given that B is going to die is

$$P(Af/Bd) = \frac{P(Af \cdot Bd)}{P(Bd)}$$
$$= \frac{P(Bd/Af)P(Af)}{P(Bd)} = \frac{1 \times \frac{1}{3}}{\frac{2}{3}} = \frac{1}{2}. \quad \text{(P.4)}$$

Similarly,

$$P(Cf/Bd) = \frac{P(Cf \cdot Bd)}{P(Bd)} = \frac{P(Bd/Cf)P(Cf)}{P(Bd)}$$
$$= \frac{1 \times 1/3}{\frac{2}{3}} = \frac{1}{2}. \quad \text{(P.5)}$$

According to this "solution", the conditional probabilities of either events Af or Cf are the same. This solution is intuitively appealing; there were equal probabilities for Af and Cf *before* A asked the question and the probabilities remain equal *after* A asked the question.

This solution, though intuitively appealing, is wrong. The reason is that we have used the event Bd as the *given* condition, instead we must use the event $W(B)$. If *given* Bd is the condition, then the probabilities in (P.4) and (P.5) are correct. However, the "information" given to A is not Bd but $W(B)$, and $W(B)$ is an event *contained* in Bd. In other words, if one knows that $W(B)$ occurred, then it follows that Bd is true. However, if Bd is true, it does not necessarily follow that $W(B)$ occurred. In terms of Venn diagrams (Figure P.2), we can see that the size (or the probability) of the event $W(B)$ (now rewritten as $W(A \to B)$, see below) is smaller than that of Bd. Note that Bd is the complementary event to Bf. We had $P(Bd) = \frac{2}{3}$, but using the theorem of total probability, we can write the event $W(B)$ as:

$$W(B) = W(B) \cdot Af + W(B) \cdot Bf + W(B) \cdot Cf \quad \text{(P.6)}$$

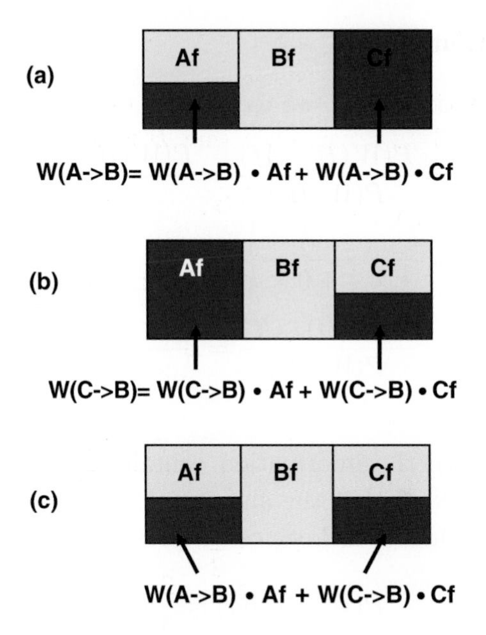

Figure P.2. Various Venn diagrams for the solution of the three prisoners problem.

and the probability of $W(B)$ is:

$$
\begin{aligned}
P(W(B)) &= P(W(B)/Af)P(Af) + P(W(B)/Bf)P(Bf) \\
&\quad + P(W(B)/Cf)P(Cf) \\
&= \frac{1}{2} \times \frac{1}{3} + 0 \times \frac{1}{3} + 1 \times \frac{1}{3} = \frac{1}{2}.
\end{aligned} \tag{P.7}
$$

Note that in (P.7), we capitalize on the fact that the warden is indifferent or unbiased towards *A or C*; if *Af* has occurred, i.e., A is going to be freed, then the warden can point at either *B or C*. The choice he makes is with probability $\frac{1}{2}$. On the other hand, if *Cf* occurred, then the warden does not have a choice but to point at B (as a result of A's question). Recognizing that the event $W(B)$ is smaller than *Bd*, in the sense that $W(B) \subset Bd$), it follows that $W(B)$ contains more information than *Bd*. Therefore, we must use $W(B)$ in the solution of the problem.

The second solution

Instead of (P.4) and (P.5), we write

$$P(Af/W(B)) = \frac{P(W(B)\cdot Af)}{P(W(B))} = \frac{P(W(B)/Af)P(Af)}{P(W(B))}$$
$$= \frac{\frac{1}{2}\times\frac{1}{3}}{\frac{1}{2}\times\frac{1}{3}+0\times\frac{1}{3}+1\times\frac{1}{3}} = \frac{1}{3}, \quad (P.8)$$
$$P(Cf/W(B)) = \frac{P(W(B)\cdot Cf)}{P(W(B))} = \frac{1\times\frac{1}{3}}{\frac{1}{2}\times\frac{1}{3}+0\times\frac{1}{3}+1\times\frac{1}{3}} = \frac{2}{3}. \quad (P.9)$$

Thus, if we use the information contained in $W(B)$, we get a different result. Note that the amount of information contained in $W(B)$ is larger than in Bd, we can express this as

$$P(W(B)/Bd) = \frac{2}{3} < 1,$$
$$P(Bd/W(B)) = 1.$$

Note that neither the "information" nor the "amount of information" that we refer to here is the kind of information used in information theory (see also the last example at the end of this appendix).

Thus, in solving the problem intuitively, we tend to use the *given* information Bd. This is less than the available information to A which is $W(B)$. Therefore, the correct solution is (P.8) and (P.9). In other words, by switching names, A can *double* his chances of survival.

A more general problem but easier to solve

The solution to the problem of the three prisoners is not easily accepted: it runs against our intuition which tells us that if the probabilities of the two events Af and Cf were initially equal, the equality of the probabilities must be maintained.

A simple generalization of the problem should convince the skeptic that the information given by the warden indeed changes the relative probabilities.

Consider the case of 100 prisoners named by the numbers "1", "2",... "100". It is known that only one prisoner is going to get freed with probability 1/100. Suppose that prisoner "1" asked the warden, "who out of the remaining 99 prisoners will be executed"? The warden points to the following: "2", "3",..., (exclude "67")...,"100", i.e. the warden tells "1" that all the 98 prisoners, except the "67", will be executed.

Clearly, by switching names with "67", the prisoner "1" increases his chances of survival from 1/100 to 99/100.

Note that initially, prisoner "1" knows that he has a 1/100 chance of surviving. He also knows that one of the remaining 99 prisoners has a 99/100 chances of survival. By acquiring the information on the 98 prisoners who will be executed, prisoner "1" still has a 1/100 chance of survival, but the chances of survival of *one* of the remaining 99 prisoners is now "concentrated" on one prisoner named "67." The latter has now a 99/100 chance of survival, therefore "1" should pay any price to switch names with "67." An illustration for ten prisoners is shown in Figure P.3b.

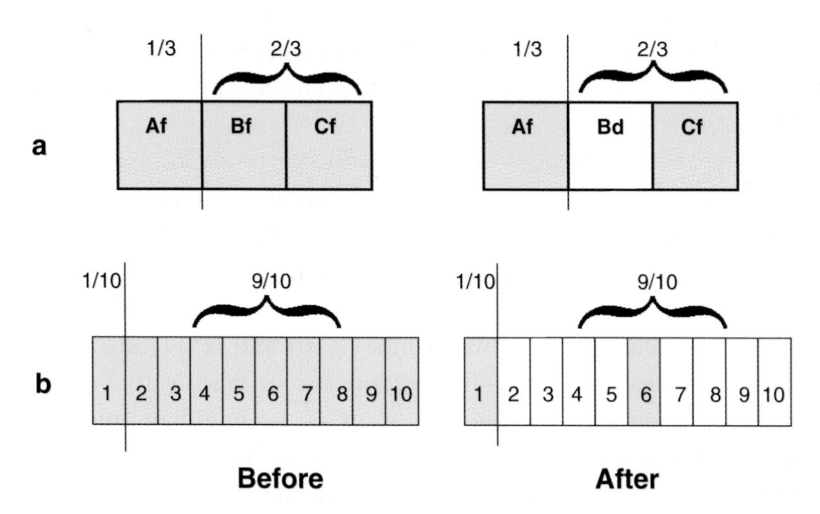

Figure P.3. The probabilities before and after the warden answers the question.
(a) The warden pointed at B to be executed. The probability 2/3 is now "concentrated" at C.
(b) The warden pointed at: 2,3,4,5,7,8,9,10 to be executed. The probability 9/10 is now "concentrated" at "6".

An apparent paradox

A student who followed the arguments of the solution to this problem asked: suppose that C asks the *same* question and the warden answers with the same answer, i.e., the warden points at B to be executed. Therefore, given $W(B)$, we should conclude, by the same arguments as above, that it is in C's advantage to switch names with A. But we have just concluded that it is in A's advantage to switch names with C.

It sounds as if given the *same* information $W(B)$, we reach two conflicting conclusions. If A asks the warden, then $P(Af/W(B)) = 1/3$ and $P(Cf/W(B)) = 2/3$. But if C asks the warden then $P(Cf/W(B)) = 1/3$ and $P(Af/W(B)) = 2/3$. How come we arrive at different conclusions based on the *same* information given to either A or to C?

The apparent paradox is a result of our reference to $W(B)$ as the *same* information, when given to either A or to C. In order to remove the apparent paradox, we should redefine $W(B)$ more precisely: instead of (P.3), we should define

$$W(A \to B) = \{\text{The warden points at } B \text{ to be doomed as a result of } A\text{'s question}\}. \tag{P.10}$$

Similarly,

$$W(C \to B) = \{\text{The warden points at } B \text{ to be doomed as a result of } C\text{'s question}\}. \tag{P.11}$$

Now it is clear that the two events (P.10) and (P.11) are not the same. The reason is that the warden *knows* who is to be freed; therefore, when answering either A *or* C, he might or might not point to the same prisoner. In terms of Venn diagrams, we can write the events (P.10) and (P.11) as (see Figure P.2)

$$\begin{aligned} W(A \to B) &= W(A \to B)\cdot Af + W(A \to B)\cdot Bf \\ &\quad + W(A \to B)\cdot Cf, \\ P(W(A \to B)) &= \frac{1}{2}\times\frac{1}{3} + 0\times\frac{1}{3} + 1\times\frac{1}{3} = \frac{1}{2}, \end{aligned} \tag{P.12}$$

$$\begin{aligned} W(C \to B) &= W(C \to B) \cdot Af + W(C \to B) \cdot Bf \\ &\quad + W(C \to B) \cdot Cf, \\ P(W(C \to B)) &= 1 \times \frac{1}{3} + 0 \times \frac{1}{3} + \frac{1}{2} \times \frac{1}{3} = \frac{1}{2} \end{aligned} \tag{P.13}$$

Note that the events $W(A \to B)$ and $W(C \to B)$ are *different* (see Figure P.2), but their probabilities (size of the area in the Venn diagram) are the same.

Since the warden knows who is going to be freed, the information given to A is different from the information given to C. Therefore, the conditional probabilities will be different depending on who receives the information. It is true though that the warden answers to both A and C by using the *same words*, i.e., "B is doomed to die," but the significance of this information is different if given to A as a result of A's question or as a result of B's question. The solution we reached above is that if $W(A \to B)$ is true, then it is in A's advantage to exchange names with C. If, on the other hand, C receives the *same* information, i.e., if C knows $W(A \to B)$, then it is still in A's advantage to exchange names. It is only when C asks the warden and gets the *different* information $W(C \to B)$, that it is in C's advantage to exchange names with A. Thus, in this problem, we have four different conditional probabilities:

(i) A asks and receives $W(A \to B)$:

$$P(Af/W(A \to B)) = \frac{P(W(A \to B)/Af)P(Af)}{P(W(A \to B))} = \frac{\frac{1}{2} \times \frac{1}{3}}{\frac{1}{2}} = \frac{1}{3}, \tag{P.14}$$

and

(ii) A asks, and C receives the answer $W(A \to B)$:

$$P(Cf/W(A \to B)) = \frac{P(W(A \to B)/Cf)P(Cf)}{P(W(A \to B))} = \frac{2}{3}. \tag{P.15}$$

(iii) C asks and receives $W(C \to B)$:

$$P(Cf/W(C \to B)) = \frac{P(W(C \to B)/Cf)P(Cf)}{P(W(C \to B))} = \frac{\frac{1}{2} \times \frac{1}{3}}{\frac{1}{2}} = \frac{1}{3} \tag{P.16}$$

and

(iv) C asks, and A receives the answer $W(C \to B)$:

$$P(Af/W(C \to B)) = \frac{P(W(C \to B)/Af)P(Af)}{P(W(C \to B))} = \frac{1 \times \frac{1}{3}}{\frac{1}{2}} = \frac{2}{3}. \tag{P.17}$$

Thus, we see that if either A or C receives the *same* information, then the conclusion is the same and there is no paradox.[16]

To complete the list of possible cases, we should add two more cases:

(v) No one asks any question. In this case, the probabilities are equal as in (P.1).

(vi) If both A and C ask the warden the same question and the warden points at B. Both A and C know the two answers. In this case, the known information is $W(A \to B) \bigcap (C \to B)$ and the probabilities are:

$$\begin{aligned} P(Af/W(A \to B) \cdot W(C \to B)) &= \frac{P(W(A \to B) \cdot W(C \to B)/Af)P(Af)}{P(W(A \to B) \cdot W(C \to B))} \\ &= \frac{\frac{1}{2} \times \frac{1}{3}}{\frac{1}{3}} = \frac{1}{2}, \end{aligned} \tag{P.18}$$

and similarly for

$$P(Cf/W(A \to B) \cdot W(C \to B)) = 1/2.$$

Note that:

$$\begin{aligned} W(A \to B) \cdot W(C \to B) &= W(A \to B) \cdot W(C \to B) \cdot Af \\ &\quad + W(A \to B) \cdot W(C \to B) \cdot Cf. \end{aligned} \tag{P.19}$$

Thus, the size (probability) of the intersection event, is 1/3 (see Figure P.2c). In this case, both A and C receive the *same* information. Hence, no one will have an advantage by switching their names.

[16] Sometimes, this apparent paradox is used to reject the solution of this problem. How could it be that given the *same* information (meaning $W(A \to B)$ leads to different conclusions? The answer is that the *same* information does lead to the *same* conclusion.

Exercise: In the solution given in (P.8) and (P.9), we have assumed that the warden is completely indifferent in the sense that if he knows Af, then he chooses to point at either B *or* C with equal probabilities. Equivalently, he tosses a fair coin to make his decision. Suppose that A asks the same question and he knows that the coin the warden uses (if indeed he uses one) is unbalanced. Therefore, if the warden needs to make a decision between B and C (i.e., when he knows Af), he tosses the coin with probabilities

$$\begin{aligned} &P(W(A \to B)/Af) = x, \\ &P(W(A \to C)/Af) = 1 - x. \end{aligned} \tag{P.20}$$

How does this new information affects A's decision to switch or not to switch?

As we have noted earlier, all the "information" used in this problem is not the kind of information used in information theory. However, one can define Shannon's information measure for this problem as follows

Initially, we have three equally likely events; hence,

$$H = -\sum_{i=1}^{3} p_i \log_2 p_i = \log_2 3 \cong 1.585. \tag{P.21}$$

When A asks the warden and receives his answer, $[W(A \to B)]$, there are only two possibilities; the corresponding MI is

$$H = -\frac{1}{3}\log_2\frac{1}{3} - \frac{2}{3}\log_2\frac{2}{3} \equiv 0.918. \tag{P.22}$$

In the more general problem, with 100 prisoners, we start with

$$H = -\sum_{i=1}^{100} p_i \log_2 p_i = \log_2 100 \cong 6.644 \tag{P.23}$$

and after prisoner "1" receives information from the warden, H reduces to

$$H = -\frac{1}{100}\log_2\frac{1}{100} - \frac{99}{100}\log_2\frac{99}{100} \cong 0.080. \tag{P.24}$$

Note that the reduction in the MI is much larger in the more general problem, and the reduction in MI is larger, the larger the number of prisoners.

Exercise: Solve the problem of 1,000 prisoners and calculate the MI before and after prisoner "1" receives the information from the warden.

Final suggestion for a teasing thought

Suppose that there are 10 prisoners as in the case of Figure P.3b. Again, prisoner "1" asks the warden the same question. But, the warden points not at all 8 prisoners that are doomed but only at k prisoners to be doomed ($k = 1, 2, \ldots, 8$). The question is again the probability of "1" surviving relative to the probability of survival of one the remaining $9 - k$ prisoners, about which no information is available. Clearly, the larger k, the more "information" is given to "1". Also, the reduction of the missing "information" in the problem is larger, the larger k is. Note, however, that the term "information" is used in different senses in the previous sentence.

References

1. Thermodynamics and Statistical Thermodynamics

Amit, D. J. and Verbin, Y. (1995), *Statistical Physics, An Introductory Course*, World Scientific, Singapore.
Baierlein, R. (1999), *Thermal Physics*, Cambridge University Press, New York.
Ben-Naim, A. (2006), *Molecular Theory of Solutions*, Oxford University Press, Oxford.
Bent, H. A. (1965), *The Second Law*, Oxford University Press, New York.
Bowley, R. and Sanchez, M. (1999), *Introductory Statistical Mechanics*, Oxford University Press, Oxford.
Bridgman, P. W. (1953), *The Nature of Thermodynamics*, Harper and Brothers, New York.
Chandler, D. (1987), *Introduction to Modern Statistical Mechanics*, Oxford University Press, Oxford.
Cohen, E. G. D. (ed) (1962), *Fundamental Problems in Statistical Mechanics*, North Holland, Amsterdam.
Cowan, B. (2005), *Topics in Statistical Mechanics*, Imperial College Press, London.
de Boer, J. and Uhlenbeck, G. E. (1962), *Studies in Statistical Mechanics*, Vol. I, North Holland, Amsterdam.
Denbigh, K. (1981), *The Principles of Chemical Equilibrium with Applications in Chemistry and Chemical Engineering*, Cambridge University Press, Cambridge, New York.
Dickerson, R. E. (1969), *Molecular Thermodynamics*, Benjamin/Cummings, London.
Dorlas, T. C. (1999), *Statistical Mechanics, Fundamentals and Model Solutions*, Institute of Physics Publishing, Bristol.
Dugdale, J. S. (1996), *Entropy and Its Physical Meaning*, Taylor and Francis, London.

Fast, J. D. (1962), *Entropy: The Significance of the Concept of Entropy and Its Applications in Science and Technology,* Philips Technical Library.

Fisher, I. Z. (1964), *Statistical Theory of Liquids*, Translated by T. M. Switz, The University of Chicago Press, Chicago.

Fowler, R. H. and Guggenheim, E. A. (1939), *Statistical Thermodynamics*, Cambridge University Press, Cambridge.

Friedman, H. L. (1985), *A Course in Statistical Mechanics*, Prentice Hall, New Jersey.

Frisch, H. L. and Lebowitz, J. L. (1964), *The Equilibrium Theory of Classical Fluids*, W.A. Benjamin, New York.

Garrod, C. (1995), *Statistical Mechanics and Thermodynamics*, Oxford University Press, Oxford.

Gasser, R. P. H. and Richards, W. G. (1995), *An Introduction to Statistical Thermodynamics*, World Scientific, Singapore.

Gibbs, J. W. (1906), Scientific papers, Vol. I, *Thermodynamics*, Longmans Green, New York, p. 166.

Glazer, A. M. and Wark, J. S. (2001), *Statistical Mechanics: A Survival Guide*, Oxford University Press, Oxford.

Goodisman, J. (1997), *Statistical Mechanics for Chemists*, John Wiley and Sons, New York.

Guggenheim, E. A. (1952), *Mixtures*, Oxford University Press, Oxford.

Hill, T. L. (1956), *Statistical Mechanics, Principles and Selected Applications*, McGraw-Hill, New York.

Hill, T. L. (1960), *Introduction to Statistical Thermodynamics*, Addison-Wesley, Reading, Massachusetts.

Hobson, A. (1971), *Concepts of Statistical Mechanics*, Gordon and Breach, New York.

Huang, K. (1963), *Statistical Mechanics*, John Wiley and Sons, New York.

Isihara, A. (1971), *Statistical Physics*, Academic Press, New York.

Jackson, E. A. (1968), *Equilibrium Statistical Mechanics*, Dover Publications, New York.

Katz, A. (1967), *Principles of Statistical Mechanics: The Informational Theory Approach*, W. H. Freeman, London.

Khinchin, A. I. (1949), *Mathematical Foundations of Statistical Mechanics*, Dover Publications, New York.

Khinchin, A. I. (1960), *Mathematical Foundations of Quantum Statistics*, Graylock Press, Albany, New York.

Kitter, C. (1958), *Elementary Statistical Physics*, John Wiley and Sons, New York.

Knuth, E. L. (1966), *Introduction to Statistical Thermodynamics*, McGraw-Hill, New York.

Kubo, R. (1965), *Statistical Mechanics*, North-Holland, Amsterdam.

Kubo, R., Toda, M. and Hashitsume, N. (1985), *Statistical Physics II*, Springer-Verlag, Berlin.

Landau, L. D. and Lifshitz, E. M. (1958), *Statistical Physics*, Pergamon Press, London.

Landsberg, P. T. (1961), *Thermodynamics with Quantum Statistical Illustrations*, Interscience Publications, New York.

Landsberg, P. T. (1978), *Thermodynamics and Statistical Mechanics*, Dover Publications, New York.

Lavenda, B. H. (1991), *Statistical Physics, A Probability Approach*, John Wiley, New York.

Lee, J. C. (2002), *Thermal Physics, Entropy and Free Energy*, World Scientific, Singapore.

Levine, R. D. and Tribus, M (eds) (1979), *The Maximum Entropy Principle*, MIT Press, Cambridge, MA.

Lindsay, R. B. (1941), *Introduction to Physical Statistics*, Dover Publications, New York.

Lucas, K. (1991), *Applied Statistical Thermodynamics*, Springer-Verlag, Berlin.

Mandl, F. (1971), *Statistical Physics*, 2nd edition, John Wiley and Sons, New York.

Martinas, K., Ropolyi, L. and Szegedi, P. (1991), *Thermodynamics: History and Philosophy. Facts, Trends, Debates*, World Scientific, Singapore.

Maczek, A. (1998), *Statistical Thermodynamics*, Oxford University Press, Oxford.

Mazenko, G. F. (2000), *Equilibrium Statistical Mechanics*, John Wiley and Sons, New York.

McGlashan, M. L. (1979), *Chemical Thermodynamics*, Academic Press, New York.

Mayer, J. E. and Mayer, M. G. (1977), *Statistical Mechanics*, John Wiley and Sons, New York.

McQuarrie, D. A. (1973), *Statistical Mechanics*, Harper and Row, New York.

Mohling, F. (1982), *Statistical Mechanics, Methods and Applications*, John Wiley and Sons, New York.

Munster, A. (1969), *Statistical Thermodynamics*, Vol. I, Springer-Verlag, Berlin.

Munster, A. (1974), *Statistical Thermodynamics*, Vol. II, Springer-Verlag, Berlin.

Penrose, O. (1979), Foundations of statistical mechanics, *Rep. Prog. Phys.* **42**, 1937.

Planck, M. (1913), *Vorlesungen uber die Theorie der Warmestrahlung.* (An English translation is reprinted as *The Theory of Heat Radiation* by Dover Publications, New York, 1991).

Planck, M. (1945), *Treatise on Thermodynamics*, Dover Publications, New York.

Plischke, M. and Bergersen, B. (1989), *Equilibrium Statistical Physics*, World Scientific Singapore.

Reichl, L. E. (1998), *A Modern Course in Statistical Physics*, 2nd edition, John Wiley and Sons, New York.

Reif, F. (1965), *Fundamentals of Statistical and Thermal Physics*, Mc-Graw Hill Book Company, New York.

Robertson, H. S. (1993), *Statistical Thermodynamics*, PTR Prentice Hall, New Jersey.

Ruelle, D. (1969), *Statistical Mechanics,* W. A. Benjamin, New York.

Rushbrooke, G. S. (1949), *Introduction to Statistical Mechanics*, Oxford University Press, London.

Schrodinger, E. (1952), *Statistical Thermodynamics: A Course of Seminar Lectures*, Cambridge University Press, Cambridge.

Temperley, H. N. V., Rowlinson, J. S. and Rushbrook, G. S. (1968), *Physics of Simple Liquids*, North-Holland, Amsterdam.

Ter Haar, D. (1954), *Elements of Statistical Mechanics*, Rhinehart and Co., New York.

Toda, M., Kubo, R., and Saito, N. (1983), *Statistical Physics I: Equilibrium Statistical Mechanics*, Springer-Verlag, Berlin.

Tolman, R. C. (1938), *Principles of Statistical Mechanics,* Clarendon Press, Oxford.

Walecka, J. D. (2000), *Fundamentals of Statistical Mechanics Manuscript and Notes of Felix Bloch*, World Scientific, Singapore.

Wannier, G. H. (1966), *Statistical Physics*, John Wiley and Sons, New York.

Whalen, J. W. (1991), *Molecular Thermodynamics: A Statistical Approach*, John Wiley and Sons, New York.

Widom, B. (2002), *Statistical Mechanics: A Concise Introduction for Chemists*, Cambridge University Press, Cambridge.

Yvon, J. (1969), *Correlations and Entropy in Classical Statistical Mechanics*, Pergamon Press.

2. Probability and Information Theory

Ambegaokar, V. (1996), *Reasoning about Luck: Probability and Its Uses in Physics*, Cambridge University Press, Cambridge.

Ash, R. B. (1965), *Information Theory*, Dover Publications, New York.

Baierlein, R. (1971), *Atoms and Information Theory: An Introduction to Statistical Mechanics*, W. H. Freeman and Co., San Francisco.

Bevensee, R. M. (1993), *Maximum Entropy Solutions to Scientific Problems*, PTR Prentice-Hall, New Jersey.

Brillouin, L. (1956), *Science and Information Theory*, Academic Press, New York.

Carnap, R. (1953), What is probability?, *Scientific American* **189**(3), 128–138.

Cover, T. M. and Thomas, J. A. (1991), *Elements of Information Theory*, John Wiley and Sons, New York.

D'Agostini, G. (2003), *Bayesian Reasoning in Data Analysis: A Critical Introduction*, World Scientific, Singapore.

Feinstein, A. (1958), *Foundations of Information Theory*, McGraw-Hill, New York.

Feller, W. (1957), *An Introduction to Probability Theory and Its Applications*, Vol. I, Wiley, New York.

Gatlin, L. L. (1972), *Information Theory and the Living System*, Columbia University Press, New York.

Georgescu-Roegen, N. (1971), *The Entropy Law and the Economic Press*, Harvard University Press, Cambridge, MA.

Gnedenko, B. V. (1962), *The Theory of Probability*, translated by B. D. Seckler, Chelsea, New York.

Goldman, S. (1954), *Information Theory*, Prentice-Hall, New York.

Gray, R. M. (1990), *Entropy and Information Theory.*

Hald, A. (1990), *A History of Probability and Statistics and Their Applications before 1750*, John Wiley and Sons, New York.

Hartley, R. V. L. (1927), Transmission of information, *Bell Syst. Tech.* J. **1**, 535–563.

Jaynes, E. T. (1957a), Information theory and statistical mechanics, *Phys. Rev.* **106**, 620.

Jaynes, E. T. (1957b), Information theory and statistical mechanics II, *Phys. Rev.* **108**, 171.

Jaynes, E. T. (1965), Gibbs vs. Boltzmann entropies, *Am. J. Phys.* **33**, 391.

Jaynes, E. T. (1973), The well-posed problem, Chapter 8 in Jaynes (1983).

Jaynes, E. T. (1983), in *Papers on Probability, Statistics, and Statistical Physics*, ed. R. D. Rosenkrantz, D. Reidel, London.

Jeffreys, H. (1939), *Theory of Probability*, Clarendon Press, Oxford.

Jones, D. S. (1979), *Elementary Information Theory*, Clarendon Press, Oxford.

Kahre, J. (2002), *The Mathematical Theory of Information*, Kluwer, Boston.

Kerrich, J. E. (1964), *An Experimental Introduction to the Theory of Probability*, Witwatersrand University Press, Johannesburg.

Khinchin, A. I. (1957), *Mathematical Foundation of Information Theory*, Dover Publications, New York.

Leff, H. S. and Rex, A. F. (eds) (1990), *Maxwell's Demon: Entropy, Information, Computing*, Adam Hilger, Bristol.

Leff, H. S. and Rex, A. F. (eds) (2003), *Maxwell's Demon 2: Entropy, Classical and Quantum Information, Computing*, IOP Publishing, London.

Mackay, D. J. C. (2003), *Information Theory, Inference and Learning Algorithms*, Cambridge University Press, Cambridge, MA.

Moran, P. A. P. (1984), *An Introduction. Probability Theory.* Clarendon Press, Oxford.

Mosteller, F. (1965), *Fifty Challenging Problems in Probability with Solutions*, Dover Publications, New York.

Nickerson, R. S. (2004), *Cognition and Chance: The Psychology of Probabilistic Reasoning*, Lawrence Erlbaum Associates, London.

Papoulis, A. (1990), *Probability and Statistics*, Prentice Hall, New Jersey.

Pierce, J. R. (1980), *An Introduction to Information Theory, Symbols, Signals and Noise*, 2nd revised edition, Dover Publications, New York.

Reza, F. M. (1961), *An Introduction to Information Theory*, McGraw-Hill, New York.

Rowlinson, J. S. (1970), Probability, information and entropy, *Nature* **225**, 1196.

Shannon, C. E., (1948), A mathematical theory of communication, *Bell Syst. Tech. J.* **27**, 379, 623.

Sklar, L. (1993), *Physics and Chance: Philosophical Issues in the Foundations of Statistical Mechanics*, Cambridge, University Press, New York.

Weaver, W. (1963), *Lady Luck: The Theory of Probability*, Dover Publications, New York.

Weber, B. H., Depew, D. J. and Smith, J. D., (1988), *Entropy, Information and Evolution: New Perspectives on Physical and Biological Evolutions*, MIT Press, Cambridge, MA.

Yaglom, A. M. and Yaglom, I. M. (1983), *Probability and Information*, translated from Russian by V. K. Jain, D. Reidel, Boston.

3. General Literature

von Baeyer, H. C. (2004), *Information, The New Language of Science*, Harvard University Press, Cambridge, MA.

Born, M. (1960), *The Classical Mechanics of Atoms*, Ungar, New York.

Broda,
E. (1983), *Ludwig Boltzmann, Man-Physicist-Philosopher.* Ox Bow Press, Woodbridge, Connecticut.

Brush, S. G. (1976), *The Kind of Motion We Call Heat*, North-Holland.

Brush, S. G. (1983), *Statistical Physics and the Atomic Theory of Matter from Boyle and Newton to Landau and Onsager*, Princeton University Press, New Jersey.

Cercignani, C. (2003), *Ludwig Boltzmann: The Man who Trusted Atoms*, Oxford University Press, London.

Cooper, L. N. (1969), *An Introduction to the Meaning and the Structure of Physics*, Harper and Row, New York.

Denbigh, K. G. (1989), Note on entropy, disorder and disorganization, *Br. J. Philos. Sci.* **40**, 323.

Feynman, R. (1996), *Feynman Lectures on Computation*, Addison-Wesley, Reading.

Frank, H. S. (1945), Free volume and entropy in condensed systems, *J. Chem. Phys.* **13**, 478–492.

Gell-Mann, M. (1994), *The Quark and the Jaguar*, Little Brown, London.

Greene, B. (1999), *The Elegant Universe*, Norton, New York.

Jaynes, E. T. (1992), The Gibbs paradox, in *Maximum Entropy and Bayesian Methods*, eds C. R. Smith, G. J. Erickson and P. O. Neudorfer, Kluwer, Dordrecht.

Lebowitz, J. L. (1993), Boltzmann's entropy and time's arrow, *Phys. Today*, September.

Mazo, R. M. (2002), *Brownian Motion, Fluctuations, Dynamics and Applications*, Clarendon Press, Oxford.

Prigogine, I. (1997), *The End of Certainty, Time, Chaos, and the New Laws of Nature*, The Free Press, New York.

Rigden, J. S. (2005), *Einstein 1905. The Standard of Greatness*, Harvard University Press, Cambridge.

Schrodinger, E. (1945), *What Is Life*, Cambridge University Press, Cambridge.

4. Cited References

Barrow, J. D. and Webb, J. K. (2005), Inconstant constants, *Sci. Am.* **292**, 32.

Ben-Naim, A. (1987), *Am. J Phys.* **55**, 725.

Ben-Naim, A. (1987), *Am. J Phys.* **55**, 1105.

Brillouin, L. (1956), *Science and Information*, Academic Press, New York.

Carnot, S. (1872), Réflexions sur la puissance motrice du teu ef sur les machines propres à développer catte puissance, *Annales scientifiques de lÉcole Normale Supérieure Sér*, 2(1), 393–457.

Denbigh, K. (1966), *Principles of Chemical Equilibrium*, Cambridge University Press, London.

Denbigh, K. (1981), How subjective is entropy? *Chem. Br.* **17**, 168–185.

Denbigh, K. G. and Denbigh, J. S. (1985), *Entropy in Relation to Incomplete Knowledge*, Cambridge University Press, Cambridge.

Dias, P. N. C. and Shimony, A. (1981), *Adv. Appl. Math.* **2**, 172.

Eliel, E. L. (1962), *Stereochemistry of Carbon Compounds*, McGraw-Hill, New York.

Falk, R. (1979), in *Proceedings of the Third International Conference for Psychology and Mathematical Education*, Warwick, UK.

Falk, R. (1993), *Understanding Probability and Statistics*, A. Peters, Wellesly, MA.

Friedman, K. and Shimony, A. (1971), *J. Stat. Phys.* **3**, 381.

Gibbs, J. W. (1906), Scientific papers, *Thermodynamics*, Longmans Green, vol. I, New York.

Jacques, J., Collet, A. and Willen, S. H. (1981), *Enantiomers, Racemates and Resolutions*, Wiley, New York.

Kirkwood, J. G. (1933), *Phys. Rev.* **44**, 31.

Leff, H. S. (1996), *Am. J. Phys.* **64**, 1261.

Leff, H. S. (1999), *Am. J. Phys.* **67**, 1114.

Leff, H. S. (2007), *Foundations of Physics* (in press).

Lesk, A. M. (1980), *J. Phys. A: Math. Gen.* **13**, L111.

Lewis, G. N. (1930), The symmetry of time in physics, *Science* **71**, 569.

McGlashan, M. L. (1966), The use and misuse of the laws of thermodynamics, *J. Chem. Educ.* **43**, 226.

McGlashan, M. L. (1979), *Chemical Thermodynamics*, Academic Press, New York.

Munowitz, M. (2005), *Knowing: The Nature of Physical Law*, Oxford University Press, Oxford.

Noyes, R. M. (1961), *J. Chem. Phys.* **34**, 1983.

Rothstein, J. (1951), *Science* **114**, 171.

Szilard, L. (1929), *Z. Phys.* **53**, 840–856.

Tribus, M. and McIrvine, E. C. (1971), Energy and information, *Sci. Am.* **225** 179–188.

Tribus, M. (1961), *Thermostatics and Thermodynamics*, Van Nostrand, Princeton NJ.

Tribus, M. (1978) in *Levine and Tribus (1979).*

Wigner, E. (1932), *Phys. Rev.* **40**, 749.

Wright, P. G. (1970), Entropy and disorder, *Contemp. Phys.* **11**, 581.

Index